Forest Vulnerability and Climate Risk Assessment:

Dynamics, Disturbances, and Resilience

Said Moukrim
https://orcid.org/0000-0002-3172-3776
Faculty of Science, Mohammed V University, Rabat, Morocco

Said Lahssini
https://orcid.org/0000-0003-1986-8240
National School of Forest Engineers, Morocco

IGI Global
Scientific Publishing
Publishing Tomorrow's Research Today

Vice President of Editorial Melissa Wagner
Director of Acquisitions Mikaela Felty
Director of Book Development Jocelynn Hessler
Production Manager Mike Brehm
Cover Design Jose Rosado

Published in the United States of America by
IGI Global Scientific Publishing
701 East Chocolate Avenue
Hershey, PA, 17033, USA
Tel: 717-533-8845 | Fax: 717-533-7115
Website: https://www.igi-global.com E-mail: cust@igi-global.com

Library of Congress Cataloging-in-Publication Data

(CIP Data Pending)
ISBN13: 979-8-2600-0783-9
Isbn13Softcover: 979-8-2600-0784-6
EISBN13: 979-8-2600-0785-3
British Cataloguing in Publication Data
A Cataloguing in Publication record for this book is available from the British Library.

Editorial Advisory Board

Table of Contents

Section 3
Climate Impacts and Environmental Processes

Detailed Table of Contents

scenarios. Although new suitable areas emerge for some species (<12.5%), these gains remain insufficient to offset projected losses, signaling increased fragmentation and ecosystem degradation. This chapter emphasizes the urgent need to integrate climate projections into conservation planning, adaptive management and restoration strategies to enhance resilience and safeguard ecosystem services across this region.

Rajaa Timzioura, Polydisciplinary Faculty, University Sultan Moulay Slimane, Morocco
Sara Ezzine, Polydisciplinary Faculty, University Sultan Moulay Slimane, Morocco
Abderrahim Ettaqy, Polydisciplinary Faculty, University Sultan Moulay Slimane, Morocco
Abdelhakim Taha, Polydisciplinary Faculty, University Sultan Moulay Slimane, Morocco
Mohamed El Mderssa, Polydisciplinary Faculty, University Sultan Moulay Slimane, Morocco
Younes Abbas, Polydisciplinary Faculty, University Sultan Moulay Slimane, Morocco

The vulnerability of Mediterranean forests results from the combined effects of climatic stressors and human pressures. In Morocco, recurrent droughts, wildfires, and rapid land-use change interact with complex governance challenges. This chapter assesses the condition and trends of major forest types such as argan, cedar, cork oak, thuya, and juniper, while placing Morocco in a regional context that includes Algeria, Tunisia, and Spain. Vulnerability is analyzed through an integrated framework linking exposure, sensitivity, and adaptive capacity using ecological and institutional indicators. Despite advances in remote sensing, modeling, and genetic-based restoration, progress is limited by fragmented information systems and weak incorporation of local knowledge. Strengthening resilience requires better interinstitutional coordination, adaptive silviculture, improved data governance, and innovative financing. A strategic roadmap outlines priorities for multi-scale monitoring and a Mediterranean forest resilience toolbox.

Morocco's forest ecosystems, spanning 5.8 million hectares, face many threats that compromise their ecological and socio-economic functions. This chapter employs the DPSIR framework combined with multi-criteria decision analysis to systematically evaluate and prioritize threats affecting Moroccan forests. Principal component analysis reveals water stress as the central node connecting all pressures through complex feedback loops and synergistic interactions. Critical threats are water stress, deforestation, timber overexploitation, and overgrazing. Major threats include forest fires, no-timber products overexploitation, and parasitic attacks. Minor threats include intersectoral coordination, scientific-management gaps and pollution. Those threats operate through self-reinforcing degradation spirals, causing massive dieback, regeneration failure, and land degradation. These pressures trigger cascading impacts on biodiversity, ecosystem services, and rural livelihoods. To address root causes, break degradation and strengthen adaptive capacity, integrated strategies are needed.

In the Kroumirie region of north-western Tunisia, populations of cork oak (Quercus suber) are suffering from drought. This decline in cork oak forests requires a re-evaluation of sustainable forest management techniques. A plant association is defined as a group of plants with a specific floristic composition, uniform physiognomy and growth under uniform site conditions. In this context, the floristic composition would include Quercus suber and one or more companion plants intended to support its persistence. As part of this initiative, the present study, aims to test the impact of such plant associations on the growth and hydraulic conductance PLC of cork oak. This chapter details the role and functioning of symbiotic associations in Mediterranean ecosystems. The study was conducted in a mixed forest of cork oak and zeen oak in the Kroumirie region of north-western Tunisia (Jendouba Governorate).

Section 2
Disturbances and Risk Dynamics

Focuses on disturbance regimes that increasingly shape the dynamics of Mediterranean forest ecosystems

This chapter introduces AI4Fire, an operational and explainable spatio-temporal machine learning framework for daily wildfire risk prediction in Northern Morocco. The system integrates large-scale geospatial data, advanced spatio-temporal feature engineering, high-performance computing, and explainable artificial intelligence to generate high-resolution risk and interpretability maps over more than one million grid points. AI4Fire combines meteorological dynamics, heatwave indicators, vegetation status, topography, socio-environmental factors, and historical wildfire pressure within a scalable Big Data architecture enabling near real-time inference. Validation over the 2021-2023 fire seasons demonstrates strong and stable predictive performance, with more than 75% of wildfire events occurring in areas classified as very high risk. By coupling accuracy, scalability, and transparency, AI4Fire provides an effective decision-support tool for proactive wildfire prevention and sustainable forest management in Mediterranean ecosystems.

Wildfires are becoming more intense across North Africa, yet research on this topic in Morocco remains limited and uneven despite rising climatic pressure and increasing fire recurrence. This chapter delivers the first bibliometric analysis on Moroccan wildfire studies, examining publication trends, collaboration networks and thematic areas. The most notable progress concerns satellite monitoring, burn-severity evaluation and spatial modelling, while governance, socio-economic vulnerability, institutional coordination, community participation and post-fire restoration are still rarely addressed. Compared with other Mediterranean countries that have developed more integrated and prevention-oriented wildfire systems, Morocco continues to face shortcomings in scientific cooperation, regional integration and links between research and policy. Improving data interoperability and coordinated research strategies will be crucial to support climate-resilient wildfire management in North Africa.

Section 3
Climate Impacts and Environmental Processes

Examines environmental processes that influence forest vulnerability under changing climatic conditions

Hicham Mharzi Alaoui, Department of Climate and Environnemental Risks, ANEF, Morocco
Majdouline Belhouari, Department of Climate and Environnemental Risks, ANEF, Morocco
Zineb Loukili, National Agency for Water and Forest, Morocco
Youness Bouziani, National Agency for Water and Forest, Morocco
Fouad Assali, National Agency for Water and Forest, Morocco
Hicham Hajji, IAV Hassan II, Morocco
Nora Hammouda, Rural Development Directorate, Morocco

Soil erosion is a major environmental challenge in Mediterranean mountainous watersheds, especially under increasing climatic variability and anthropogenic pressure. This chapter presents an integrated and dynamic framework for soil erosion assessment and prioritization, combining RUSLE-based modeling, field-based validation, and SMART multi-criteria decision analysis, applied to the N'Fis watershed (Morocco), upstream of the Lalla Takerkoust Dam. Results reveal strong spatial heterogeneity, with soil loss rates ranging from <3 to >22.7 $t \cdot ha^{-1} \cdot yr^{-1}$. Field surveys highlight marked local contrasts related to erosion dynamics, vegetation degradation, land use, and infrastructure exposure. SMART-based multi-criteria analysis integrates biophysical and socio-economic factors into a unified prioritization framework, enabling refined identification of priority intervention zones. The proposed approach offers a robust, scalable, and transferable decision-support tool for sustainable watershed management, dam protection, and climate adaptation in semi-arid and mountainous regions.

Climate change is an increasing global threat to biodiversity and ecosystem functioning, affecting species distributions, ecological interactions, and habitat conditions. In Morocco, rising temperatures and increasing aridity pose major challenges to ecosystem conservation and adaptation. This study assesses the national-scale impacts of climate change on Moroccan ecosystems by integrating artificial intelligence and cloud computing to analyze the evolution of Emberger's bioclimatic zones. The results indicate a marked aridification of bioclimatic zones, with projected losses of approximately 85% and 36% of humid and sub-humid zones, respectively, while arid and semi-arid zones are expected to expand by about 15% and 10%. These changes reflect heterogeneous vulnerabilities among bioclimatic zones and sub-zones, ranging from intra-zone thermal shifts to transitions toward new bioclimatic regimes. These findings support adaptive ecosystem management, biodiversity conservation, and the sustainability of ecosystem services under future climate scenarios.

Foreword

The forests of the Mediterranean basin have fascinated naturalists, ecologists, and geographers for centuries. Their biodiversity, shaped by millennia of climatic oscillation, geological upheaval, and intimate coevolution with human activities, makes them among the most scientifically rich and historically significant terrestrial ecosystems on the planet. Yet it is the forests of the southern shore of the Mediterranean, stretching across Morocco, Algeria, Tunisia, and Libya, that have remained most persistently underrepresented in the international scientific literature. These sub-Mediterranean forests, dominated by cedar, cork oak, holm oak, argan, and Aleppo pine, prevail within ecological and socioeconomic conditions that differ profoundly from those of their northern counterparts. They face challenges which, in the context of accelerating climate change and mounting human pressure, demand a science as rigorous and distinctive as the ecosystems themselves.

For several decades, researchers and foresters working in North Africa have accumulated a body of knowledge on the vulnerability of these forests to drought, fire, land degradation, and the compounding effects of human pressure. Evidence of progressive ecosystem destabilization, from the dieback of Atlas cedar under prolonged water stress, to the erosion of mixed-oak landscapes under shifting disturbance regimes, to the accelerating reconfiguration of bioclimatic zones across Morocco and the wider Maghreb, has been documented with increasing precision. Yet much of this knowledge has circulated in national reports, in French-language journals, and through practitioner networks, rarely reaching the international scientific community with the visibility it deserves. This volume, edited by Said Lahssini and Said Moukrim, addresses that gap directly.

The volume is organized around three thematic axes that reflect the natural progression of vulnerability science. The first section examines the responses of forest ecosystems to climate change, drawing on projected dynamics modelling, comparative vulnerability assessments across the Southern Mediterranean, and case studies from Morocco and Tunisia. The second section turns to the disturbance regimes that are increasingly reshaping Mediterranean forest dynamics, with a sustained focus on wildfire. The third section broadens the lens to the environmental processes that modulate forest vulnerability under changing climatic conditions.

This volume is the second of two companion works. While the first, *Dynamics, Challenges, and Perspectives for Sustainable Forest Management*, addresses forest management frameworks, governance structures, ecosystem services, and technological innovation, the present volume concentrates on the ecological and climatic threats that make the urgency of that management so acute. The two volumes are designed to be read in dialogue: the vulnerability assessments and risk analyses assembled here provide the diagnostic foundation upon which the management responses examined in the companion volume must be built. Together, they offer a comprehensive scientific reference for sub-Mediterranean forest ecosystems, integrating biophysical understanding of forest functioning and risk with the institutional and technical dimensions of long-term stewardship.

I have had the privilege of working alongside Mediterranean forest scientists for much of my career, and I am convinced that this volume makes a timely and necessary contribution. The questions it addresses, how vulnerable are these forests, how will their disturbance regimes evolve under future climate scenarios, what environmental processes amplify or attenuate that vulnerability, and what tools can translate scientific understanding into operational risk management, are among the most consequential in contemporary forest ecology. I hope this work will serve as a foundational reference for the scientists, managers, and policymakers responsible for these ecosystems, and that it will contribute to building the knowledge base on which their long-term resilience depends. The forests of the Southern Mediterranean have endured a great deal over the millennia. With the scientific knowledge assembled in this volume, and the foresight it is designed to cultivate, there is good reason to believe they can endure what is coming.

Riccardo Valentini

University of Tuscia, Italy

Riccardo Valentini is a Full Professor of Forest Ecology at the University of Tuscia. He was lead author of the third and fifth reports of the IPCC (Intergovernmental Panel on Climate Change) on climate impacts and adaptation and author of the IPCC special report "Land and Climate". Member of the European Academy of Sciences and Medal of the Academy of Sciences of the "XL" for Physics and Natural Sciences. He dedicated his research to understand forest role on global biogeochemical cycles with a special focus on Mediterranean forests.

Preface

Mediterranean forest ecosystems are increasingly recognized as among the most vulnerable to global environmental change. Situated at the climatic interface between humid temperate regions and expanding arid zones, these ecosystems are particularly sensitive to rising temperatures, declining precipitation, and increasing climatic variability. Projections consistently indicate that the Mediterranean basin is one of the most climate-vulnerable regions on Earth, with mean annual temperatures expected to increase by 2 to 4°C and precipitation to decline by 10 to 30% by the end of the century under moderate to high emission scenarios (Giorgi & Lionello, 2008; IPCC, 2021). In recent decades, these environmental pressures have intensified, profoundly affecting the structure, functioning, and spatial distribution of forest ecosystems throughout the basin. A transformation now documented at a pace and spatial extent that places current dynamics outside the range of historical variability (Miranda et al., 2023; Moukrim et al., 2020; 2022).

In North Africa, these challenges are especially pronounced. Forest landscapes are exposed to recurrent drought events, progressive aridification, soil degradation, and increasing disturbance risks. Drought-induced tree mortality, already affecting key species such as *Cedrus atlantica, Quercus suber,* and *Pinus halepensis* across portions of their native ranges, is generating pressures that threaten the long-term viability of these ecosystems (Allen et al., 2010). Wildfire, whose frequency and severity are being exacerbated by climate-driven fuel desiccation and landscape changes, is increasingly reshaping forest structure and post-fire recovery dynamics across the region (Turco et al., 2018). The combination of vegetation cover loss and intense rainfall events further accelerates soil erosion and watershed degradation, compounding the threats to both forest productivity and downstream hydrological services (Borrelli et al., 2017). These biophysical processes do not act independently: anthropogenic pressures (overgrazing, land-use change, and demographic expansion) interact with climatic stressors to create complex vulnerability patterns. Where these pressures intersect, the resilience of forest systems is progressively eroded, increasing the probability of regime shifts toward degraded, lower-productivity states (Folke et al., 2004).

Understanding these dynamics has become a central priority for forest science and environmental policy. Assessing vulnerability requires an integrated perspective that combines ecological knowledge, climate projections, disturbance analysis, and spatial modelling. Advances in geospatial technologies, environmental modelling, and artificial intelligence are now providing powerful tools to analyse these complex interactions and to anticipate future transformations of forest ecosystems; from species distribution models that map potential habitat loss under future climate scenarios, to machine learning frameworks that generate real-time wildfire risk predictions, to spatial big data approaches that track shifts in bioclimatic zones at regional scales.

The chapters presented in this volume contribute to this effort by examining the vulnerability of sub-Mediterranean forest ecosystems and the environmental processes shaping their future trajectories. Together, they explore how climate change, disturbance regimes, and landscape dynamics interact to influence forest resilience and how scientific tools can support more adaptive, anticipatory, and climate-resilient management responses across north Africa and the broader Mediterranean region. This volume is the second of two companion works: while the first volume focuses on forest management paradigms, governance, ecosystem services, and technological innovation for practitioners, the present work concentrates on the scientific foundations of vulnerability assessment, disturbance ecology, and climate risk analysis.

ORGANIZATION OF THE BOOK

The book is organized into three thematic sections addressing complementary dimensions of forest vulnerability and climate risk in sub-Mediterranean ecosystems.

Section I: Forest Vulnerability and Climate Change

This section examines the responses of forest ecosystems to climate change and analyzes the vulnerability of forest species and landscapes across the Southern Mediterranean, from species-level distributional shifts to landscape-scale resilience dynamics.

The first chapter opens the volume by assessing the vulnerability of several emblematic forest species in North Africa through species distribution modeling. By projecting future distributions under different climate scenarios, the study highlights potential habitat contractions and identifies areas of particular vulnerability across the principal forest types of Morocco and the broader Maghreb region.

The second chapter provides a comparative synthesis of forest vulnerability assessments conducted across the Southern Mediterranean. It discusses the conceptual frameworks and methodological approaches used to evaluate vulnerability and identifies key ecological and institutional factors influencing forest resilience.

The third chapter examines the principal environmental and anthropogenic pressures affecting forest resilience in Morocco. Using an integrated analytical framework, it identifies the main drivers of forest degradation and explores their interactions within complex socio-ecological systems.

The fourth chapter investigates the resilience of Mediterranean mixed-oak forest landscapes facing increasing drought stress. The study analyzes ecological processes that influence the adaptive capacity of these ecosystems and discusses management strategies that could enhance their resilience.

Section II: Disturbances and Risk Dynamics

This section focuses on disturbance regimes that increasingly shape the dynamics of Mediterranean forest ecosystems, with particular attention to wildfire as both an ecological process and a governance challenge.

The first chapter in this section presents an operational framework based on machine learning techniques for predicting wildfire risk at high spatial and temporal resolution. By integrating large geospatial datasets and explainable artificial intelligence approaches, the study demonstrates how data-driven models can support proactive fire prevention strategies.

The following chapter analyzes the evolution of wildfire research and governance frameworks in Morocco within the broader Mediterranean context. It examines scientific production, collaboration networks, and institutional arrangements related to wildfire management, identifying key gaps and proposing directions for more integrated and resilient fire governance.

Section III: Climate Impacts and Environmental Processes

The final section examines environmental processes that influence forest vulnerability under changing climatic conditions, linking biophysical dynamics at the watershed and regional scales to the broader challenge of climate adaptation.

The first chapter of this section develops a spatial modeling approach for assessing soil erosion dynamics in mountainous forest watersheds. By combining modeling techniques with field validation and decision-support analysis, the study provides tools for identifying priority areas for watershed management.

The concluding chapter explores the use of spatial big data and artificial intelligence to analyze the evolution of bioclimatic zones under future climate scenarios.

The results highlight major shifts in climatic conditions that could significantly reshape ecosystem distributions and forest dynamics in Morocco, offering a spatially explicit prospective view for forest planners and policymakers.

Taken together, the eight contributions assembled in this volume advance our understanding of the mechanisms driving forest vulnerability, disturbance dynamics, and climate-driven environmental change in sub-Mediterranean ecosystems. By integrating ecological analysis, environmental modelling, and advanced geospatial technologies, from species distribution models to explainable machine learning frameworks and spatial big data platforms, the volume aims to provide both a rigorous scientific foundation and a set of operational analytical tools for those engaged in the development of more adaptive and climate-resilient forest management strategies in North Africa and the wider Mediterranean region.

Said Moukrim

Faculty of Science, Mohammed V University in Rabat, Morocco

Said Lahssini

National School of Forest Engineers, Salé, Morocco

REFERENCES

Allen, C. D., Macalady, A. K., Chenchouni, H., Bachelet, D., McDowell, N., Vennetier, M., Kitzberger, T., Rigling, A., Breshears, D. D., Hogg, E. H. T., Gonzalez, P., Fensham, R., Zhang, Z., Castro, J., Demidova, N., Lim, J.-H., Allard, G., Running, S. W., Semerci, A., & Cobb, N. (2010). A global overview of drought and heat-induced tree mortality reveals emerging climate change risks for forests. *Forest Ecology and Management*, *259*(4), 660–684. DOI: 10.1016/j.foreco.2009.09.001

Borrelli, P., Robinson, D. A., Fleischer, L. R., Lugato, E., Ballabio, C., Alewell, C., Meusburger, K., Modugno, S., Schütt, B., Ferro, V., Bagarello, V., Oost, K. V., Montanarella, L., & Panagos, P. (2017). An assessment of the global impact of 21st century land use change on soil erosion. *Nature Communications*, *8*(1), 2013. DOI: 10.1038/s41467-017-02142-7 PMID: 29222506

Folke, C., Carpenter, S., Walker, B., Scheffer, M., Elmqvist, T., Gunderson, L., & Holling, C. S. (2004). Regime shifts, resilience, and biodiversity in ecosystem management. *Annual Review of Ecology, Evolution, and Systematics*, *35*(1), 557–581. DOI: 10.1146/annurev.ecolsys.35.021103.105711

Giorgi, F., & Lionello, P. (2008). Climate change projections for the Mediterranean region. *Global and Planetary Change*, *63*(2–3), 90–104. DOI: 10.1016/j.gloplacha.2007.09.005

IPCC. (2021). *Climate Change 2021: The Physical Science Basis. Contribution of Working Group I to the Sixth Assessment Report of the Intergovernmental Panel on Climate Change*. Cambridge University Press., DOI: 10.1017/9781009157896

Miranda, A., Lara, A., Altamirano, A., Di Bella, C., González, M. E., & Camarero, J. J. (2023). Widespread synchronous decline of Mediterranean-type forest driven by accelerated aridity. *Nature Plants*, *9*(11), 1810–1817. DOI: 10.1038/s41477-023-01541-7 PMID: 37845335

Moukrim, S., Lahssini, S., Rhazi, M., Menzou, K., El Madihi, M., Rifai, N., Bouziani, Y., Azedou, A., Boukhris, I., & Rhazi, L. (2022). Climate change impact on potential distribution of an endemic species Abies marocana Trabut. *Ekologia (Bratislava)*, *41*(4), 329–339. DOI: 10.2478/eko-2022-0034

Moukrim, S., Lahssini, S., Rifai, N., Menzou, K., Mharzi-Alaoui, H., Labbaci, A., Rhazi, M., Wahby, I. W., El Madihi, M., & Rhazi, L. (2020). Modélisation de la distribution potentielle de Cedrus atlantica Manetti au Maroc et impacts du changement climatique. *Bois et Forêts des Tropiques*, *344*, 3–16. DOI: 10.19182/bft2020.344.a31888

Turco, M., Rosa-Cánovas, J. J., Bedia, J., Jerez, S., Montávez, J. P., Provenzale, A., & Llasat, M. C. (2018). Exacerbated fires in Mediterranean Europe due to anthropogenic warming projected with non-stationary climate-fire models. *Nature Communications*, *9*(1), 3821. DOI: 10.1038/s41467-018-06358-z PMID: 30279564

Acknowledgment

The completion of this collective volume stands as a testament to the dedication, expertise, and collaborative spirit of a community of scholars united by a common commitment to advancing knowledge in this field.

The editors wish to express their profound gratitude to all who contributed to this undertaking. Above all, our deepest appreciation is owed to the chapter authors, whose intellectual rigor, scholarly generosity, and sustained engagement made this work possible. Their contributions reflect not only scientific excellence but also a genuine willingness to share knowledge for the benefit of the wider academic community.

The editors also wish to pay tribute to the reviewers, whose exacting and constructive assessments were essential in upholding the standards of quality, coherence, and scientific integrity that a publication of this nature demands. Many of the authors graciously assumed the additional responsibility of peer review, a dual contribution that speaks to their exemplary commitment to the collective endeavor.

We extend our sincere appreciation to colleagues from various academic departments and research institutions for their support in disseminating the call for contributions and for the encouragement they offered throughout the editorial process.

The editors further wish to acknowledge, with particular gratitude, the *CEMED* "*Centre des Experts "Mastery" pour les Études et le Développement*", whose institutional support and steadfast accompaniment throughout this project have been invaluable. Their engagement reflects a commendable commitment to the promotion of research and scholarly excellence.

It is our hope that this volume will serve as a meaningful contribution to its field, and a worthy reflection of the collective effort that brought it into being.

Said Moukrim
Faculty of Science, Mohammed V University in Rabat, Morocco

Said Lahssini
National School of Forest Engineers, Salé, Morocco

Section 1
Forest Vulnerability and Climate Change

Chapter 1

Vulnerability and Projected Dynamics of North African Forest Ecosystems Under Climate Change

Said Moukrim
http://orcid.org/0000-0002-3172-3776
Research Center of Plant and Microbial Biotechnologies, Biodiversity, and Environment. Faculty of Sciences. Mohammed V University in Rabat, Morocco

Said Lahssini
http://orcid.org/0000-0003-1986-8240
National School of Forest Engineers, Morocco

Kamal Menzou
http://orcid.org/0009-0005-1125-1854
National School of Forest Engineers, Morocco

Abdelkader Benabbou
Agence Nationale des Eaux et Forêts, Morocco

Abdessadeq Boudjaj
Faculty of Sciences, Mohammed V University in Rabat, Morocco

Mohamed Ben Bammou
Faculty of Sciences, Mohammed V University in Rabat, Morocco

Mohammed El Madihi
http://orcid.org/0000-0002-2425-5396
Faculty of Sciences, Chouaib Doukkali University, El Jadida, Morocco

Laila Rhazi
Faculty of Sciences, Mohammed V University in Rabat, Morocco

DOI: 10.4018/979-8-2600-0783-9.ch001

ABSTRACT

North African forest ecosystems provide essential services while facing unprecedented pressures from climate change and anthropogenic activities. This chapter examines the vulnerability of seven emblematic Moroccan forest species using species distribution modeling (MaxEnt) to project current and future (2070) distributions under optimistic and pessimistic scenarios. The authors identify key bioclimatic drivers and reveal extensive habitat contractions ranging from 15% to 100%. Critically vulnerable species face complete habitat elimination under severe warming, while widespread species experience catastrophic losses exceeding 94% under pessimistic scenarios. Although new suitable areas emerge for some species (<12.5%), these gains remain insufficient to offset projected losses, signaling increased fragmentation and ecosystem degradation. This chapter emphasizes the urgent need to integrate climate projections into conservation planning, adaptive management and restoration strategies to enhance resilience and safeguard ecosystem services across this region.

INTRODUCTION

Climate change is recognized as one of the greatest challenges facing natural ecosystems and human societies (IPCC, 2021; Stocker et al., 2013). Over recent decades, scientific evidence has shown that anthropogenic greenhouse gas emissions are the dominant driver of observed warming, which is expected to accelerate in the coming decades (IPCC, 2021; Pielke et al., 2022). Global mean temperature is projected to increase between 0.3 and 4.8 °C by the end of the century depending on emission scenarios, with associated shifts in rainfall regimes, increased frequency of extreme events, and greater variability in hydrological cycles (Hallegatte et al., 2015; IPCC, 2023). These changes are already affecting ecosystems worldwide, and forests are among the most exposed and vulnerable ecosystems because of their strong dependence on climatic factors for growth, regeneration, and long-term resilience.

The Mediterranean Basin has been identified as one of the main "hotspots" of climate change due to its geographic position at the interface between temperate and arid zones (Diffenbaugh et al., 2007). This region is expected to undergo above-average warming compared to the global mean, combined with reduced precipitation and more frequent droughts and heatwaves (Lionello, 2012). Recent research underscores that Mediterranean forests rank among the most vulnerable forest systems globally, due to their sensitivity to rising temperatures, increasing drought stress, and intensifying disturbance regimes including fire and biotic outbreaks. Climate models and empirical assessments show that large proportions of Mediterranean forests face very high climate exposure and risk under future warming scenarios, with compounded effects of drought, wildfire, and extreme climatic events accelerating forest decline and reducing ecosystem resilience. Such vulnerability reflects not only climate exposure but also low adaptive capacity of many Mediterranean tree species and forest types, placing these ecosystems at heightened risk of structural change and functional loss (Hidalgo-Triana et al., 2023; Miranda et al., 2023). This profound vulnerability is further amplified by the region's history of fragmentation, shallow soils, and exposure to recurrent disturbances such as fire and overexploitation (Mharzi Alaoui et al., 2017). Furthermore, the Mediterranean space has been subject to millennia of human occupation and land-use pressures, which have shaped its ecosystems into inherently fragile systems (Blondel, 2006). Climate change adds an additional layer of stress that threatens to disrupt their structure, composition, and functions, with cascading effects on biodiversity and ecosystem services (Cheddadi et al., 2006).

Forests in the southern Mediterranean, particularly in North Africa, illustrate these challenges in an acute way. North African forests cover relatively small areas compared to total landmass, but harboring a remarkable variety of ecosystems that are strongly influenced by millennia of human activity. They are marked by pronounced climatic gradients (Medail & Quezel, 1999) and providing vital ecological and socio-economic services (FAO, 2016). They act as reservoirs of biodiversity, many hosting endemic taxa restricted to Mediterranean mountain ranges (Benabid, 2000). They regulate hydrological cycles, protect soils from erosion, sequester carbon, conserve biodiversity, and sustain rural livelihoods by providing timber, fuelwood, cork, forage, and non-wood forest products (Benabou, Moukrim, Laaribya, et al., 2022). They also buffer against desertification processes, which are intensifying under warming and declining rainfall. This multifunctionality underscores both their ecological value and the socio-political importance of ensuring their long-term resilience. Despite this importance, these ecosystems face severe pressures: climatic stress through increased temperatures, recurrent droughts, and rising fire risk; anthropogenic stress through overgrazing, land clearing, and uncontrolled harvesting; and socio-economic stress linked to poverty, population growth, and limited management capacity. Together, these drivers create conditions of heightened vulnerability and render North African forest ecosystems highly vulnerable to climate change and place them among the most threatened globally (Medail & Quezel, 1999). The dynamics of these ecosystems reveal negative trends. Forest cover across the Maghreb has been declining or degrading under combined pressures of overexploitation, land conversion, overgrazing, and recurrent fires. Climate change compounds these anthropogenic drivers, accelerating processes of dieback, regeneration failure, and range shifts. Multiple studies highlight that Mediterranean-type forests are particularly sensitive to changes in water availability, given their historical adaptation to marked seasonality and recurrent summer droughts (Aussenac, 1973).

However, the magnitude of recent climatic shifts appears to exceed the adaptive capacity of many species, threatening to disrupt long-established equilibria.

In this context, Morocco offers a particularly illustrative case within the South-Mediterranean domain, both because of its bioclimatic diversity, from arid lowlands to humid montane belts, and its forests harbor emblematic Mediterranean and endemic species. With more than 9 million hectares of forests and woodlands, representing roughly 12% of the national territory, Morocco hosts a large diversity of forest which play a key role in biodiversity conservation and contribute directly to the well-being of rural populations (Serbouti et al., 2023). This diversity renders Morocco a microcosm of South-Mediterranean Forest dynamics, while simultaneously exposing its ecosystems to a broad spectrum of climate-related stressors.

Climate change affects forests at multiple levels, from physiology to biogeography. Trees face increased evapotranspiration, water stress, and reduced growth rates, while phenological shifts in flowering and fruiting alter reproduction and regeneration cycles. Species distributions are projected to shift upward in altitude or northward in latitude, raising risks of local extinction when migration is blocked by topography or human land use. At the ecosystem level, dieback events such as those observed in cedar forests, reduced regeneration in cork oak stands, and degradation of argan and holm oak woodlands illustrate the loss of ecosystem integrity and the decline of essential services such as carbon storage, soil protection, and hydrological regulation (ElMderssa et al., 2025; Rhanem, 2011; Varela, 2008). These ecological impacts have direct repercussions on rural communities, exacerbating exposure to desertification, reducing the availability of forest products, and eroding cultural and recreational values linked to forests. These examples reflect the broader dynamics of forest vulnerability in Morocco and across the southern Mediterranean.

Given the complexity of these interactions, an integrative approach is required to assess and anticipate the vulnerability of Mediterranean and North African forests under climate change. This research aims to contribute to this effort by synthesizing knowledge on the projected dynamics of key South-Mediterranean Forest ecosystems, with Morocco as a central case study. A set of model species was selected to illustrate contrasting ecological niches and socio-economic importance: *Argania spinosa* as a keystone of arid and semi-arid ecosystems, *Cedrus atlantica* as a high-altitude species sensitive to warming and drought, *Abies marocana* as a restricted endemic at risk of extinction, *Quercus suber* as a socio-economically valuable lowland species, *Quercus rotundifolia* as the most widespread and ecologically plastic Mediterranean oak, *Pistacia atlantica* as a drought-tolerant pre-forest species characteristic of arid and semi-arid landscapes, and *Tetraclinis articulata* as a semi-arid conifer with unique coppicing ability, forming mixed pre-forest stands and representing one of the most widespread woodland ecosystems in southern Mediterranean Morocco. Together, these species encompass a wide range of ecological strategies, biogeographic patterns, and sensitivities to climatic stress, and thus serve as robust models for understanding both the resilience and vulnerability of South Mediterranean forests under environmental change.

To address this objective, the study adopts a methodological framework based on species distribution modeling (SDM), which has emerged as a robust tool for projecting potential shifts in habitat suitability under different climate scenarios. By linking ecological traits with climatic projections, species distribution models provide insights into the potential trajectories of studied forests under climate change and offer a scientific basis for conservation and management strategies.

MATERIALS AND METHODS

Study area and ecology of model species

The study was conducted in the territory of the Kingdom of Morocco. This country is located at the northwestern corner of the African continent within 21° and 36° North and 1° and 17° West, between the Atlantic Ocean in the west, the Sahara Desert in the south, and the Mediterranean Sea in the north. This unique geographic position places Morocco at the interface of major biogeographic regions, while its complex orography, shaped by the Rif and Atlas Mountain chains, creates a rich mosaic of ecological conditions. The diversity of climates, from humid Mediterranean zones to semi-arid and arid landscapes, makes Morocco one of the most representative territories of North African ecosystems. The country hosts four of the world's 14 terrestrial biomes and includes several critical and endangered ecoregions identified in the WWF Global 200 list (Dinerstein et al., 2017). With approximately 4,200 vascular plant taxa and a high level of endemism (19.2%) (Fennane & Ibn Tattou, 2012), Morocco is considered as the most biologically diverse country of the South-Mediterranean Basin (Medail & Quezel, 1999). This exceptional ecological heterogeneity justifies its selection as a case study, since it encapsulates the main dynamics and vulnerabilities that characterize forest ecosystems of the southern Mediterranean.

Within this framework, seven emblematic forest tree species were selected as models because of their ecological, economic, and cultural significance, as well as their distribution across contrasting environments. Together, they reflect the range of challenges posed by climate change to North African forests. The argan tree (*Argania spinosa*), endemic to Morocco and is a xero-thermophilous species belonging to the Sapotaceae family (Fennane & Ibn Tattou, 2012), forms characteristic presteppic woodlands in the central-west and southwest of the country, particularly in the Souss plain and the western Anti-Atlas. As the only representative of its family in the region (De Candolle, 1844), it is a species of outstanding ecological, cultural, and socio-economic value. Beyond its role in maintaining fragile arid ecosystems, it underpins the livelihoods of millions of rural inhabitants through its multifunctional products, particularly argan oil, which has gained worldwide recognition. The Atlas cedar (*Cedrus atlantica*), a mountain conifer endemic to the Maghreb, occupies a privileged ecological niche in the Middle Atlas, Rif (Muller et al., 2022), and parts of the High Atlas, usually between 1,400 and 2,400 meters (Benabid, 2000). It represents one of the most emblematic forest ecosystems of North Africa (Boudy, 1950), combining ecological functions such as soil protection and water regulation with socio-cultural symbolism, as well as recreational and landscape value. The Moroccan fir (*Abies marocana*), a narrow endemic restricted to the Rif Mountains, belongs to the Pinaceae family (Fennane & Ibn Tattou, 2012) and is considered a Tertiary relict species of exceptional biogeographic importance. With a distribution limited to less than 4,500 hectares, it constitutes one of the rarest and most remarkable conifers of the Mediterranean basin. The cork oak (*Quercus suber*), widely distributed across western Mediterranean countries, attains in Morocco some of its most extensive lowland formations in the Maâmora forest (Benabid, 2000; Natividade, 1956). As a keystone species, it provides economic benificts through cork production and ecological services by preserving biodiversity, sequestering carbon, and soil stabilization. The holm oak (*Quercus rotundifolia*), Morocco's most widespread forest species covering about 1.4 million hectares, belongs to the Fagaceae family. Its ecological flexibility allows it to establish itself on a broad spectrum of bioclimatic gradients, from semi-arid plains to humid montane areas, to provide the structural background of many landscapes. It is usually associated with pastoral systems and traditional land uses, which emphasizes its ecological flexibility and central role in Mediterranean forest ecology. The Atlas pistachio (*Pistacia atlantica*), a drought-tolerant deciduous tree of the Anacardiaceae family, reaching up to 20 meters in height, represents one of the most characteristic woody species of the arid and semi-arid regions of North Africa. In Morocco, the subspecies *atlantica* persists mainly in the Oriental high plateaus and parts of the Middle Atlas foothills, where it historically formed extensive pre-forest climax communities, often covering approximately 607,900 ha. This drought-tolerant species tolerates extremely low annual rainfall (150 mm or less) and plays a key ecological role in structuring open woodland and steppe-forest mosaics, often in mixture with *Tetraclinis articulata* (Benabid & Fennane, 1994). Finally, the Barbary thuja (*Tetraclinis articulata*), endemic to the western Mediterranean, currently occupies approximately 607,900 ha in Morocco, primarily in semi-arid temperate and warm bioclimatic zones. This coniferous species is distributed across the Mediterranean slope of the Rif, the Oriental, the central Middle Atlas, valleys of the Central Plateau and western Meseta, as well as the High Atlas and Anti-Atlas, typically below 1,400 m elevation

(Benabid, 2000). Unique among Moroccan conifers, it exhibits remarkable coppicing ability and natural regeneration capacity under favorable ecological conditions. It frequently forms mixed stands with various species supporting a rich understory flora and diverse arthropod fauna.

These seven species were thus selected to stay on as model taxa since they collectively exemplify a broad spectrum of ecological conditions, from xeric to humid cold environments, and exhibit different adaptive strategies. They also play a central role in sustaining local livelihoods, from the provision of timber, cork, fuelwood, and non-timber production to cultural, recreational, and regulating ecosystem services. In addition, the fact that they occur across a broad range of different bioclimatic environments makes them suitable for investigation of differential climate change responses in south-Mediterranean forest ecosystems. By focusing on this sample group of taxa, the research aims to highlight the variety of responses to climatic stress to be expected from North African forests.

Taken together, these ecosystems illustrate the multiple pressures affecting Moroccan forest ecosystems. Climate change acts as a major driver, intensifying drought, heat stress, and wildfire risk, while human activities such as overgrazing, wood harvesting, and land conversion exacerbate their vulnerability (Mharzi Alaoui et al., 2017; Rifai et al., 2018). This convergence of climatic and anthropogenic threats compromise forest ecosystems resilience, undermines their capacity to regenerate, and threatens both biodiversity and the ecosystem services on which rural communities depend.

Modeling approach

Species Distribution Modeling (SDM) represents a widely used approaches in ecological forecasting, conservation planning, risk assessment, biogeography, and species management (Franklin, 2009; Guisan et al., 2017). It establishes statistical relationships between known occurrences of a species and the associated environmental conditions, then predict its potential distribution across unsampled areas and under future scenarios (Franklin, 2009; Phillips et al., 2006). This approach has been extensively applied to evaluate potential range shifts under climate change (Franklin, 2009; Guisan et al., 2017),, with techniques ranging from correlative to mechanistic models, and from classical statistical approaches to machine learning methods (Franklin, 2009; Guisan et al., 2017). The maximum entropy algorithm (MaxEnt) was adopted by using the "MaxEnt 3.3. 3k" algorithm for training SDM, which has been demonstrated as one of the most performing methods for presence-only data (Merow et al., 2013; Phillips et al., 2006). MaxEnt requires relatively few occurrence records to build reliable models, can accommodate both continuous and categorical predictors, and shows resilience to spatial errors in location data (Elith

& Graham, 2009). Additionally, it provides measures of variable importance and model performance, making it particularly suitable for assessing species vulnerability to climate change (Merow et al., 2013). Such modeling is particularly relevant in the Mediterranean context, where sharp environmental gradients and strong anthropogenic pressures interact to shape species distributions. Applied to Morocco, this framework allows not only an assessment of the vulnerability of these emblematic species but also a broader understanding of how climate change may reconfigure forest ecosystems across the southern Mediterranean.

However, SDM-based projections are inherently subject to multiple sources of uncertainty that must be acknowledged to avoid overly deterministic interpretations. These uncertainties arise from limitations in occurrence data quality and completeness, assumptions of species–climate equilibrium, potential omission of biotic interactions and dispersal constraints, and the correlative nature of the modeling approach. Consequently, the resulting maps should be interpreted as probabilistic indications of potential suitability under given assumptions, rather than exact predictions of future distributions.

Datasets used

The majority of SDM applications rely on biological data (precise geographic coordinates of species occurrences) combined with environmental variables. Occurrence data were primarily extracted from the National Forest Inventory (NFI, 2005), and complemented by additional information derived from forest management studies and field surveys. These complementary sources were essential to verify and refine the inventory, especially in areas where species are reported in the literature but not systematically mapped. To reduce sampling bias and spatial autocorrelation, a single occurrence point was retained per grid cell of the environmental dataset (Guisan & Zimmermann, 2000). Environmental data included nineteen bioclimatic variables considered biologically meaningful to define the eco-physiological needs of plant species. These variables indicate mean, seasonal, and extreme monthly of climatic conditions and were generated from WorldClim v2.1 dataset (Fick & Hijmans, 2017). This dataset provides interpolated climate surfaces at 30 seconds spatial resolution ($\approx 1\ km^2$ cell size). Elevation was also included as a key biophysical predictor. Future climate conditions were derived from an ensemble of 18 General Circulation Models (GCMs) used by the IPCC. Two Representative Concentration Pathways (RCP2.6 and RCP8.5), corresponding to optimistic and pessimistic greenhouse gas trajectories (Kriegler et al., 2012), were used for long-term projections (2070s). The average maps of GCMs for each each scenario were calculated to limit model-specific uncertainties (Gleckler et al., 2008; Knutti et al., 2010; Pierce et al.,

2009). This ensemble approach reduces variance and provides a more robust basis for predicting potential shifts in species distributions under climate change.

Model training, validation and data analysis

Default parameters of MaxEnt algorithm were applied for background selection, number of iterations, and feature types. Occurrence points were randomly divided into two subsets: one used for model training and the other for model testing, ensuring independent evaluation. Model performance was assessed through cross-validation and statistical metrics. The area under the receiver operating characteristic curve (AUC) and the True Skill Statistic (TSS) were calculated to quantify the ability of the model to discriminate suitable versus unsuitable areas (Allouche et al., 2006; Hanley & McNeil, 1982). The "equal training sensitivity and specificity" threshold was adopted to generate binary presence–absence maps, as it provides a balanced compromise between omission and commission errors. To identify the relevant predictors of species distribution, relative contribution, permutation importance and response curves of each predictor were examined, and the Jackknife test was applied (Miller, 1974).

To analyze spatial dynamics under climate change, suitability maps were generated for both current and future conditions. Habitat changes (stable, lost, gain of suitable areas and unsuitable areas) were quantified under RCP2.6 and RCP8.5 for the 2070s. All spatial processing, mapping, and zonal statistics were performed with R software and Quantum GIS (QGIS).

RESULTS AND DISCUSSION

Model performance and variables' contribution

This study capitalizes on our previous work on key North African forest species (Benabou, Moukrim, Lahssini, et al., 2022; Moukrim et al., 2019, 2020, 2022) and it is further complemented by new results, especially for holm oak, providing a broader understanding and a more representative overview of forest ecosystems across the region. Analysis of the predictive models developed with the MaxEnt algorithm showed a high predictive performance across all studied species. Both the AUC and TSS indicated excellent accuracy in distinguishing suitable from unsuitable habitats. The consistently high values of these metrics confirm the robustness of the models and their reliability for predicting the current and future distributions of the selected tree species. The assessment of variable importance, based on relative contribution, permutation importance, and response curves, revealed that only a restricted set of environmental variables provided the most useful information in explaining the distribution patterns. The Jackknife test confirmed the influence of these variables by demonstrating the highest training gain when used in isolation, as well as the strongest decrease in model performance when omitted. The most influential predictors for those species were: i) Precipitation of the coldest quarter, temperature seasonality, and precipitation of the wettest month for *Argania spinosa*; ii) Precipitation of the wettest quarter and precipitation of the coldest quarter for *Abies marocana*; iii) Precipitation of the coldest quarter, precipitation of the wettest quarter, and temperature annual range for *Quercus suber*; iv) Precipitation of the driest quarter, precipitation of the wettest month, mean diurnal range, and altitude for *Cedrus atlantica*; v) Annual precipitation and altitude were the most decisive predictors for *Quercus rotundifolia*; vi) Precipitation of coldest quarter, precipitation of wettest month and temperature seasonality for *Pistacia atlantica*; vii) Precipitation of coldest and warmest quarters, and precipitation of wettest month for *Tetraclinis articulata*.

Other variables, while included in the models, contributed only marginally and did not significantly improve predictive performance. The dominance of precipitation-related variables highlights the strong climatic dependence of these north African species, where seasonal and inter-annual rainfall regimes exert a primary control on their ecological niches. Likewise, the importance of altitude in explaining cedar and holm oak distributions reflects the influence of topography on climate.

Potential current and future distributions

The training of the different models allowed us to predict and map the potential ranges of the studied species under both current and future (optimistic and pessimistic scenarios for 2070) climatic conditions. The predicted current distributions largely confirmed the ecological affinity of the species with the territories where they are actually present, as reported in the National Forest Inventory (NFI, 2005), forest management studies, and the existing literature. However, the modeled suitable areas were generally broader than the actual occurrence ranges (Table 1). For instance, while *Quercus rotundifolia* currently occupies approximately 1.4 million ha, its potential suitable area reaches over 2 million ha. Similarly, *Tetraclinis articulata* shows the largest discrepancy, with a current distribution of 545,450 ha compared to a potential suitable area exceeding 3.1 million ha, and *Pistacia atlantica* displays 607,900 ha of current distribution against 750,969 ha of potential suitable habitat. This suggests that additional regions of Morocco are bioclimatically suitable for those species, even though they have not colonized them. Such discrepancies may reflect anthropogenic constraints, including land use allocation (agriculture, urbanization) and the legal framework regulating forest domains, but also biotic interactions, dispersal barriers, and historical factors that limit natural expansion (Moukrim et al., 2025). The overlap between predicted suitable areas and actual occurrence points further demonstrated strong consistency, with most known occurrences falling within the zones of highest suitability. These results highlight the potential of SDM to guide further ecological research, including paleoecological investigations, to better understand historical dynamics and past distributions (e.g., Atlas cedar in the southwestern High Atlas; (Reille, 1976)).

Future projections indicate that the distribution of suitable habitats is highly site-specific, species-specific and strongly dependent on the climate scenario. Under both RCPs, suitable habitats are projected to shrink, with more pronounced reductions under the pessimistic scenario (Table 1). Under the optimistic scenario (RCP2.6), projected suitable areas range from 27.3% (*Q. rotundifolia*) to 89.9% (*C. atlantica*) of current potential distribution, while the pessimistic scenario (RCP8.5) reveals dramatic reductions, with suitable areas declining to only 5.8% for *Q. rotundifolia*, 6.9% for *C. atlantica*, and complete habitat loss (0%) for both *A. marocana* and *Q. suber* in Maamora forest. Notably, *T. articulata* and *P. atlantica* show intermediate vulnerability, retaining 31.3% and 25.7% of their potential suitable areas, respectively, under RCP8.5. Suitable areas are expected to decline significantly, partly compensated by an expansion of moderately suitable zones. This pattern suggests a shift from optimal to suboptimal habitats. Moreover, projected changes point to both gains and losses in suitable areas, with overall contraction and increasing fragmentation as the climate scenario shifts from optimistic to pessimistic. These projections are consistent with historical evidence of range contractions since the mid-20th century (Boudy, 1950) and align with global patterns of latitudinal and altitudinal shifts in species distributions under climate change (Hellmann et al., 2008; Parmesan & Yohe, 2003).

Climate change impact on potential distributions

By overlaying the predicted suitable areas for the studied species under future scenarios (optimistic and pessimistic, projected to 2070) with those derived from current climatic conditions, it was possible to assess the likely impacts of climate change on their potential ranges (Figure 1). The analysis indicates predominantly negative, and only rarely positive, impacts on the extent of suitable habitats (Table 1). Under the optimistic scenario (RCP2.6), habitat losses range from 15.4% for *C. atlantica* to 73% for *Q. rotundifolia*, while the pessimistic scenario (RCP8.5) reveals catastrophic contractions, with losses reaching 94% for *C. atlantica* and *Q. rotundifolia*, and complete habitat elimination (100%) for *A. marocana* and *Q. suber*. In particular, the future distributions of suitable areas appear to be strongly dependent on the intensity of climate change, with the precise location and magnitude of these changes varying across the territory.

A closer examination of the projections highlights clear spatial shifts, with significant contractions of currently suitable areas and, to a much lesser extent, the emergence of new favorable zones. For instance, under RCP2.6, stable suitable areas represent 84.6% for *C. atlantica*, 50.9% for *T. articulata*, and 44.7% for A. spinosa, with minimal habitat gains (new areas ranging from 0% to 5.3%). Under the severe RCP8.5 scenario, stable areas collapse dramatically to 21.7% for *T. articulata*, 19.3% for *P. atlantica*, and only 6% for *C. atlantica*. Notably, T. articulata shows the highest potential for new suitable areas under RCP8.5 (+9.4%, representing 295,500 ha), followed by *A. spinosa* (+12.5%, 327,139 ha), though these gains remain insufficient to compensate for massive losses of 78.3% and 72.1%, respectively. However, these new areas are limited and unlikely to offset the anticipated losses (Figure 1 and Table 1).

The ability of species to persist in stable suitable areas or to colonize new areas will largely depend on their ecological and genetic adaptability, particularly phenotypic plasticity and genetic resilience that may allow some populations to persist despite unfavorable climatic shift, as well as their dispersal capacity, and the extent to which anthropogenic pressures or land-use changes facilitate or constrain their establishment. Beyond direct climatic effects, these projections likely underestimate real vulnerability, as they do not explicitly account for synergistic stressors such as the interaction between rising temperatures and the increasing frequency of forest fires, prolonged droughts, or pathogen outbreaks. Such interacting pressures can amplify ecosystem stress, accelerate population decline, and trigger ecosystem collapse well before critical climatic thresholds are reached.

For some species with restricted ranges, particularly endemics, the projected loss of suitable habitat raises concerns about local extinctions and the long-term erosion of ecosystem resilience. This risk is further exacerbated under scenarios where climate-induced habitat contraction coincides with disturbance-driven mortality, reducing recovery potential and adaptive responses. The complete projected disappearance of suitable habitat for *A. marocana* and *Q. suber* under RCP8.5 exemplifies the critical vulnerability of narrow endemics and economically valuable species, respectively.

Figure 1. Habitat suitability changes by 2070 under climate change: optimistic (RCP2.6) and pessimistic (RCP8.5) scenarios. a, b Argania spinosa; c, d Cedrus atlantica; e, f Abies marocana; g, h Quercus suber; i, j Quercus rotundifolia; k, l Pistacia atlantica; m, n Tetraclinis articulata

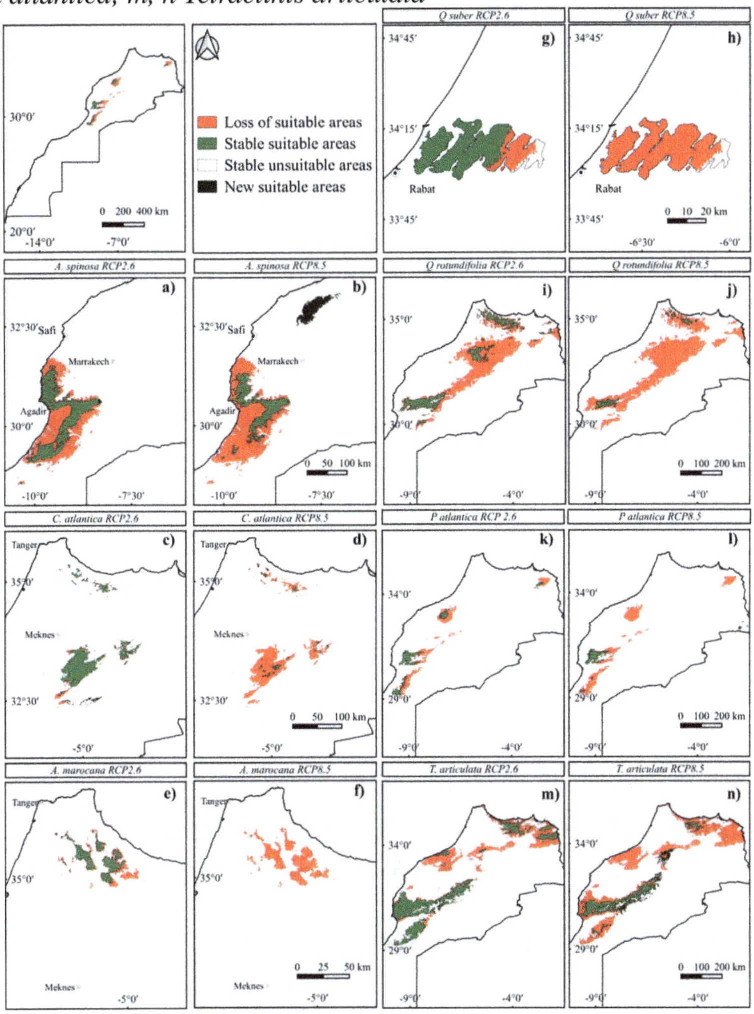

Table 1. Current and future suitable areas (ha) and predicted habitat suitability change for studied species

	Current distribution		Futur distribution 2070							
	Actuel Occurrence	Potential distribution	Optimistic climate scenarios RCP2.6				Pessimistic climate scenarios RCP8.5			
			Total	Stable	New	Loss	Total	Stable	New	Loss
Current study										
Quercus rotundifolia	1 415 000	2 020 827	551 659 (27,3)	544 430 (26,9)	7 229 (+0,4)	1 476 397 (-73)	117 812 (5,8)	112 405 (5,6)	5 407 (+0,3)	1 908 421 (-94,4)
Pistacia atlantica	607 900	750 969	307 113 (40,9)	270 789 (36,1)	36 323 (+4,8)	480 180 (-63,9)	192 654 (25,7)	144 779 (19,3)	47 875 (+6,4)	606 190 (-80,7)
Tetraclinis articulata	545 450	3 146 928	1 680 160 (53,4)	1 600 600 (50,9)	79 560 (+2,5)	1 546 327 (-49,1)	979 371 (31,3)	683 871 (21,7)	295 500 (+9,4)	2 463 057 (-78,3)
Previous study										
Argania spinosa (Moukrim et al., 2019)	871 210	2 616 717	1 178 522 (45)	1 169 674 (44,7)	8 848 (+0,3)	1 447 043 (- 55,3)	1 058 284 (40,4)	731 145 (27,9)	327 139 (+12,5)	1 885 572 (- 72,1)
Cedrus atlantica (Moukrim et al., 2020)	133 000	662 653	595 739 (89,9)	560 742 (84,6)	34 997 (+5,3)	101 911 (-15,4)	45 978 (6,9)	39 579 (6)	6 399 (+1)	623 074 (-94)
Abies marocana (Moukrim et al., 2022)	4 500	130 460	76 740 (58,9)	76 740 (58,9)	0 (0)	53 630 (-41,1)	0 (0)	0 (0)	0 (0)	130 460 (-100)
Quercus suber (Maamora forest) (Benabou et al., 2022)	70 000	121 727 (92,2% of the forest)	94 015 (77,23)	94 015 (77,23)	0 (0)	27 712 (-22,77)	0 (0)	0 (0)	0 (0)	121 727 (-100)

Notes: For future habitat suitability, the percent changes from current climate predictions are provided in parenthesis.

Vulnerability of Southern Mediterranean Forest Ecosystems to Climate Change and Management Implications

Differential Species Vulnerability and Projected Impacts

Southern Mediterranean Forest ecosystems, including those of Morocco, are already functioning at the edge of their ecological tolerance. Increasing aridity and recurrent anthropogenic pressures make them among the most vulnerable systems in the Mediterranean basin (Giorgi & Lionello, 2008). Species distribution models of seven emblematic North African species reveal significant contractions of suitable habitats under future scenarios, with strong dependence on warming severity. Under the optimistic scenario (RCP2.6), habitat losses range from 15.4% for *Cedrus atlantica* to 73% for *Quercus rotundifolia*, while the pessimistic scenario (RCP8.5) projects catastrophic contractions of 72.1% to 100%, with two species

(*Abies marocana* and *Quercus suber*) facing complete elimination of climatically suitable habitat. While some new climatically favorable areas may emerge (less than 12.5% of potential distribution), these gains remain insufficient to compensate for widespread losses, particularly for species with narrow ranges or restricted dispersal capacities (Engler et al., 2011).

The projected impacts vary among species, reflecting differences in ecological niches, geographic ranges, and adaptive capacities. Critically vulnerable species face near-total or complete habitat loss under severe scenarios. *Abies marocana*, already restricted to less than 4,500 ha in the Rif Mountains, is projected to lose 41.1% of suitable habitat under RCP2.6 and experience complete habitat elimination under RCP8.5, with no compensatory gains in either scenario. This signals imminent extinction risk for one of the Mediterranean's rarest conifers and a Tertiary relict of exceptional biogeographic value. Similarly, *Quercus suber* in the Maamora forest, despite its socio-economic importance, is projected to retain 77.2% of suitable habitat under RCP2.6 but suffer total habitat loss in this area under RCP8.5, threatening collapse of one of Morocco's most valuable cork-producing ecosystems in Maamora forest.

Highly vulnerable species experience dramatic range contractions exceeding 90% under pessimistic scenarios. *Cedrus atlantica*, relatively resilient under RCP2.6 (retaining 89.9% of suitable habitat with 5.3% (34,997 ha) new areas), collapses to only 6.9% suitable habitat under RCP8.5, representing a 94% loss. *Quercus rotundifolia*, currently Morocco's most widespread forest species, faces a 73% reduction under RCP2.6 and a devastating 94.4% loss under RCP8.5, with minimal gains (<0.4%) insufficient to offset losses. This massive contraction threatens the structural foundation of Mediterranean forest landscapes and associated pastoral systems.

Moderately vulnerable species show intermediate responses. *Pistacia atlantica* retains 40.9% and 25.7% of suitable habitat under RCP2.6 and RCP8.5 respectively, with modest gains of 36,323 ha and 47,875 ha. Tetraclinis articulata demonstrates the highest retention capacity, maintaining 53.4% under RCP2.6 (1,600,600 ha stable) and 31.3% under RCP8.5, with the most significant new habitat opportunities (9.4%, representing 295,500 ha). *Argania spinosa* retains 45% under RCP2.6 and 40.4% under RCP8.5, with the highest expansion potential (12.5%, representing 327,139 ha). However, even for these adaptable species, losses (49.1% to 80.7%) far exceed gains. The stark difference between scenarios; transforming manageable shifts into catastrophic collapses, underscores the critical importance of global climate mitigation efforts (Scholes & Biggs, 2005).

Cascading Ecological and Socio-Economic Consequences

These projected habitat changes generate profound cascading effects extending beyond individual species to reshape entire ecosystems and threaten human well-being. Habitat contractions and spatial shifts accentuate fragmentation, erode ecological connectivity, and increase vulnerability to pests, diseases, and fires (Dale et al., 2001). Loss of dominant or keystone species fundamentally alters ecosystem structure, with potential regime shifts toward degraded or non-forested states. For *Q. rotundifolia*, massive range contractions imply transformation of extensive Mediterranean landscapes into scrubland mosaics, altering fire regimes, increasing erosion, reducing carbon sequestration, and diminishing ecosystem resilience. Complete projected disappearance of *A. marocana* under RCP8.5 represents not only species extinction but potential collapse of associated montane ecosystems, including specialized flora and fauna dependent on unique microclimatic conditions. The limited emergence of new suitable areas creates additional challenges. Spatial dislocation, with losses in current core areas and gains in distant or fragmented patches, creates migration barriers most tree species cannot overcome within projected climate change timeframes. Dispersal limitations, soil constraints, competing land uses, and anthropogenic barriers further constrain natural colonization, increasing extinction risk and threatening protected area network integrity (Hannah et al., 2007).

Beyond ecological impacts, forest service degradation directly threatens rural communities dependent on fuelwood, grazing, timber, cork, and non-timber forest products. Complete projected loss of *Q. suber* habitat in Maamora forest under RCP8.5 threatens one of Morocco's most economically valuable ecosystems. Cork production, sustaining thousands of jobs in harvesting, processing, and export, faces potential collapse. The forest also provides critical regulating services, groundwater recharge, erosion control, carbon storage, whose loss would generate regional hydrological and land degradation consequences. Similarly, *A. spinosa* supports approximately 2.2 million people through argan oil production, a globally recognized high-value product. Despite relatively higher adaptive capacity (Timzioura et al., 2025), habitat losses of 72.1% still represent dramatic reductions threatening traditional argan-based economies and cultural heritage of this UNESCO-recognized biosphere reserve.

Weakening of regulating functions, particularly hydrological control, erosion prevention, and carbon storage, amplifies land degradation and desertification risks. Forest cover loss reduces infiltration, increases runoff, and accelerates erosion, creating feedback loops degrading land productivity. In semi-arid and arid environments already water-stressed, loss of forest-mediated hydrological regulation threatens agricultural systems, urban water supplies, and ecosystem stability. Moreover, degraded forests become more fire-susceptible, creating positive feedbacks between climate change, forest degradation, and fire risk that further reduce resilience, accelerate biodiversity loss, and release stored carbon.

Adaptive Management and Conservation Strategies

Confronted with these challenges, conventional forest management strategies based on historical distributions and past climatic stability appear increasingly inadequate. Climate change introduces uncertainties requiring adaptive, forward-looking, and proactive approaches (Moukrim et al., 2019). Species distribution modeling, as developed in this work, offers critical decision-support tools extending beyond occurrence records or forest inventories, providing spatially explicit guidance for anticipating climate impacts and designing adaptive responses. In this context, the translation of modeled outcomes into geographically targeted management actions is essential to support evidence-based Mediterranean forest governance.

Priority conservation efforts should focus on protecting stable areas, those regions projected to remain climatically suitable across scenarios. For *C. atlantica*, intensive management should target the stable areas under RCP2.6 and particularly the core areas stable even under RCP8.5, representing the most resilient populations. For *A. marocana*, the projected suitable under RCP2.6 should be designated strict conservation zones with enhanced protection, genetic monitoring, and ex-situ conservation programs through seed banks, botanical gardens, and clonal archives ensuring genetic diversity preservation. Given the projected losses of suitable habitat under different scenarios, these areas should be prioritized as sources for assisted migration and as key nodes within integrated, spatially explicit ex-situ in-situ conservation network.

Restoration priorities should target both degraded current habitats and emerging future suitable areas. For *T. articulata* and *A. spinosa*, showing notable habitat gains under RCP8.5, proactive restoration and assisted migration can establish resilient populations in advance of climate-driven range shifts. Such anticipatory interventions require careful site selection prioritizing ecological connectivity, favorable soil conditions, low land-use conflict, and community engagement. Spatial prioritization of these interventions, informed by suitability maps, can help identify potential assisted migration corridors and restoration hotspots, thereby optimizing limited conservation resources. Adaptive silviculture, adjusting stand densities, favoring natural regeneration from locally adapted seed sources, promoting species diversity, can reduce water stress and enhance tree resilience while maintaining genetic diversity and ecosystem functioning.

Ecosystem-based adaptation emphasizes landscape-scale interventions enhancing connectivity, maintaining heterogeneity, and reducing non-climatic stressors. Establishing ecological corridors between current core areas and projected future suitable zones facilitates natural migration and gene flow. These corridors, when explicitly mapped and integrated into land-use planning, can function as climate adaptation pathways at regional scales. Protecting riparian zones, north-facing slopes, and high-elevation areas likely to serve as microclimatic refugia provides buffers against regional warming. Reducing anthropogenic pressures, represented by over-grazing, unsustainable harvesting and land conversion, enhances adaptive capacity. Diversification strategies, including mixed-species stands and agroforestry systems, reduce dependence on vulnerable species while maintaining livelihood benefits and forest cover. Together, these spatially explicit measures transform modeling outputs into actionable decision-support tools, bridging the gap between climate projections and practical forest management under Mediterranean conditions.

Policy Integration and Regional Cooperation

Building resilience requires integrating climate projections into forest policy at national and regional levels, shifting from reactive crisis management to proactive anticipatory governance. Morocco's forest strategies should explicitly incorporate species distribution projections to guide reforestation programs, prioritize climate-resilient species, allocate resources to high-priority conservation zones, and update subsidy schemes rewarding adaptive practices. Multi-sectoral coordination between forest, water, agriculture, and energy sectors is essential given interconnections between climate-driven forest degradation and water availability, erosion, and carbon sequestration.

Community engagement is critical for securing local support. Rural communities must be partners in designing and implementing strategies. Benefit-sharing mechanisms, alternative livelihood programs (ecotourism, sustainable non-timber forest product enterprises), and participatory decision-making align conservation objectives with community needs. For regions dependent on *Q. suber* cork or *A. spinosa* argan oil, anticipatory livelihood diversification reduces socio-economic vulnerability.

Regional cooperation across the Maghreb is essential for shared species forming common ecological and cultural heritage. Transboundary conservation initiatives, harmonized policies, genetic resource sharing through regional seed banks, and knowledge exchange platforms enhance conservation effectiveness and build capacity. International partnerships can mobilize technical expertise and climate finance for large-scale adaptation initiatives.

CONCLUSION

This study provides spatially explicit evidence of the profound vulnerability of southern Mediterranean forest ecosystems to climate change, revealing a future marked by widespread habitat contractions, ecological disruption, and socio-economic stress. Species distribution modeling of seven emblematic North African species projects habitat losses ranging from 15% to 100% depending on species and climate scenario, with critically vulnerable taxa such as *Abies marocana* facing complete elimination of suitable habitat under severe warming (RCP8.5). Even widespread species such as *Quercus rotundifolia* and *Cedrus atlantica* experience catastrophic contractions exceeding 94%. The limited emergence of new favorable zones (less than 12.5% of potential distribution) offers insufficient compensation, underscoring a future trajectory of range collapse, fragmentation, and ecosystem degradation. These projections translate into cascading consequences extending beyond species range shifts. The loss of forest cover threatens essential ecosystem services; carbon sequestration, water regulation, soil protection; underpinning regional stability and human well-being. Millions of rural inhabitant's dependents on cork production, argan oil, fuelwood, and grazing face livelihood disruption, while weakening hydrological functions amplifies desertification risks across water-stressed landscapes.

This diagnosis also illuminates pathways for proactive response. The stark difference between optimistic and pessimistic scenarios demonstrates that the future remains open, contingent upon immediate climate mitigation and local adaptation efforts. Species distribution modeling offers decision-support tools that can guide conservation prioritization, identify climate refugia, and rationalize restoration investments. By integrating these projections into forest policy, promoting adaptive silviculture, enhancing landscape connectivity, and fostering regional cooperation, it is possible to build resilience and secure ecosystem services even under changing climatic conditions. The resilience of southern Mediterranean forests ultimately depends on the capacity and willingness of managers, policymakers, and communities to act decisively on scientific evidence. The Moroccan case illustrates both the magnitude of the climate challenge and the opportunity to transform forest governance into anticipatory, climate-informed stewardship.

REFERENCES

Allouche, O., Tsoar, A., & Kadmon, R. (2006). Assessing the accuracy of species distribution models : Prevalence, kappa and the true skill statistic (TSS): Assessing the accuracy of distribution models. *Journal of Applied Ecology*, *43*(6), 1223–1232. DOI: 10.1111/j.1365-2664.2006.01214.x

Aussenac, G. (1973). Climat, microclimat et production ligneuse. *Annales des Sciences Forestieres*, *30*(3), 239–258. https://www.afs-journal.org/articles/forest/pdf/1973/03/AFS_0003-4312_1973_30_3_ART0003.pdf. DOI: 10.1051/forest/19730303

Benabid, A. (2000). *Flore et écosystèmes du Maroc : Evaluation et préservation de la biodiversité*. Ibis Press & Kalila Wa Dimna. https://agris.fao.org/agris-search/search.do?recordID=XF2015030117

Benabou, A., Moukrim, S., Laaribya, S., Aafi, A., Chkhichekh, A., El Maadidi, T., & El Aboudi, A. (2022). Mapping Ecosystem Services of Forest Stands : Case Study of Maamora, Morocco. *Geography, Environment, Sustainability*, *15*(1), 141–149. DOI: 10.24057/2071-9388-2021-047

Benabou, A., Moukrim, S., Lahssini, S., El Aboudi, A., Menzou, K., Elmalki, M., El Madihi, M., & Rhazi, L. (2022). Impact of the climate change on potential distribution of Quercus suber in the conditions of North Africa. *Biosystems Diversity*, *30*(3), 289–294. DOI: 10.15421/012231

Boudy, P. (1950). *Economie forestière Nord-africaine-Tome 2 : Monographies et traitements des essences forestières*. Edition Larose.

Cheddadi, R., François, L., Bar-Hen, A., Huang, K., Demarteau, M., Roujjati, A., Baali, A., Damnati, B., Taieb, M., & Laurent, J.-M. (2006). Impact du climat sur quelques espèces méditerranéennes : Du passé au futur. *Réponses adaptatives au changement global : résultats et prospective*, 91-96.

Dale, V. H., Joyce, L. A., McNulty, S., Neilson, R. P., Ayres, M. P., Flannigan, M. D., Hanson, P. J., Irland, L. C., Lugo, A. E., Peterson, C. J., Simberloff, D., Swanson, F. J., Stocks, B. J., & Wotton, B. M. (2001). Climate Change and Forest Disturbances : Climate change can affect forests by altering the frequency, intensity, duration, and timing of fire, drought, introduced species, insect and pathogen outbreaks, hurricanes, windstorms, ice storms, or landslides. *Bioscience*, *51*(9), 723–734. DOI: 10.1641/0006-3568(2001)051[0723:CCAFD]2.0.CO;2

Diffenbaugh, N. S., Pal, J. S., Giorgi, F., & Gao, X. (2007). Heat stress intensification in the Mediterranean climate change hotspot. *Geophysical Research Letters*, *34*(11), 2007GL030000. Advance online publication. DOI: 10.1029/2007GL030000

Dinerstein, E., Olson, D., Joshi, A., Vynne, C., Burgess, N. D., Wikramanayake, E., Hahn, N., Palminteri, S., Hedao, P., Noss, R., Hansen, M., Locke, H., Ellis, E. C., Jones, B., Barber, C. V., Hayes, R., Kormos, C., Martin, V., Crist, E., & Saleem, M. (2017). An Ecoregion-Based Approach to Protecting Half the Terrestrial Realm. *Bioscience*, *67*(6), 534–545. DOI: 10.1093/biosci/bix014 PMID: 28608869

Elith, J., & Graham, C. H. (2009). Do they? How do they? WHY do they differ? On finding reasons for differing performances of species distribution models. *Ecography*, *32*(1), 66–77. DOI: 10.1111/j.1600-0587.2008.05505.x

ElMderssa, M., Atlas, S., Ettaqy, A., Mouridi, Z. E., Abbas, Y., & Ibijbijen, J. (2025). Comparative evaluation of three methods for soil organic carbon quantification in Morocco's Middle Atlas forests. *Discover Soil*, *2*(1), 98. Advance online publication. DOI: 10.1007/s44378-025-00127-3

Engler, R., Randin, C. F., Thuiller, W., Dullinger, S., Zimmermann, N. E., Araújo, M. B., Pearman, P. B., Le Lay, G., Piedallu, C., Albert, C. H., Choler, P., Coldea, G., De LAMO, X., Dirnböck, T., Gégout, J.-C., Gómez-García, D., Grytnes, J.-A., Heegaard, E., Høistad, F., … Guisan, A. (2011). 21st century climate change threatens mountain flora unequally across Europe. *Global Change Biology, 17*(7), 2330-2341. DOI: 10.1111/j.1365-2486.2010.02393.x

FAO. (2016). *Global Forest Resources Assessment*. https://www.fao.org/3/a-i4808e.pdf

Fennane, M., & Ibn Tattou, M. (2012). Statistiques et commentaires sur l'inventaire actuel de la flore vasculaire du Maroc. *Bulletin de l'Institut Scientifique, Rabat, section Sciences de la Vie, 34*(1), 1-9.

Fick, S. E., & Hijmans, R. J. (2017). WorldClim 2 : New 1-km spatial resolution climate surfaces for global land areas. *International Journal of Climatology*, *37*(12), 4302–4315. DOI: 10.1002/joc.5086

Franklin, J. (2009). *Mapping species distributions : Spatial inference and prediction*. Cambridge University Press., DOI: 10.1017/CBO9780511810602

Giorgi, F., & Lionello, P. (2008). Climate change projections for the Mediterranean region. *Global and Planetary Change, Mediterranean climate: trends, variability and change, 63*(2), 90-104. DOI: 10.1016/j.gloplacha.2007.09.005

Gleckler, P. J., Taylor, K. E., & Doutriaux, C. (2008). Performance metrics for climate models. *Journal of Geophysical Research*, *113*(D6), 2007JD008972. Advance online publication. DOI: 10.1029/2007JD008972

Guisan, A., Thuiller, W., & Zimmermann, N. E. (2017). *Habitat Suitability and Distribution Models With Applications in R*. Cambridge University Press., DOI: 10.1017/9781139028271

Guisan, A., & Zimmermann, N. E. (2000). Predictive habitat distribution models in ecology. *Ecological Modelling, 135*(2), 147–186. DOI: 10.1016/S0304-3800(00)00354-9

Hallegatte, S., Fay, M., Bangalore, M., Kane, T., & Bonzanigo, L. (2015). *Shock Waves : Managing the Impacts of Climate Change on Poverty*. World Bank Publications., DOI: 10.1596/978-1-4648-0673-5

Hanley, J. A., & McNeil, B. J. (1982). The meaning and use of the area under a receiver operating characteristic (ROC) curve. *Radiology, 143*(1), 29–36. DOI: 10.1148/radiology.143.1.7063747 PMID: 7063747

Hannah, L., Midgley, G., Andelman, S., Araújo, M., Hughes, G., Martinez-Meyer, E., Pearson, R., & Williams, P. (2007). Protected area needs in a changing climate. *Frontiers in Ecology and the Environment, 5*(3), 131–138. DOI: 10.1890/1540-9295(2007)5[131:PANIAC]2.0.CO;2

Hellmann, J. J., Byers, J. E., Bierwagen, B. G., & Dukes, J. S. (2008). Five potential consequences of climate change for invasive species. *Conservation Biology, 22*(3), 534–543. DOI: 10.1111/j.1523-1739.2008.00951.x PMID: 18577082

Hidalgo-Triana, N., Solakis, A., Casimiro-Soriguer, F., Choe, H., Navarro, T., Pérez-Latorre, A. V., & Thorne, J. H. (2023). The high climate vulnerability of western Mediterranean forests. *The Science of the Total Environment, 895*, 164983. DOI: 10.1016/j.scitotenv.2023.164983 PMID: 37353024

IPCC. (2021). Summary for Policymakers. In Allan, R. P., Cassou, C., Chen, D., Cherchi, A., Connors, L., Doblas-Reyes, F. J., Douville, H., Driouech, F., Edwards, T. L., Fischer, E., Flato, G. M., Forster, P., AchutaRao, K. M., Adhikary, B., Aldrian, E., & Armour, K. (Eds.), *Climate Change 2021 : The Physical Science Basis. Contribution of Working Group I to the Sixth Assessment Report of the Intergovernmental Panel on Climate Change* (p. 33). Cambridge University Press., DOI: 10.1017/9781009157896.001

IPCC. (2023). *Climate Change 2022 – Impacts, Adaptation and Vulnerability : Working Group II Contribution to the Sixth Assessment Report of the Intergovernmental Panel on Climate Change* (1re éd.). Cambridge University Press. DOI: 10.1017/9781009325844

Knutti, R., Abramowitz, G., Eyring, V., Gleckler, P. J., Hewitson, B., & Mearns, L. (2010). Good practice guidance paper on assessing and combining multi model climate projections. In T. F. Stocker, D. Qin, G.-K. Plattner, M. Tignor, & P. M. Midgley (Éds.), *Meeting Report of the IPCC Expert meeting on assessing and combining multi model climate projections* (p. 1-11). IPCC Working Group I Technical Support Unit, University of Bern. https://academicjournals.org/journal/JENE/article-abstract/C1CDB822968

Kriegler, E., O'Neill, B. C., Hallegatte, S., Kram, T., Lempert, R. J., Moss, R. H., & Wilbanks, T. (2012). The need for and use of socio-economic scenarios for climate change analysis : A new approach based on shared socio-economic pathways. *Global Environmental Change*, *22*(4), 807–822. DOI: 10.1016/j.gloenvcha.2012.05.005

Lionello, P. (Ed.). (2012). *The climate of the Mediterranean region : From the past to the future* (1st ed.). Elsevier.

Medail, F., & Quezel, P. (1999). Biodiversity hotspots in the Mediterranean Basin : Setting global conservation priorities. *Conservation Biology*, *13*(6), 1510–1513. DOI: 10.1046/j.1523-1739.1999.98467.x

Merow, C., Smith, M. J., & Silander, J. A.Jr. (2013). A practical guide to MaxEnt for modeling species' distributions : What it does, and why inputs and settings matter. *Ecography*, *36*(10), 1058–1069. DOI: 10.1111/j.1600-0587.2013.07872.x

Mharzi Alaoui, H., Assali, F., Rouchdi, M., Tahiri, D., Lahssini, S., Aafi, A., & Moukrim, S. (2017). Cartographie de l'intensité du feu et de la sensibilité aux feux de cimes en forêts méditerranéennes. Cas de la province de Chefchaouen au nord du Maroc. [Terre et Vie]. *Revue d'Écologie*, *72*(4), 387–409. DOI: 10.3406/revec.2017.1900

Miller, R. G. (1974). The jackknife-a review. *Biometrika*, *61*(1), 1–15. DOI: 10.1093/biomet/61.1.1

Miranda, A., Syphard, A. D., Berdugo, M., Carrasco, J., Gómez-González, S., Ovalle, J. F., Delpiano, C. A., Vargas, S., Squeo, F. A., Miranda, M. D., Dobbs, C., Mentler, R., Lara, A., & Garreaud, R. (2023). Widespread synchronous decline of Mediterranean-type forest driven by accelerated aridity. *Nature Plants*, *9*(11), 1810–1817. DOI: 10.1038/s41477-023-01541-7 PMID: 37845335

Moukrim, S., Lahssini, S., Rhazi, M., Menzou, K., El Madihi, M., Rifai, N., Bouziani, Y., Azedou, A., Boukhris, I., & Rhazi, L. (2022). Climate Change Impact on Potential Distribution of an Endemic Species *Abies marocana* Trabut. *Ekologia (Bratislava)*, *41*(4), 329–339. DOI: 10.2478/eko-2022-0034

Moukrim, S., Lahssini, S., Rhazi, M., Mharzi, A. H., Benabou, A., Wahby, I., El Madihi, M., Arahou, M., & Rhazi, L. (2019). Climate change impacts on potential distribution of multipurpose agro-forestry species : Argania spinosa (L.) Skeels as case study. *Agroforestry Systems*, *93*(4), 1209–1219. DOI: 10.1007/s10457-018-0232-8

Moukrim, S., Lahssini, S., Rifai, N., Menzou, K., Mharzi-Alaoui, H., Labbaci, A., Rhazi, M., Wahby, I. W., El Madihi, M., & Rhazi, L. (2020). Modélisation de la distribution potentielle de Cedrus atlantica Manetti au Maroc et impacts du changement climatique. *Bois et Forêts des Tropiques*, *344*(2), 3–16. DOI: 10.19182/bft2020.344.a31888

Moukrim, S., Tellal, J., Lahssini, S., Boudjaj, A., Rhazi, M., Madihi, M. E., Bammou, M. B., Moukrim, M., Saber, E.-R., Wahby, I., Rhazi, L., & Grillas, P. (2025). Long-Time Dynamics in a North African Coastal Ramsar Site, the Sidi Boughaba Lake. *Ekologia (Bratislava)*, *44*(2), 147–159. DOI: 10.2478/eko-2025-0017

Muller, S. D., Daoud-Bouattour, A., Fauquette, S., Bottollier-Curtet, M., Rifai, N., Robles, M., Saber, E.-R., El Madihi, M., Moukrim, S., & Rhazi, L. (2022). Holocene history of peatland communities of central Rif (Northern Morocco). *Geobios*, *70*, 35–53. DOI: 10.1016/j.geobios.2021.12.001

Natividade, J. V. (1956). *Subériculture, édition française de l'ouvrage portugais «Subericultura»*. Ecole Nationale des Eaux et Forêts.

NFI. (2005). *National Forest Inventory*. High Commission for Water, Forests and Combating Desertification.

Parmesan, C., & Yohe, G. (2003). A globally coherent fingerprint of climate change impacts across natural systems. *Nature*, *421*(6918), 37–42. DOI: 10.1038/nature01286 PMID: 12511946

Phillips, S. J., Anderson, R. P., & Schapire, R. E. (2006). Maximum entropy modeling of species geographic distributions. *Ecological Modelling*, *190*(3-4), 231–259. DOI: 10.1016/j.ecolmodel.2005.03.026

Pielke, R. J.Jr, Burgess, M. G., & Ritchie, J. (2022). Plausible 2005–2050 emissions scenarios project between 2 °C and 3 °C of warming by 2100. *Environmental Research Letters*, *17*(2), 024027. DOI: 10.1088/1748-9326/ac4ebf

Pierce, D. W., Barnett, T. P., Santer, B. D., & Gleckler, P. J. (2009). Selecting global climate models for regional climate change studies. *Proceedings of the National Academy of Sciences of the United States of America*, *106*(21), 8441–8446. DOI: 10.1073/pnas.0900094106 PMID: 19439652

Reille, M. (1976). Analyse pollinique de sédiments postglaciaires dans le Moyen Atlas et le Haut Atlas marocains : Premiers résultats. *Ecologia Mediterranea*, *2*(1), 154–172. DOI: 10.3406/ecmed.1976.925

Rhanem, M. (2011). Aridification du climat régional et remontée de la limite inférieure du cèdre de l'Atlas (Cedrus atlantica Manetti) aux confins de la plaine de Midelt (Maroc). *Physio-Géo. Géographie physique et environnement, 5*(1), 143-165. DOI: 10.4000/physio-geo.1983

Rifai, N., Khattabi, A., Moukrim, S., Arahou, M., & Rhazi, L. (2018). Évaluation de la dynamique de l'occupation du sol dans la zone humide RAMSAR de Tahaddart (nord-ouest du Maroc). *Revue d'Ecologie (Terre et Vie), 73*(2), 142-152. https://hal.science/hal-03532629/document

Scholes, R. J., & Biggs, R. (2005). A biodiversity intactness index. *Nature*, *434*(7029), 45–49. DOI: 10.1038/nature03289 PMID: 15744293

Serbouti, S., Ettaqy, A., Boukcim, H., Mderssa, M. E., El Ghachtouli, N., & Abbas, Y. (2023). Forests and woodlands in Morocco : Review of historical evolution, services, priorities for conservation measures and future research. *International Forestry Review*, *25*(1), 121–145. DOI: 10.1505/146554823836838745

Stocker, T. F., & Qin, D. Plattner, G.-K., Alexander, L. V., Allen, Simon K., Bindoff, N. L., Bréon, F.-M., Church, J. A., Cubasch, U., Emori, S., & others. (2013). Technical summary. In T.F. Stocker, D. Qin, G.-K. Plattner, M. Tignor, S.K. Allen, J. Boschung, A. Nauels, Y. Xia, V. Bex, & P. M. Midgley (Éds.), *Climate change 2013 : The physical science basis. Contribution of Working Group I to the Fifth Assessment Report of the Intergovernmental Panel on Climate Change* (p. 33-115). Cambridge University Press.

Timzioura, R., Ezzine, S., Benomar, L., Lamhamedi, M. S., Ettaqy, A., Zine El Abidine, A., Zaher, H., Khasa, D. P., Pepin, S., & Abbas, Y. (2025). Bibliometric Analysis of Argan (Argania spinosa (L.) Skeels) Research : Scientific Trends and Strategic Directions for Climate-Resilient Ecosystem Management. *Forests*, *16*(6), 892. DOI: 10.3390/f16060892

Varela, M.-C. (2008). Dépérissement des peuplements de chêne-liège et changement climatique. *forêt méditerranéenne, XXIX*(2), 209-212.

Chapter 2
A Comparative Synthesis of Forest Vulnerability Assessments in the Southern Mediterranean With a Focus on Morocco and Regional Perspectives

Rajaa Timzioura
http://orcid.org/0009-0008-6233-1119
Polydisciplinary Faculty, University Sultan Moulay Slimane, Morocco

Sara Ezzine
Polydisciplinary Faculty, University Sultan Moulay Slimane, Morocco

Abderrahim Ettaqy
http://orcid.org/0000-0003-4662-5770
Polydisciplinary Faculty, University Sultan Moulay Slimane, Morocco

Abdelhakim Taha
Polydisciplinary Faculty, University Sultan Moulay Slimane, Morocco

Mohamed El Mderssa
http://orcid.org/0000-0003-1643-7218
Polydisciplinary Faculty, University Sultan Moulay Slimane, Morocco

Younes Abbas
http://orcid.org/0009-0003-8681-2172
Polydisciplinary Faculty, University Sultan Moulay Slimane, Morocco

ABSTRACT

The vulnerability of Mediterranean forests results from the combined effects of climatic stressors and human pressures. In Morocco, recurrent droughts, wildfires,

DOI: 10.4018/979-8-2600-0783-9.ch002

and rapid land-use change interact with complex governance challenges. This chapter assesses the condition and trends of major forest types such as argan, cedar, cork oak, thuya, and juniper, while placing Morocco in a regional context that includes Algeria, Tunisia, and Spain. Vulnerability is analyzed through an integrated framework linking exposure, sensitivity, and adaptive capacity using ecological and institutional indicators. Despite advances in remote sensing, modeling, and genetic-based restoration, progress is limited by fragmented information systems and weak incorporation of local knowledge. Strengthening resilience requires better interinstitutional coordination, adaptive silviculture, improved data governance, and innovative financing. A strategic roadmap outlines priorities for multi-scale monitoring and a Mediterranean forest resilience toolbox.

INTRODUCTION

Located at the intersection of Africa, Asia, and Europe, the Mediterranean Basin is a global biodiversity hotspot, harboring nearly 500 native tree species, including more than 130 endemic taxa across its botanical territories (Fady et al., 2025). Traditionally recognized as a global biodiversity hotspot (Médail and Diadema, 2006), it has increasingly emerged as one of the world's most vulnerable regions to climate change (Fady et al., 2025). This dual status stems from the convergence of multiple stressors: intensifying human activities, the rising frequency and severity of wildfires, accelerating habitat fragmentation, and the expansion of hotter and drier climatic regimes (Ilmen and Benjelloun, 2013). The northward displacement of arid and semi-arid bioclimatic zones, often described as a climate shift, illustrates the ongoing reorganization of Mediterranean ecosystems and foreshadows major transformations in the coming decades.

Despite these mounting pressures, Mediterranean forest species have historically shown remarkable resilience, relying on combinations of physiological acclimation, phenotypic plasticity, and intraspecific variability to endure environmental changes (Martin-StPaul et al., 2017). Species such as holm oak (*Quercus ilex*) and cork oak (*Q. suber*) exemplify this adaptive capacity (Gimeno et al., 2009; Ramírez-Valiente et al., 2010). A visible expression of these adjustments is the ongoing redistribution of species along thermal and elevational gradients, directly linked to regional climatic displacement (Valladares et al., 2014).

In Morocco, such patterns are manifest in the upward or latitudinal advance of the Atlas cedar (*Cedrus atlantica*) into areas historically dominated by Barbary thuya (*Tetraclinis articulata*), which in turn are increasingly colonized by holm oak (Ilmen and Benjelloun, 2013). This observed migration is likely a response to the deterioration of optimal climatic conditions for certain species, particularly increasing temperatures and reduced water availability, prompting populations to shift toward northern regions where these climatic constraints are less severe.

Morocco provides a particularly instructive setting for examining these challenges. Its forest landscapes, ranging from the humid cedar forests of the Middle Atlas and Rif to coastal cork oak woodlands and the arid argan (*Argania spinosa*) formations of the southwest, combine high ecological diversity with substantial socio-economic importance (ANEF, 2022; Alasli, 2019). However, these ecosystems are increasingly exposed to prolonged drought episodes, recurrent wildfires, habitat fragmentation, and escalating land-use conflicts (Serbouti et al., 2023; Francos et al., 2024). Collectively, these pressures are already reshaping species distributions, regeneration patterns, and the delivery of key ecosystem services, raising significant concerns for national conservation and sustainable management goals. A comprehensive understanding of the factors underlying the fragility of Moroccan forests is essential and should be framed within an integrated approach based on the Know–React–Prevent triptych (Ilmen and Benjelloun, 2013).

To synthesize these complex interactions, a conceptual framework is proposed (Figure 1), illustrating how ecological stressors and human pressures interact with governance fragmentation to amplify forest vulnerability in Morocco. The diagram highlights a self-reinforcing vicious circle driven by weak enforcement and institutional fragmentation (red arrows), while also identifying key resilience levers based on coordinated action across science, institutions, and society (green arrows), ultimately contributing to reduced forest vulnerability and enhanced ecosystem resilience.

Figure 1. Conceptual framework illustrating the vulnerability of Moroccan Mediterranean forests

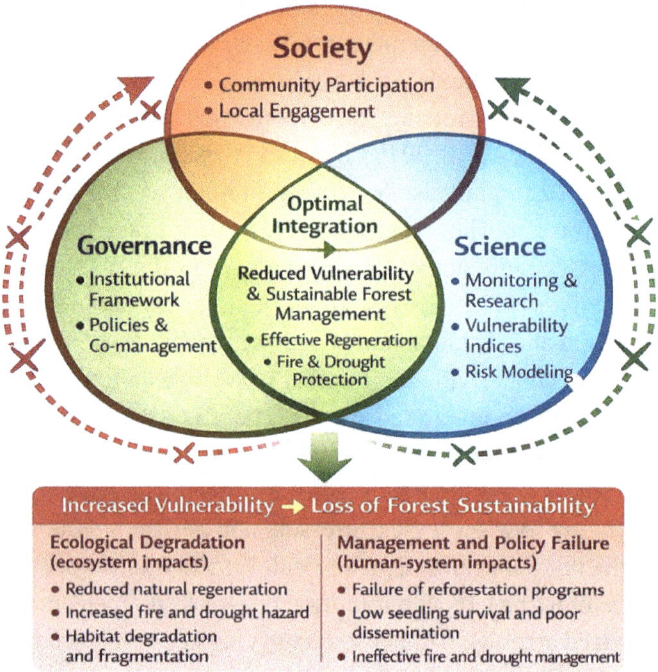

This chapter delivers a national-scale assessment of forest vulnerability in Morocco, based on an integrative review of ecological, institutional, and management indicators, and places it within a broader Southern Mediterranean context. The analysis is organized into three interconnected sections: a synthesis of the current state and trajectories of Morocco's major forest ecosystems, with comparisons to similar conditions in Algeria, Tunisia, and Spain; a critical appraisal of institutional instruments and management frameworks, with particular attention to the "Forests of Morocco 2020–2030" strategy; and the outline of an operational governance framework supported by remote sensing, Geographic Information Systems (GIS), ecological modelling, and machine learning for monitoring, prioritization, and early-warning systems.

MOROCCAN CONTEXT

Morocco's forest ecosystems are shaped by an exceptional diversity of climates, landforms, and biogeographical influences. Situated between the Mediterranean and Atlantic domains and affected by Saharan incursions, the country displays one of the most complex ecological mosaics of the southern Mediterranean (Benabid, 1985; Maliha et al., 2008). Four major mountain ranges, the Rif, Middle Atlas, High Atlas, and Anti-Atlas, act as ecological corridors and barriers, generating strong topoclimatic contrasts and creating a genuine "ecological staircase" from south to north (Ilmen and Benjelloun, 2013; Gharnit et al., 2023). Across this staircase, a succession of bioclimatic belts ranges from Saharan arid conditions to humid montane zones (Emberger, 1955), supporting a high diversity of forest types and plant communities.

Ecological Gradients

With 9 million hectares of forests, pre-forest formations, and steppes, equivalent to 12.7% of the national territory, Morocco holds a prominent place in the Mediterranean biogeographical region (Benabid, 1985; ANEF, 2022). Forest ecosystems predominantly occupy semi-arid, subhumid, and humid climatic conditions (Alaoui et al., 2020) and exhibit strong structural and floristic differentiation aligned with environmental gradients. This spatial variability fuels long-standing debates over the delineation of bioclimatic zones (Emberger, 1939) and altitudinal vegetation belts (Ozenda, 1975; Achhal et al., 1979), concepts that have been used to describe the hierarchical organization of Moroccan vegetation (Benabid, 1985).

This diversity results in the coexistence of contrasting forest formations, from xeric argan (*Argania spinosa*) and thuya (*Tetraclinis articulata*) woodlands in the south to oak, cedar, and fir forests in the more humid mountain belts, each serving as indicators of ecosystem resilience and climatic sensitivity (Benabid & Fennane, 1994). In comparison with neighboring Maghreb countries, Morocco also contributes substantially to regional biodiversity. It supports 112 indigenous tree species from 30 families and 59 genera, including several endemic taxa (Fady et al., 2025); which corresponds to an endemism rate of approximately 3.6%, highlighting the country's ecological significance within the Mediterranean basin.

Anthropogenic Pressures

Although Morocco's ecological gradients confer significant biogeographical richness, they are increasingly challenged by the combined impacts of climate change and anthropogenic pressures. Among these, overexploitation of forest resources, unsustainable grazing, agricultural expansion, informal extraction of timber and non-timber products, and rural demographic pressures are the most prominent drivers of degradation (Alaoui et al., 2020; Serbouti et al., 2023). In particular, grazing pressure in Moroccan forest ecosystems often exceeds sustainable thresholds, with forage extraction rates estimated to be *two to three times higher than the ecologi-*

cal carrying capacity in several regions, leading to poor natural regeneration, soil compaction, and accelerated land degradation (CESE, 2022). As in most Mediterranean landscapes, Moroccan forests are intimately tied to rural livelihoods, serving as critical sources of fuelwood, fodder, medicinal and aromatic plants, and other provisioning ecosystem services (Boubekraoui et al., 2023).

The cumulative effect of these pressures has led to significant forest decline, loss of soil fertility, and the deterioration of hydrological regulation services, with direct consequences for the socio-economic stability of rural populations (Laaribya, 2011). Morocco's forest area has been declining at an average rate of around 17 000 hectares per year, reflecting cumulative drivers including overgrazing, agricultural expansion, and unsustainable wood extraction (CESE, 2022). Several emblematic species now face substantial reductions in their natural ranges, including *Argania spinosa* (Moukrim et al., 2019a; Hakam et al., 2024), *Cedrus atlantica* (Linares et al., 2011; Abel-Schaad et al., 2018), and multiple *Quercus* species (Alaoui et al., 2020). These declines underscore the vulnerability of Moroccan forests to both climatic and human-induced degradation.

Legal and institutional frameworks

Historical foundations and evolution

Morocco's forest governance is rooted in a long legal history dating back to the early 20th century. Under the French and Spanish protectorates (from 1912), communal forest practices, formerly governed by customary systems such as *agdal*, *terroir*, and *jemàa*, were replaced by centralized forest administration (Benzyane, 2007; Aubert, 2013). The foundational Dahir of 1917 nationalized the forests and introduced formal regulations aimed at exploitation control, soil protection, and timber production (Aubert, 2010; Court of Auditors, 2018). Parallel regulations were issued in the Spanish zone from 1918 to 1922.

After independence in 1956, Morocco inherited this colonial framework and progressively adapted it through several legislative milestones, including the Dahir of 1969 on soil restoration and the 1976 Dahir promoting community participation in forest management (Benzyane, 2007; Ellatifi, 2012). From the 1970s onward, forestry policy underwent a paradigm shift, gradually integrating pastoral considerations, reforestation objectives, and ultimately, participatory and ecosystem-based management approaches (CESE, 2022; Benzid & Madani, 2025).

Integration into global and regional environmental agendas

Since the Rio Earth Summit in 1992, Morocco has aligned its national forestry strategies with major international conventions, including the Convention on Biological Diversity (CBD), UNFCCC, Kyoto Protocol, UNCCD, and the Ramsar Convention (Benzyane, 2007; Ghanam, 2003). This alignment facilitated the adoption of significant national planning instruments such as the National Forest Inventory (1996), the Master Reforestation Plan (1997), and the National Watershed Management Plan (1996), each integrating biodiversity, climate, and land restoration considerations (CESE, 2022).

Persistent constraints and governance challenges

Despite these advances, the Moroccan forestry code remains anchored in principles established by the Dahirs of 1917 and 1921 (Aubert, 2013). Experts highlight the need for deeper decentralization, enhanced legal clarity, and more effective integration of local communities in forest decision-making processes (Benzyane et al., 2002; Naggar, 2020; Serbouti et al., 2023). Central to current debates is the question of shared responsibility in forest governance, particularly the redefinition of roles between the State, forest users, civil society, and territorial institutions.

The Dahir of 10 October 1917 still governs Morocco's forests, placing them under state ownership with strict administrative and policing rules. It recognizes certain local use rights (grazing, non-timber products) but does not grant land ownership to rural communities. Despite post-independence reforms promoting participation, state predominance and unclear boundaries have caused conflicts over access and land use, particularly in the Rif.

National Strategies (ANEF, Forests of Morocco 2020–2030)

The launch of the "Forests of Morocco 2020–2030" strategy marked a turning point in national forest governance. This forward-looking framework recognizes that sustainable forest management must simultaneously address ecological integrity, socio-economic development, and institutional modernization (Naggar, 2020; CESE, 2022). Inspired by long-standing recommendations from Moroccan and international scholars advocating participatory and decentralized management models (Benzyane et al., 2002; Aubert, 2010), the strategy positions local communities as central actors in co-managing forest resources.

The strategy is organized around five structural pillars, development, sustainability, participation, productivity, and biodiversity, and is accompanied by significant institutional reforms. The creation of the National Agency for Water and Forests

(ANEF) embodies this modernization effort, replacing previous structures with a streamlined and autonomous institution responsible for implementing forest policy. ANEF's operational mandate includes strengthening partnerships with international organizations (e.g., FAO), promoting public–private partnerships, and developing technical guidelines for the management and restoration of key forest species such as cedar, holm oak, cork oak, and argan tree (CESE, 2022; FAO, 2024).

STRATEGIC INSIGHTS ON MOROCCAN FOREST GOVERNANCE AND SUSTAINABILITY

Morocco illustrates both the challenges and opportunities associated with the management of Mediterranean forest ecosystems. Despite structural, institutional, and environmental constraints, the country has demonstrated sustained political commitment to forest conservation through national initiatives and international partnerships, although effectiveness varies across regions and sectors. Innovation—technological, digital, social, institutional, and financial—remains unevenly integrated, with relative strengths in governance reform and public–private partnerships, but weaknesses in technology adoption, digitalization, and social innovation.

Evaluations of the "Forests of Morocco 2020–2030" strategy reveal a gap between stated ambitions and observed outcomes, as deforestation rates more than doubled between 2020 and 2025. These challenges are further compounded by socio-economic and demographic pressures, including population growth, rural poverty, and recurrent wildfires, while technical and organizational capacity at the community level remains limited.

Moroccan research nevertheless contributes actively to the development of strategies for conservation, restoration, assisted regeneration, and the migration of key forest species such as *Argania spinosa*, *Cedrus atlantica*, and *Abies maroccana*, thereby providing a scientific basis for climate-resilient forest management (Moukrim et al., 2019a; 2020; 2022). Overall, the long-term success of Moroccan forest governance will depend on a systemic approach that integrates institutional reform, technological innovation, scientific knowledge, and community engagement, offering lessons that are also relevant for other Mediterranean countries.

STATE AND TRAJECTORIES OF MAJOR MOROCCAN FOREST ECOSYSTEMS

Figure 2 illustrates the spatial distribution of the major forest ecosystems in Morocco, providing a geographical framework for the ecosystem-specific analyses presented below.

Figure 2. Spatial distribution of the major forest ecosystems in Morocco (NFI, 2018)

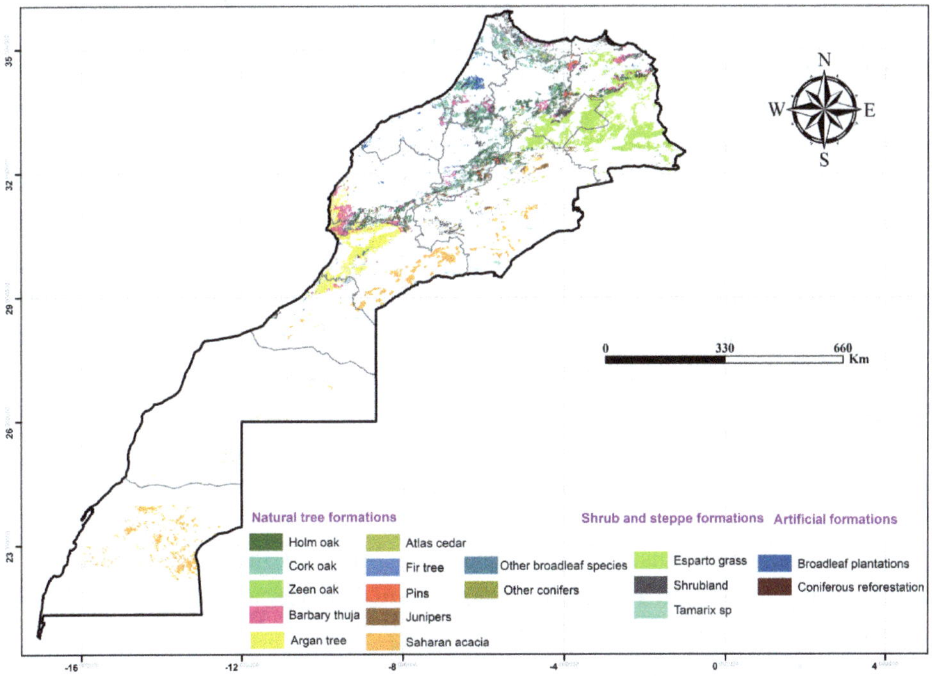

Argan Woodlands (*Argania spinosa L. Skeels*)

Endemic to Morocco, the argan tree (*Argania spinosa L. Skeels*) is one of the country's most emblematic woody species and a cornerstone of its dryland forest landscapes (Emberger, 1925; M'hirit et al., 1998). Although its distribution is concentrated in the southwest, recent analyses confirm that the species extends beyond this core range, occupying a broader set of ecological contexts than previously recognized (Timzioura et al., 2025).

Argan woodlands deliver a wide spectrum of ecosystem services: they help stabilize soils, conserve biodiversity, and buffer microclimates (M'hirit et al.,

1998; Laaribya et al., 2017). Socio-economically, they underpin local livelihoods through the production of argan oil, an activity largely structured around women's cooperatives and territorial value chains (Benchekroun & Buttoud, 1989; Lybbert et al., 2011; Karmoude et al., 2025). In recognition of their ecological and cultural significance, argan forests were designated a UNESCO Biosphere Reserve in 1998.

Despite their xerophytic and thermophilic adaptations (Laaribya et al., 2017), argan landscapes are now subject to intense degradation. Overexploitation of wood and fruit, unsustainable nut collection practices, overgrazing, and rapid land conversion interact with climate warming, recurrent drought, and advancing aridity. Together, these drivers erode the long-term resilience of this agro-silvopastoral system (Laaribya et al., 2017; Moukrim et al., 2019a; Timzioura et al., 2025; Karmoude et al., 2025). In this sense, argan woodlands epitomize the broader degradation trajectories affecting Moroccan and Mediterranean dry forests.

Recent research increasingly relies on remote sensing and ecological niche modelling to quantify degradation trends and anticipate future dynamics of *Argania spinosa*. Quantitative metrics indicate that more than 32% of currently suitable habitats may become unsuitable under future climate scenarios (Moukrim et al., 2019a), with a projected 47–65% reduction in highly suitable areas by mid-century and up to 81% by the end of the century (Hakam et al., 2024). Land-cover analyses further reveal substantial forest cover losses, reaching 1,065 ha·yr^{-1} in Essaouira and increasing to 1,533 ha·yr^{-1} in Tiznit, highlighting strong spatial heterogeneity in degradation rates (Sinsin et al., 2020). The resulting picture is consistent: ongoing habitat loss, fine-scale deforestation, and progressive decline driven primarily by grazing pressure and unsustainable fruit harvest, coupled with a projected reduction in climatically suitable areas and a likely northward displacement of optimal habitats under continued aridification (Timzioura et al., 2025).

Operational responses remain limited. Assisted migration and large-scale proactive relocation of argan provenances remain limited in practice (Timzioura et al., 2025). Nonetheless, recent work has begun to bridge the gap between restoration concepts and field-level implementation. Studies by Lamhamedi et al. (2015) and Zine El Abidine et al. (2020) have optimized nursery production systems, identified provenance–site matching rules, and evaluated the performance of planted material over time. These integrated efforts, combining genetic conservation, ecological restoration, and local development, are central to enhancing the adaptive capacity of argan ecosystems while improving the livelihoods of highly vulnerable rural communities (Moukrim et al., 2019a; Timzioura et al., 2025).

Atlas Cedar Forests (*Cedrus atlantica Manetti*)

Like the argan tree, *Cedrus atlantica* occupies a strategic ecological and socio-economic position in Moroccan forestry. This montane conifer underpins the functioning of numerous high-altitude ecosystems, notably through its role in regulating hydrological regimes, stabilizing soils, and maintaining microclimatic conditions (M'Hirit et al., 1994). Economically, Atlas cedar remains a key timber resource, supplying the vast majority of Morocco's high-quality saw timber (Cheddadi et al., 2017; Arar et al., 2020; El Mderssa et al., 2022; Hamida and Kouba, 2025; Boussalim and Dallahi, 2025). Several cedar landscapes are included in UNESCO Biosphere Reserves, highlighting their global conservation importance (Cheddadi et al., 2017).

Over recent decades, however, the species has undergone a marked contraction of its natural range (Abel-Schaad et al., 2018). This regression stems from the cumulative effect of severe droughts, progressive aridification, land-use change, and chronic human pressures (Moukrim et al., 2020). Distribution models and field observations consistently reveal declining habitat suitability, particularly in eastern populations, heightened drought sensitivity in aging stands, and the persistence of small, relict populations within microrefugia that retain cooler and moister conditions (Cheddadi et al., 2009; Cheddadi et al., 2017; Arar et al., 2020). Future distribution models further project substantial losses in climatically suitable habitats, ranging from –21.7% under the optimistic RCP 2.6 scenario by 2050 to –93.1% under the pessimistic RCP 8.5 scenario by 2070 (Moukrim et al., 2020).

Although *C. atlantica* exhibits some drought tolerance compared with other conifers, it is highly exposed to the ongoing shift towards hotter, drier climates. Recurrent multi-year droughts, rising temperatures, and declining atmospheric humidity are already translating into growth decline, increased mortality, and regeneration failure in many stands (Cheddadi et al., 2017; Abel-Schaad et al., 2018; Hamida and Kouba, 2025). Recent work by Boussalim and Dallahi (2025) provides detailed insights: regeneration is constrained by overgrazing, harsh high-altitude site conditions, and climatic extremes, which together reduce seedling establishment and survival.

In several degraded cedar stands, holm oak (*Quercus ilex*) is expanding and gradually replacing cedar as the dominant canopy species (Hamida and Kouba, 2025). This shift signals a potential reconfiguration of mountain forest composition under future climate scenarios. Predictive modelling suggests that the persistence of *C. atlantica* will depend heavily on maintaining sufficient atmospheric moisture and conserving humid microhabitats that function as climatic refugia (Emberger, 1938; Arar et al., 2020).

Cork Oak Systems (*Quercus suber* L.)

Cork oak (*Quercus suber L.*) forms one of the flagship forest systems of the western Mediterranean, and Morocco's Maâmora Forest is among the most emblematic lowland cork oak landscapes in the region (Fennane and Ibn Tattou, 2012). The Maâmora is characterized by extensive *Q. suber* stands, which occupy more than half of its forest area (Benabid, 2000; Aafi et al., 2005; Belghazi et al., 2011; Lahssini et al., 2015; FAO, 2017).

In recent decades, this system has been exposed to multiple, interacting stressors. Overgrazing, land clearing, overharvesting of timber and acorns, recurrent fires, and attacks by the pathogenic fungus *Hypoxylon mediterraneum* have driven pronounced decline in many sectors of the forest (Belghazi et al., 2011; FAO, 2017; Benabou et al., 2022). Degradation patterns are spatially heterogeneous: some stands exhibit severe canopy dieback and altered structure, whereas others retain relatively stable and functional socio-ecological dynamics.

Species distribution models indicate that *Q. suber* is highly sensitive to future climate scenarios, with projected reductions in suitable habitats under continued warming and drying (Benabou et al., 2022). At the same time, natural regeneration is often weak or absent due to grazing, soil compaction, and competition from shrubs and herbaceous species. These constraints pose significant challenges for the long-term conservation and sustainable management of cork oak systems (FAO, 2017).

Research on Maâmora forest has yielded several recommendations to improve reforestation and restoration outcomes. Belghazi et al. (2011) advocate direct acorn sowing combined with adequate soil preparation (e.g. deep ripping), supplemental irrigation in the first years, and improved nursery practices, while discouraging the replacement of cork oak with exotic species such as eucalyptus on unsuitable sites. Zine El Abidine et al. (2020) emphasize that the deep, sandy soils of Maâmora, combined with intensifying drought, limit the success of planting nursery seedlings with shallow root systems. They argue for more drought-adapted planting material and site-specific silvicultural strategies. These insights point to the need for climate-smart reforestation approaches that integrate ecological constraints with the socio-economic realities of local communities (Serbouti et al., 2023; Timzioura et al., 2025).

Holm oak (*Quercus ilex L. subsp. rotundifolia*), which frequently co-occurs with or replaces cedar and cork oak in transitional zones, plays a complementary role in these landscapes. While it is the most widespread woody formation in Morocco and contributes substantially to fuelwood supply and soil protection, it is itself affected by overgrazing, pruning, and recurrent droughts that reduce regeneration and increase physiological stress (Aafi et al., 2005; Lahssini et al., 2015). In many sectors, its upward expansion into declining cedar belts does not guarantee long-term ecosystem stability under continued climate warming.

Thuya Woodlands (*Tetraclinis articulata (Vahl) Masters*)

Thuya (*Tetraclinis articulata*), endemic to the western Mediterranean, is a dominant component of Morocco's semi-arid Atlantic and Central Plateau regions (Benabid, 2000). It is among the most drought-tolerant native trees, contributing crucially to slope stabilization, erosion control, and the preservation of residual forest cover in degraded landscapes.

However, many thuya stands are highly fragmented and degraded. Excessive fuelwood cutting, charcoal production, overgrazing, and limited natural regeneration have sharply reduced stand density. Seed biology studies reveal low and variable germination and marked differences in early growth among provenances, reflecting both genetic differentiation and environmental filtering (El Khaddari et al., 2019). This combination of ecophysiological robustness and socio-ecological fragility makes *T. articulata* a textbook case of "structural vulnerability": the species is intrinsically adapted to drought, but its populations are weakened by chronic human pressure and recruitment failure.

Climate projections suggest that thuya habitats may contract further as aridification intensifies (FAO, 2017). To avoid continued decline, management responses must couple assisted regeneration and provenance-based planting with community-based governance arrangements. Diversified reforestation using native xerophilous species, including thuya, could support both ecosystem stability and local livelihoods if embedded in participatory, benefit-sharing frameworks (Lamhamedi et al., 2015).

COMPARED REGIONAL DYNAMICS

Algeria

In Algeria, Mediterranean forests dominated by pines and other conifers have long been recognized as highly fire-prone, especially under hot, dry summer conditions (Quézel and Médail, 2003). Between 1979 and 2010, nearly 1.16 million hectares of forest were affected by approximately 41,600 fires, with substantial impacts on pine and cedar stands (Belgherbi et al., 2018; Ayache et al., 2020; Ghefar et al., 2024). These recurrent disturbances, compounded by intensifying droughts and soil degradation, have severely weakened natural regeneration and increased the risk of shrub encroachment and irreversible vegetation shifts.

Algerian case studies highlight the same structural drivers observed in Morocco: climatic aridification, rural dependence on forest resources, limited enforcement of land-use regulations, and fragmented governance systems. Together, these factors contribute to a pronounced vulnerability of Algerian Mediterranean forests to large-scale, high-severity fires and long-term landscape degradation.

Tunisia

In Tunisia, sclerophyllous and semi-deciduous forests show clear signs of decline due to increasing drought, higher temperatures, and reduced water resources (Touhami et al., 2023). Studies report severe dieback in cork oak forests, especially in Bellif, where drought has reduced growth and increased vulnerability (Ben Jamâa et al., 2005). Drought has also altered plant cover dynamics in arid areas of southern Tunisia (Ben Salem et al., 2007), while phenological changes in cork oak in Kroumirie reflect increased climate stress (Ben Yahia et al., 2016). Changes in rainfall patterns (Berndtsson & Niemczynowicz, 1986) and warming-driven shifts in forest pest dynamics add further pressure on these ecosystems (Bourougaaoui et al., 2021).

This climatic aridification is closely associated with increased fire incidence and forest loss. In 2021 alone, Tunisia lost nearly 280 km^2 of forest, approximately 26% of the total forest area lost between 2008 and 2021 (Touhami et al., 2023). Achour et al. (2018) report a three-fold increase in forest loss between 2007–2010 (32 km^2) and 2011–2014 (90 km^2), corresponding to an average annual increase of over 3 km^2. These figures confirm the structural nature of forest degradation in the country and highlight the need for robust fire-risk management, drought adaptation, and restoration policies.

Spain

On the northern shore of the western Mediterranean, Spain provides a complementary perspective on similar drivers operating within different socio-economic and institutional contexts. Spanish Mediterranean forests, dominated by *Pinus* spp., *Quercus* spp. (including *Q. ilex subsp. ilex* and *Q. suber*), and mixed shrublands, have experienced major land-use changes over the past century, leading to extensive forest expansion on abandoned agricultural land but also to increased fuel loads and higher fire risk (e.g. Pausas and Fernández-Muñoz, 2012).

Recent decades have seen a marked rise in large, high-severity wildfires in eastern and southern Spain, linked to climate warming, longer fire seasons, fuel accumulation, and human ignition sources (Pausas and Keeley, 2014; FAO, 2017). Drought-induced mortality events have been documented in both conifer and oak species, particularly during severe multi-year droughts, with consequences for carbon storage, biodiversity, and hydrological regulation.

Regional Conclusions

Across Morocco, Algeria, Tunisia, and Spain, southern and western Mediterranean forests are converging toward similar trajectories, characterized by increasing exposure to drought and heat, more frequent and intense wildfires, persistent anthropogenic pressure, and structural weaknesses in governance and rural development. These trends threaten not only forest cover and biodiversity but also critical ecosystem services that support water security, soil protection, carbon sequestration, and rural livelihoods (FAO, 2017; FAO, 2024; Gauquelin et al., 2018; Peñuelas and Sardans, 2021).

The Mediterranean Forest Research Agenda 2030 (Bou Dagher-Kharrat et al., 2022) provides a coherent strategic response to these challenges, structured around four complementary priorities: forest resilience under climate-related disturbances; conservation and sustainable management of biodiversity and forest genetic resources; multifunctional forest management balancing trade-offs and synergies among ecosystem services; and social, commercial, and policy innovation to support a sustainable forest bioeconomy.

These priorities are articulated within a "Mediterranean strategic triangle" linking research, innovation, and education as mutually reinforcing pillars for translating scientific advances into operational forest management. The trajectories observed in Morocco, Algeria, Tunisia, and Spain closely align with this framework and highlight the urgency of coordinated regional action. Strengthening cooperation among research institutions, forest administrations, local authorities, and civil society will be essential to enhancing the ecological and socio-economic resilience of southern Mediterranean forests in the context of accelerating global change.

VULNERABILITY AND GOVERNANCE ANALYSIS

Exposure, Sensitivity and Adaptive Capacity

Assessing forest vulnerability requires more than a conceptual definition; it demands the translation of abstract frameworks into operational indicators measurable in the field. Since the Third Assessment Report, the IPCC (2001, 2007) has popularized a three-component structure, exposure, sensitivity, and adaptive capacity that has become the backbone of most vulnerability assessments. When applied to forest socio-ecological systems, these components provide a flexible but rigorous lens for examining how climatic and anthropogenic pressures interact across scales.

- *Exposure* describes the extent to which forest systems are subjected to climate-related hazards such as droughts, heatwaves, shifts in precipitation regimes, or extreme events (Weißhuhn et al., 2018). In practice, exposure is commonly quantified using long-term climatic series and derived indicators: temperature anomalies, intra- and interannual rainfall variability, drought indices, or hazard-frequency metrics. These can be obtained from meteorological networks, remote sensing products, and downscaled climate projections (Roshani et al., 2024).
- *Sensitivity* refers to how strongly a given forest ecosystem responds when exposed to these disturbances. It can be captured through indicators of species composition, taxonomic and functional diversity (e.g. Shannon, Simpson indices), and structural attributes such as canopy cover, basal area, stand density, and regeneration status (Pandey et al., 2022). High sensitivity may be associated with low diversity, dominance of drought-susceptible species, or structurally simplified stands.
- *Adaptive capacity* encompasses the ability of forest ecosystems and dependent communities to absorb disturbances, reorganize, and maintain their core functions. From an ecological perspective, this includes genetic diversity, phenotypic plasticity, regeneration rates, and the availability of microrefugia. From a social perspective, it depends on livelihood diversification, access to information and finance, strength of local institutions, and the existence of learning and cooperation mechanisms (Locatelli et al., 2008; IPCC, 2007). Operationalizing this triptych requires combining *biophysical* variables (climate, topography, species diversity, biomass, disturbance history) with *socio-economic* indicators (income, education, land tenure, local governance capacity). To integrate such heterogeneous data, many studies rely on multicriteria or composite approaches, Principal Component Analysis (PCA), Analytical Hierarchy Process (AHP), normalized indices, and weighting schemes (Sharma et al., 2017; Kumar et al., 2021). These tools allow the construction of composite vulnerability indices that are both spatially explicit and interpretable for decision-makers.

Nevertheless, important conceptual ambiguities persist. In many applications (Table 1), adaptive capacity is reduced to socio-economic variables alone, while ecological resilience (e.g. genetic diversity, regeneration potential, functional redundancy) is placed under sensitivity or overlooked entirely (Füssel, 2010). For forest socio-ecological systems, this creates a biased view of vulnerability and weakens the relevance of resulting indices for management. A key challenge, particularly at local scale, is therefore to build *context-sensitive indicator sets* that reflect the coupled nature of forests and communities and that can be measured, monitored, and updated through field work.

These approaches illustrate how exposure, sensitivity, and adaptive capacity can be translated into operational metrics at different spatial scales.

Bottom-up and socio-ecological approaches are particularly suited to Moroccan contexts, where local knowledge, land-use practices, and social vulnerabilities play a decisive role in shaping forest responses to climate hazards.

Table 1. Examples of methodological frameworks for forest vulnerability assessment and their field relevance

Approach type	Indicator domains	Typical scale	Field / management relevance	Reference
Multicriteria quantitative	Ecological, climatic, anthropogenic	Regional	PCA and index ranking of climate and pressure variables	Banerjee et al., 2022
Socio-ecological	Ecological, climatic, anthropogenic	Local	Vulnerability = Sensitivity + Exposure – Adaptive Capacity; integrates local observations	Bourgoin et al., 2022
Top-down quantitative	Climatic and ecological	National	Vulnerability index (Sensitivity – Adaptive Capacity) for policy prioritization	Cui et al., 2016
Bottom-up participatory	Multidimensional local indicators	Local	Normalized indicators derived from field surveys and local knowledge	Jha et al., 2022
Indicator-based (ahp)	Biodiversity, disturbance, forest cover, slope, biomass	Regional	Hierarchical weighting of exposure and sensitivity components	Kumar et al., 2021
Integrated multidimensional	Climatic, topographic, ecological, socio-economic	Extended regional	PCA and normalization used to design multi-metric monitoring frameworks	Sun et al., 2021
Integrated composite	Climatic, ecological, socio-economic	Local	Simple averages per component; useful in data-scarce field contexts	Roshani et al., 2024
Socio-economic vulnerability	42 socio-economic variables	Local	Social Vulnerability Index; measures community-level adaptive capacity	Cutter et al., 2003

continued on following page

Table 1. Continued

Approach type	Indicator domains	Typical scale	Field / management relevance	Reference
Semi-qualitative / multicriteria	Bioclimatic and biological indicators	Regional / National	Expert scoring of exposure, sensitivity, and adaptation options	Thorne et al., 2018
Bottom-up composite multidimensional	Exposure (temp/precip), Sensitivity (species indices), Adaptive capacity (equity, diversity, maturity)	Local / Regional	Aggregation using Manush approach; integrates detailed field metrics	Pandey et al., 2022

Institutions and Decentralization

The effectiveness of vulnerability governance depends as much on institutional arrangements and decision-making structures as on scientific diagnostics. While the previous section addressed the conceptual and methodological underpinnings of exposure, sensitivity, and adaptive capacity, this section focuses on how these dimensions are, or are not, embedded in Morocco's forest governance architecture.

Historically, Moroccan forest policy has been shaped by a top-down paradigm, inherited from colonial administration and later adapted through national reforms after independence (Benzyane, 2007; Benzid & Madani, 2025). National strategies and legal frameworks provide overall coherence but tend to overlook local socio-ecological realities, which limits their usefulness for adaptive, place-based management (Cui et al., 2016; Schultz et al., 2019). Many vulnerability assessments still prioritize biophysical indicators, while social, institutional, and economic determinants of vulnerability remain underrepresented (Füssel, 2010; Craig et al., 2017).

Decentralization has been repeatedly identified as a necessary condition for adaptive governance, yet its implementation remains partial (Craig et al., 2017). Centralized planning and sectoral silos restrict the capacity of local actors, communes, forest users, and civil society, to influence decision-making and shape management practices. As a result, conservation outcomes are heterogeneous and often misaligned with local needs and ecological dynamics (Schultz et al., 2019; Cosens et al., 2017).

Despite these limitations, Morocco has progressively constructed a suite of national instruments that constitute the backbone of evidence-based forest governance:

National Strategy for the Prevention and Control of Forest Fire (2001, updated 2020):

This strategy establishes inter-agency coordination and is operationalized through the SysFeu platform, which generates twice-daily fire-danger maps from meteorological and fuel-load indicators. Historical fire databases (since 1997), combined with satellite detection and collaboration with FAO, the US Forest Service, and EFFIS, have significantly improved risk forecasting and crisis response (FAO, 2017; Alaoui et al., 2017). However, the integration of drought and hydrological stress indicators is still incomplete, particularly in mountainous regions, and early-warning coverage remains uneven.

National Phytosanitary Surveillance Network:

Since 2008, forest health has been monitored through a permanent 8 × 8 km grid of observation plots. Annual surveys document crown defoliation, dieback, pest outbreaks, and abiotic damage, following a formalized cycle of alert, diagnosis, intervention, and post-treatment evaluation. Yet weak linkage between observed decline and climatic or socio-economic drivers reduces the predictive and strategic value of the network.

Observations from this network and related studies have repeatedly identified a limited number of major phytosanitary pressures affecting Moroccan forest ecosystems. These include bark beetle outbreaks in cedar forests, particularly *Orthotomicus erosus* and related scolytine species, often associated with drought-induced weakening of stands; defoliating insects such as *Lymantria dispar* in oak-dominated systems; and pathogenic fungi, notably *Hypoxylon mediterraneum*, contributing to cork oak decline in lowland and coastal forests. In addition, episodic outbreaks of processionary moth (*Thaumetopoea* spp.) affect pine and mixed forests, with implications for both forest health and public health. Despite the regular documentation of these infestations, weak integration between phytosanitary observations, climatic stress indicators, and socio-economic drivers currently limits the predictive and strategic potential of the surveillance network.

National Forest Inventory (IFN):

Initiated in 1990, the inventory combines aerial photographs, satellite imagery, and 3,635 sample plots to estimate forest area, volume, and stand structure. Around one million hectares are updated annually, but full inventory cycles may exceed 15 years, slowing feedback loops for adaptive management. Interoperability with climate, fire, and phytosanitary databases is limited, hindering integrated vulnerability assessments.

National Action Programme to Combat Desertification (PAN-LCD):

Adopted in 2001 and revised in 2022, this program extends vulnerability monitoring to drylands and steppe systems. It employs NDVI, land productivity trends, and other biophysical indicators and supports the creation of a National Observatory for Drought and Desertification. Implemented with partners such as UNDP, IFAD, GIZ, and the Sahara and Sahel Observatory (OSS), it strengthens the link between land-degradation monitoring and rural development. Yet its outputs are only partially connected to forest-sector planning, and socio-economic resilience metrics remain marginal. Together, these systems represent substantial progress towards evidence-informed governance, but they still operate largely in isolation. The absence of a unified ecosystem monitoring observatory that integrates climatic, biological, and socio-economic data prevents the emergence of a coherent picture of forest vulnerability. Overcoming this fragmentation, technically, institutionally, and territorially, is a prerequisite for turning data into timely adaptation decisions.

In this perspective, vulnerability governance in Morocco appears as both a technical and a political–social process. Building resilience requires not only improved modelling and monitoring, but also shared authority, negotiated access, and multi-level coordination among state agencies, territorial authorities, local users, and private actors.

Co-management and Use Conflicts

Effective forest vulnerability reduction cannot be achieved without the involvement of the communities that depend on forest ecosystems. Co-management and participatory governance models seek to create institutional arrangements where local actors, state services, and civil society share responsibilities, information, and benefits (Cash et al., 2006).

In Morocco, forest-dependent populations rely heavily on ecosystem services, fuelwood, grazing, non-timber forest products, medicinal plants, for their livelihoods (Alaoui et al., 2020). This dependency often brings local needs into tension with conservation and restoration objectives. Land tenure complexity, overlapping user rights, and the coexistence of statutory and customary norms further complicate the picture. Where resource access is contested or poorly defined, conflicts over grazing, wood extraction, and land conversion are frequent.

Co-management frameworks must therefore explicitly address:

- **Land tenure and rights**: clarity over who can use what, where, and under which conditions;

- **Conflict resolution mechanisms**: transparent, accessible procedures to mediate disputes;
- **Community capacities**: technical training, organizational support, and financial mechanisms enabling local actors to participate meaningfully (Moukrim et al., 2019b);
- **Accountability and transparency**: shared monitoring and clear rules for decision-making and benefit sharing (Bednar, 2008; Craig et al., 2017; Thorne et al., 2018).

Participatory monitoring approaches, where local knowledge and field observations are combined with scientific tools, help create feedback loops between communities and institutions. They make vulnerability assessments more grounded and improve the likelihood that adaptation options will be implemented and maintained (Schultz et al., 2019; Locatelli et al., 2008; Jha et al., 2022). In this view, multi-level governance becomes essential: national strategies must be adapted and negotiated at regional and local levels, in line with ecological processes and socio-economic realities (Table 2).

Table 2. Community and co-management dimensions in forest vulnerability approaches

Approach type	*Community and governance focus*	*Co-management / field application*	*Reference*
Bottom-up participatory	Local engagement, use of indigenous knowledge	Participatory monitoring; integration of community observations in indicators	Jha et al., 2022; Locatelli et al., 2008
Socio-ecological	Combined socio-economic and ecological metrics	Supports co-management decisions; assesses community vulnerability and adaptive capacity	Bourgoin et al., 2022
Bottom-up composite multidimensional	Integrated socio-ecological perspective	Aggregates exposure, sensitivity, and adaptive capacity into locally actionable strategies	Pandey et al., 2022
Semi-qualitative / multicriteria	Attention to land tenure and conflicts	Evaluates use rights, governance quality, and potential for adaptive co-management	Thorne et al., 2018; Bednar, 2008

In the Moroccan context, strengthening participatory forest management through local user associations, communal councils, and territorial contracts could help reconcile conservation and development objectives. Embedding vulnerability assessments within such co-management arrangements ensures that proposed measures are socially legitimate, ecologically robust, and politically feasible. As illustrated in Figure 3, reducing forest vulnerability requires the effective integration of science (monitoring and risk modeling), governance (institutional frameworks and co-management), and society (community participation and local engagement).

Moving from vulnerability assessment to vulnerability reduction therefore hinges on linking conceptual frameworks, monitoring systems, and conflict-sensitive co-management in a coherent governance model. The figure 3, highlights that when these three pillars are weakly connected, governance fragmentation emerges, leading to ecological degradation and the failure of reforestation and fire-management programs.

Overall, effective management of forest vulnerability in Morocco requires bridging: (i) conceptual frameworks and field metrics, (ii) centralized policy and decentralized practice, and (iii) scientific diagnosis and community-driven action. Integrating vulnerability analysis into adaptive, participatory governance systems is therefore a critical step toward durable reductions in forest vulnerability.

Building on this integrative framework (Figure 3), the integration of emerging technologies and data-driven approaches can enhance monitoring, inform decision-making, and support adaptive management in Moroccan forests.

Figure 3. Interdependence between society, governance, and science in forest vulnerability management

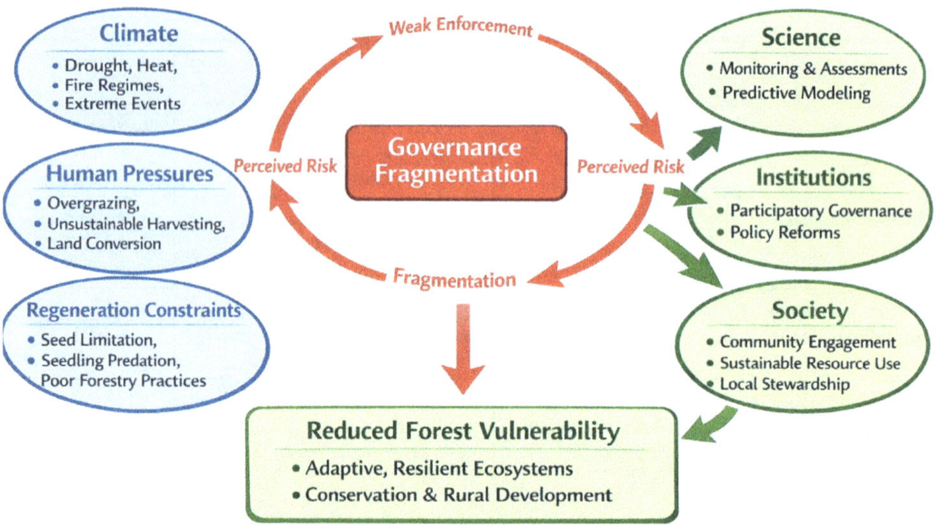

TECHNOLOGIES AND PROSPECTIVE APPROACHES

Across the southern Mediterranean, technology has become a core component of forest governance rather than a mere add-on. In Morocco, the "Forêts du Maroc 2020–2030" strategy couples a renewed forest inventory system and the digitalization of key forestry functions with increasing operational use of Earth observation, modelling, and forest genetics (MAPMDREF, 2020; CESE, 2022). Regional assessments consistently describe the Mediterranean as a climate-change hotspot where more frequent droughts and fire-conducive weather conditions make anticipatory, data-driven management indispensable (MedECC, 2019, 2020). In this context, Morocco is gradually converging toward practices seen in Spain and Italy, while sharing structural constraints with Tunisia and Algeria, creating both opportunities for replication and the need for careful institutional design (CESE, 2022; FAO, 2023).

Remote Sensing, Geographic Information Systems and Artificial Intelligence

Remote sensing and geospatial technologies now underpin forest monitoring and early-warning systems throughout the southern Mediterranean. In Morocco, these tools are being integrated into a broader modernization process under Forêts du Maroc 2020–2030, which places digitalization at the heart of institutional reform (MAPMDREF, 2020). The ANEF has begun to assemble a monitoring architecture that combines Earth Observation (EO), GIS platforms, and Machine-Learning (ML) approaches to support fire-risk analysis, vegetation monitoring, and the evaluation of reforestation outcomes (MAP/ANEF, 2024).

Time-series from Sentinel-2 and Landsat are used to map forest cover and land use, quantify burn severity, and track temporal trends in canopy condition. These EO products feed ML classifiers for land-cover mapping, fire-susceptibility modelling, and post-fire recovery assessments, in line with FAO recommendations on combining multi-sensor data with ground measurements (FAO, 2023). Morocco is also progressively deploying uncrewed aerial systems (UAS/drones) for fire reconnaissance, hotspot localization, and high-resolution monitoring of plantations and regeneration sites. These drone-based surveys fill the spatial and temporal gaps between permanent plots and satellite imagery, improving the granularity of decision-making (MAP/ANEF, 2024; CESE, 2022).

Compared with Spain and Italy, where EO-based observatories (e.g. national forest inventories linked to Copernicus services and fire-danger products) are fully institutionalized, Morocco is still in a transition phase but is moving in the same direction by embedding EO and UAS into routine workflows rather than relying solely on short-term projects. Remaining obstacles include fragmented databases, unequal regional capacities, and limited automation of periodic updates (CESE, 2022). Tunisia's National Forestry and Pasture Strategy 2015–2024 illustrates another path, linking EO-derived indicators to explicit forest-cover targets and donor-funded restoration (AfDB, 2025). However, its dependence on external finance highlights the importance of strong domestic data governance, an area where Morocco is already investing.

In Algeria, FAO's NENFIRE project has focused on reconstructing multi-decadal fire histories using Landsat archives, thereby compensating for incomplete official fire records and informing prevention strategies (Kouachi et al., 2024). Morocco, which already benefits from tighter field–satellite integration, could draw on this experience to release open, authoritative national fire and restoration layers, improving transparency and interoperability across the Maghreb.

Ecological and Climatic Modelling

Ecological and climatic modelling enables a shift from reactive to anticipatory management by translating climate trajectories into projections of species ranges, disturbance regimes, and management thresholds. In Morocco and the wider Maghreb, recent modelling efforts for emblematic species such as *Cedrus atlantica* indicate altitudinal and latitudinal shifts, habitat contraction, and changing suitability patterns under warming and drying scenarios. These projections support the design of refugia, assisted migration schemes, and ecological corridors (Laaribya et al., 2024).

Regional syntheses stress that foresight should not remain confined to isolated research projects but must be embedded systematically into planning, budgeting, and policy evaluation cycles (MedECC, 2020). Spain and France provide illustrative examples where scenario platforms, based on CMIP-class climate ensembles and downscaled projections, are formally linked to forest-management plans and regional strategies. Tunisia, despite active restoration efforts, still tends to treat modelling as a deliverable of externally funded projects rather than as a core planning input, while Algeria is only beginning to connect EO-based fire histories with forecasting tools through FAO-supported programmes (MedECC, 2019).

For Morocco, the key issue is the translation interface: how to connect global climate scenarios and regional projections with national EO datasets and silvicultural prescriptions within a unified decision chain. In practical terms, this would involve establishing a "scenario canon" for Moroccan forestry, standardized time horizons, emissions pathways, species sets, and management thresholds that must be used when drafting or revising management plans (MedECC, 2020; CESE, 2022; Laaribya et al., 2024). Such a framework would ensure consistency across agencies and regions and would anchor management choices in explicit assumptions about future climate.

Genetics and Assisted Natural Regeneration

Technological innovation in forestry also concerns forest reproductive material, provenance choice, and regeneration strategies. Accumulating evidence from Morocco and other Mediterranean countries shows that integrating Forest Genetic Resources (FGR), provenance testing, and nursery selection into restoration programs enhances drought tolerance and survival of key taxa such as *Argania spinosa* and *Quercus ilex* (Alami et al., 2024; Timzioura et al., 2025).

The policy window created by Forêts du Maroc 2020–2030 offers Morocco an opportunity to more tightly link science and operations. Aligning nursery production, assisted natural regeneration (ANR), and planting programs with site-suitability maps, certified seed sources, and genetic monitoring would transform isolated pilot projects into long-term pipelines. A critical step would be to require seed source documentation and genetic-quality criteria in management plans and procurement procedures, as part of national technical standards. This would bring Moroccan practice closer to that of northern Mediterranean countries while providing a south-rim model adaptable to Tunisia and Algeria (Alami et al., 2024; CESE, 2022).

Data Governance

The benefits of EO, modelling, and genetic approaches ultimately depend on robust *data governance*. In Morocco, forest-related information is still dispersed across agencies and projects, with heterogeneous formats and access rules. The drive by ANEF to create an integrated forest-information backbone is therefore a pivotal step. Such a backbone aims to harmonize cartographic layers, plot-level forest inventory data, biodiversity records, and socio-economic indicators under shared standards and quality-control procedures (MAP/ANEF, 2024).

FAO and the broader GEO community stress that open metadata, versioned workflows, and independent quality assurance are essential conditions for credibility and cross-border comparability, especially in regions like the Mediterranean where transboundary processes (fires, droughts, migratory pests) are common (MedECC, 2020; FAO, 2023). Spain and Italy illustrate how publishing open, auditable layers on fires, restoration areas, and land-cover dynamics can accelerate feedback between research and operations, stimulate external validation, and bolster public accountability.

Conversely, Tunisia's experience shows that donor-driven databases may become fragmented or obsolete when project cycles end, unless they are embedded within national information systems with clear mandates, budgets, and access policies. Algeria's collaboration with FAO to define national-standard fire perimeters illustrates how publishing authoritative layers can be as transformative as generating them.

For Morocco, the logical next step is to institutionalize *SIFI-type governance (Single, Integrated Forest Information systems)* with regular, versioned releases of national datasets on fires, restoration, regeneration, and land degradation. Each release should be accompanied by detailed documentation of methods, uncertainties, and *quality assurance/quality control (QA/QC)* procedures (CESE, 2022; FAO, 2023). Such a system would ensure that information survives project cycles and becomes a durable asset for both national decision-making and Mediterranean-scale analyses.

RECOMMENDATIONS AND ROADMAP

Building upon the preceding diagnosis, this roadmap consolidates a comprehensive framework to enhance the predictive, monitoring, and adaptive capacities of Morocco's forest ecosystems. It synthesizes structural reforms, technical measures, and socio-economic levers required to bridge the persistent gaps between knowledge production and management implementation.

Near-term Risk Reduction and Multi-scale Monitoring

Morocco's forest adaptation strategy must begin with an integrated and hierarchical monitoring system capable of linking local ecological indicators to national and regional decision frameworks. Current observation gaps can be addressed through a multi-scale architecture combining satellite time series (Sentinel, MODIS, Landsat) with permanent ground plots and climatic stations managed by ANEF. These datasets should feed a national open-access dashboard updated on a seasonal basis, ensuring transparency and reproducibility. At the local level, community-based observatories, supported by forest cooperatives and local authorities, must be formally integrated into the monitoring chain. Their participatory reports on fire incidence, dieback, and illegal logging enhance spatial precision and foster ownership of ecosystem protection (FAO, 2017; IRES, 2024).

Medium-term Adaptive Silviculture and Restoration

To ensure long-term resilience, reforestation and rehabilitation programs should evolve toward function-oriented, site-specific silviculture. Our analysis shows that restoration success depends on the capacity to secure soil stability, regulate water flows, and maintains species diversity rather than on the number of hectares planted.

In the Atlas forests, priority should be given to microrefugia protection, assisted regeneration, and mixed-species mosaics that reflect climatic and edaphic variability.

Climatic zoning must guide planting density, thinning regimes, and regeneration cycles in accordance with projected aridity gradients.

Genetic quality remains a decisive factor. National nurseries should systematically apply the production standards and provenance certification detailed by Lamhamedi et al. (2015), guaranteeing drought-tolerant genotypes and healthy root systems. Integrating these standards into procurement and monitoring frameworks would directly enhance survival rates and forest functionality.

Enabling Governance Reforms and Scientific Capacity

Addressing vulnerability also requires reforming the forest information and decision system. The creation of a national forest data backbone, unifying inventories, biodiversity databases, and socio-economic indicators, is crucial for evidence-based management. Such an infrastructure must adhere to interoperability, open-data, and independent quality control standards, ensuring data continuity across institutions (CESE, 2022; IRES, 2024).

Institutional reforms should further promote decentralized accountability: regional services (ZDTF, DFP, DPANEF) must be empowered to tailor interventions within national zoning frameworks, with clear obligations of reporting and evaluation.

In parallel, capacity development must bridge technical and social dimensions. Training curricula for forestry engineers should integrate remote sensing, GIS analysis, ecological modeling, and participatory planning to foster multidisciplinary expertise. This shift will transform monitoring from a descriptive exercise into an adaptive management instrument.

Financing Resilience and Valuing Ecosystem Services

Forest adaptation will not be sustainable without innovative financial mechanisms. Beyond state budgets, Morocco should activate climate-resilience funds and payment schemes for ecosystem services (PES), linking financial disbursement to measurable ecological outcomes. Transparent performance-based funding ("no data, no payment") would align national efforts with international standards for accountability and impact evaluation (CESE, 2022).

Economic valuation of non-timber ecosystem services-carbon storage, water regulation, soil protection-should be incorporated into planning tools, reinforcing the social and fiscal justification for forest investment (FAO, 2017; IRES, 2024).

A Regional Knowledge Toolbox for Mediterranean Managers

To facilitate replication and cooperation, Morocco can spearhead the development of a regional forest-resilience toolbox structured around:

1. Monitoring protocols integrating remote-sensing metrics with standardized field indicators for vegetation stress, biomass dynamics, and fire exposure;
2. Decision-support templates linking vulnerability mapping, climatic scenarios, and restoration prioritization (IRES, 2024);
3. Governance modules for open-data management, training exchange, and peer review (CESE, 2022).

Such a toolbox would enable convergence among North-African administrations, transforming isolated initiatives into a shared Mediterranean framework for adaptive forest management.

By coupling open data, adaptive planning, and social participation, Morocco can bridge the long-standing gap between strategic intent and ecological performance, positioning its forestry sector as a regional laboratory of resilience and foresight in the Mediterranean.

MEDITERRANEAN TOOLBOX FOR FOREST RESILIENCE

To facilitate regional convergence and mutual learning, Morocco is well positioned to contribute to a Mediterranean forest-resilience toolbox designed for managers and policymakers. This toolbox can be structured around five complementary components: monitoring protocols that integrate satellite indicators (e.g., NDVI, NBR, soil-moisture proxies) with standardized field metrics such as defoliation, regeneration, biomass, and fuel loads; decision-support templates linking vulnerability maps, climate scenarios, and restoration priorities with explicit thresholds and management responses (IRES, 2024); governance modules defining procedures for open-data management, peer review, and cross-country training exchanges (CESE, 2022); genetic and restoration guidelines covering provenance selection, seed systems, assisted natural regeneration, and climate-resilient planting designs; and fire- and risk-management components including common indicators, early-warning systems, and post-fire restoration protocols adapted to Mediterranean conditions.

By spearheading such a toolbox in collaboration with neighboring countries and regional platforms (FAO/Silva Mediterranea, MedECC, EFI/EFIMED), Morocco could help transform currently fragmented initiatives into a shared Mediterranean framework for adaptive forest management.

Taken together, this proposed roadmap turns vulnerability analysis into a continuous learning cycle that connects prediction, prevention, and adaptation. In the short term, it institutionalizes multi-scale monitoring and participatory early-warning systems; in the medium term, it aligns silviculture and restoration practices with climate-functional criteria and genetic quality control; and in the long term, it embeds transparency, innovative financing, and scientific excellence at the core of forest governance. By combining open data, adaptive planning, and social participation, Morocco can reduce the persistent gap between strategic ambition and ecological performance and position its forestry sector as a regional laboratory of resilience and foresight in the Mediterranean.

CONCLUSION

Forest vulnerability in Morocco encapsulates the broader tension between ecological fragility and the capacity of institutions to adapt within Mediterranean socio-ecological systems. The analysis developed in this chapter shows that vulnerability cannot be interpreted solely through the lens of climate hazards or biophysical exposure. Rather, it emerges at the intersection of how societies perceive, quantify, and govern change across space and time. By articulating field-based indicators of hazard, exposure, sensitivity, and adaptive capacity with an examination of governance structures, this work highlights the dual scientific and institutional foundations of resilience.

Empirical evidence confirms that Moroccan forest ecosystems are increasingly stressed by recurrent droughts, rising fire incidence, and cumulative anthropogenic pressures. At the same time, the country has developed a substantial suite of monitoring instruments and strategic frameworks. Yet these tools still function largely in parallel, without a truly integrated platform for cross-sectoral analysis and decision support. Overcoming this fragmentation requires governance arrangements that explicitly link knowledge production to action, ensuring that vulnerability assessments trigger timely, context-specific responses on the ground.

The transition toward resilient forest governance in Morocco hinges on moving beyond top-down, project-based logics toward adaptive and shared management. This involves three interdependent shifts:

1. *Scientific integration* – the construction of interoperable databases and an ecosystem-vulnerability observatory that connects climatic, biological, and socio-economic indicators within a coherent analytical framework;

2. *Institutional reform* – the consolidation of co-management and decentralization within transparent, accountable national structures, clarifying mandates and fostering multi-level coordination;
3. *Social inclusion* – the recognition and empowerment of local communities as full partners in risk monitoring, restoration planning, and the equitable distribution of benefits.

In the end, resilient forest governance is not only about ecological recovery; it is also about renegotiating the relationships between knowledge, authority, and territory. Morocco's evolving framework, anchored in strong public institutions yet increasingly open to participatory renewal, has the potential to serve as a reference for the southern Mediterranean. It offers a promising pathway for aligning climate adaptation, biodiversity conservation, and sustainable rural livelihoods within a single, integrated vision.

REFERENCES

Aafi, A., El Kadmiri, A. A., Benabid, A., & Rochdi, M. (2005). Richesse et diversité floristique de la suberaie de la Mamora (Maroc) [Richness and floristic diversity of the cork oak forest of Mamora (Morocco)]. *Acta Botanica Malacitana*, *30*, 127–138. DOI: 10.24310/abm.v30i0.7187

Abel-Schaad, D., Iriarte, E., López-Sáez, J. A., Pérez-Díaz, S., Sabariego Ruiz, S., Cheddadi, R., & Alba-Sánchez, F. (2018). Are *Cedrus atlantica* forests in the Rif Mountains of Morocco heading towards local extinction? *The Holocene*, *28*(6), 1023–1037. DOI: 10.1177/0959683617752842

Achhal, A., Akabli, A., Barbero, M., Benabid, A., M'Hirit, O., Peyre, C., Quézel, P., & Rivas-Martinez, S. (1979). A propos de la valeur bioclimatique et dynamique de quelques essences forestières au Maroc. *Ecologia Mediterranea*, *5*(1), 211–249. DOI: 10.3406/ecmed.1979.960

Achour, H., Toujani, A., Rzigui, T., & Faïz, S. (2018). Forest cover in Tunisia before and after the 2011 Tunisian Revolution: A spatial analysis approach. *Journal of Geovisualization and Spatial Analysis*, *2*(2), 10. DOI: 10.1007/s41651-018-0017-7

AfDB. (2025). Tunisia launches $24 million project to restore forests and revitalize rural economies. African Development Bank. https://www.afdb.org/en/news-and-events/press-releases/tunisia-launches-24-million-project-restore-forests-and-revitalize-rural-economies-83225

African Development Bank (AfDB). (2025). *Tunisia launches $24 million project to restore forests and revitalize rural economies*. African Development Bank. https://www.afdb.org/en/news-and-events/press-releases/tunisia-launches-24-million-project-restore-forests-and-revitalize-rural-economies-83225

Agence Nationale des Eaux et Forêts. (ANEF). (2022). *National Water and Forests Agency in Morocco*. http://www.eauxetforets.gov.ma/Biodiversite/GestionBiodiversite/Pages/Gestion-Biodiversite.aspx

Alami, M.. (2024). Genomic scan of endangered Moroccan endemic trees: *Argania spinosa* as a case of adaptive capacity under Mediterranean stressors. *Environmental Advances*, *16*, 100339.

Alaoui, A., Laaribya, S., & Ayan, S. (2020). The evolution of the forest cover with the effect of anthropic pressure: The case study of Sehoul cork-oak forest in Morocco, North Atlantic. *Kastamonu Üniversitesi Orman Fakültesi Dergisi*, *20*(1), 62–73. DOI: 10.17475/kastorman.705872

Alaoui, H. M., Assali, F., Rouchdi, M., Tahiri, D., Lahssini, S., Aafi, A., & Moukrim, S. (2017). Cartographie de l'intensité du feu et de la sensibilité aux feux de cimes en forêts méditerranéennes. Cas de la province de Chefchaouen au nord du Maroc. *Revue d'Écologie*, *72*(4), 387–409. DOI: 10.3406/revec.2017.1900

Alasli, M. (2019). Static risk mapping of forest fires in the case of the Province of Chefchaouen (Morocco). *Proceedings of the ICA*. Copernicus GmbH. DOI: 10.5194/ica-proc-2-2-2019

Arar, A., Nouidjem, Y., Bounar, R., Tabet, S., & Kouba, Y. (2020). Modeling of the current and future potential distribution of Atlas cedar (*Cedrus atlantica*) forests revealed shifts in the latitudinal, longitudinal and altitudinal range towards more humid conditions. *Ecological Questions*, *31*(3), 49–62. DOI: 10.12775/EQ.2020.022

Aubert, P.-M. (2010). *Action publique et société rurale dans la gestion des forêts marocaines: Changement social et efficacité environnementale* (Thèse de doctorat en sciences sociales pour l'environnement). AgroParisTech.

Aubert, P.-M. (2013). Les évolutions de la politique forestière au Maroc: Entre réappropriation du modèle forestier français et idéalisation de la tribu. *Revue Forestière Française*, *65*(4), 305–316. DOI: 10.4267/2042/53627

Ayache, A., Hellal, B., & Ayad, N.. (2020). Modelling of the changes in land cover in the forest of Sidi Bel'Abbes between 1987 and 2018 and the impact of fires on soil evolution in western Algeria. *Authorea*. DOI: 10.22541/au.159672708.89027787

Banerjee, S., Niyogi, R., Sarkar, M. S., & John, R. (2022). Assessing the vulnerability of protected areas in the eastern Himalayas based on their biological, anthropogenic, and environmental aspects. *Trees People 8*. 10.1016/j

Bednar, J. (2008). *The robust federation: Principles of design*. Cambridge University Press., https://books.google.co.ma/books?id=cba5vpiFMiMC DOI: 10.1017/CBO9780511819445

Belghazi, B., Badouzi, M., Belghazi, T., & Moujjani, S. (2011). Semis et plantations dans la forêt de chêne-liège de la Maâmora (Maroc). *Forêt Méditerranéenne*, *32*(3), 301–314.

Belgherbi, B., Benabdeli, K., & Mostefai, K. (2018). Mapping the risk of forest fires in Algeria: Application of the forest of Guetarnia in Western Algeria. *Ekologia (Bratislava)*, *37*(3), 289–300. DOI: 10.2478/eko-2018-0022

Ben Jamâa, M. L., Sghaier, T., Mnara, S., Nouri, M., & Sellemi, H. (2005). Le dépérissement du chêne-liège dans la subéraie de Béllif (Tunisie): Caractérisation et évaluation de son impact sur l'accroissement du liège. Integrated Protection in Oak Forests. *IOBC/WPRS Bulletin*, *28*, 17–24.

Ben Salem, F., Tarhouni, M., Ouled Belgacem, A., & Neffati, M. (2007). Impact of Drought on Plant Cover Dynamics in Two Natural Areas of Southern Tunisia. *Journal of Biological Sciences (Faisalabad, Pakistan)*, *7*(8), 1539–1544. DOI: 10.3923/jbs.2007.1539.1544

Ben Yahia, K., Chaar, H., Bahri, S., Mhamdi, S., Soudani, K., Khouaja, A., & Hasnaoui, B. (2016). Phenological monitoring of Cork oak in Kroumirie (northwest Tunisia). *Journal of Chemical, Biological and Physical Sciences, Section B*, *6*(4), 1258–1281.

Benabid, A. (1985). Les écosystèmes forestiers, préforestiers et présteppiques du Maroc: Diversité, répartition biogéographique et problèmes posés par leur aménagement. *Forêt Méditerranéenne*, *7*, 53–64.

Benabid, A. (2000). *Flore et écosystèmes du Maroc: Diversité, phytogéographie et conservation*. Ibis Press.

Benabid, A. (2000). *Flore et écosystèmes du Maroc: Évaluation et préservation de la biodiversité* [Flora and ecosystems of Morocco: Assessment and preservation of biodiversity]. Ibis Press; Kalila Wa Dimna.

Benabid, A., & Fennane, M. (1994). Connaissances sur la végétation du Maroc: Phytogéographie, phytosociologie et séries de végétation. *Lazaroa*, *14*, 21–97.

Benabou, A., Moukrim, S., Lahssini, S., El Aboudi, A., Menzou, K., Elmalki, M., Madihi, M., & Rhazi, L. (2022). Impact of climate change on potential distribution of *Quercus suber* in the conditions of North Africa. *Biosystems Diversity*, *30*(3), 289–294. DOI: 10.15421/012231

Benchekroun, F., & Buttoud, G. (1989). L'arganeraie dans l'économie rurale du sud-ouest marocain. *Forêt Méditerranéenne*, *11*, 127–136.

Benzid, H., & Madani, N. (2025). Protection of mediterranean forest biodiversity: The case of Morroco-current status, legislation and opportunities. *Revista Catalana de Dret Ambiental*, *16*(1), 1–34. DOI: 10.17345/rcda4066

Benzyane, M. (2007). La gestion durable des ressources forestières au Maroc: Quelle stratégie? *Forêt Méditerranéenne*, *28*(1), 47–54.

Benzyane, M., Naggar, M., & Lahlou, B. (2002). L'aménagement des forêts sud-méditerranéennes: Quelle approche? *Forêt Méditerranéenne*, *23*(3), 201–210.

Berndtsson, R., & Niemczynowicz, J. (1986). Spatial and Temporal Characteristics of Highintensive Rainfall in Northern Tunisia. *Journal of Hydrology (Amsterdam)*, *87*(3-4), 285–298. DOI: 10.1016/0022-1694(86)90020-X

Bou Dagher-Kharrat, M., de Arano, I. M., Zeki-Bašken, E., Feder, S., Adams, S., Briers, S., Fady, B., Lefèvre, F., Górriz-Mifsud, E., Mauri, E., Mavsar, R., Muys, B., Nocentini, S., Pettenella, D., Peñuelas, J., Sardans, J., Secco, L., Travaglini, D., Tripodi, G., Wunder, S., & Bozzano, M. (2022). *Plan stratégique de recherche pour la forêt méditerranéenne 2030*. Institut européen des forêts. DOI: 10.36333/rs5

Boubekraoui, H., Maouni, Y., Ghallab, A., Draoui, M., & Maouni, A. (2023). Spatio-temporal analysis and identification of deforestation hotspots in the Moroccan Western Rif. *Trees, Forests and People*, *12*, 100388. DOI: 10.1016/j.tfp.2023.100388

Bourgoin, C., Oszwald, J., Bourgoin, J., Gond, V., Blanc, L., Dessard, H., Van Phan, T., Sist, P., Laderach, P., & Reymondin, L. (2022). Assessing the ecological vulnerability of forest landscape to agricultural frontier expansion in the central highlands of Vietnam. *International Journal of Applied Earth Observation and Geoinformation*, *84*, •••. 10.1016/j

Bourougaaoui, A., Ben Jamâa, M. L., & Robinet, C. (2021). Has North Africa turned too warm for a Mediterranean forest pest because of climate change? *Climatic Change*, *165*(3-4), 46. DOI: 10.1007/s10584-021-03077-1

Boussalim, Y., & Dallahi, Y. (2025). Predictive modeling of *Cedrus atlantica* natural regeneration: Grazing intensity and bioclimatic extremes as the primary drivers in the Ain Leuh Forest, Middle Atlas of Morocco. *Modeling Earth Systems and Environment*, *11*(1), 17. DOI: 10.1007/s40808-024-02191-0

Cash, D. W., Adger, W. N., Berkes, F., Garden, P., Lebel, L., Olsson, P., Pritchard, L., & Young, O. (2006). cale and cross-scale dynamics: Governance and information in a multilevel world. *Ecology and Society*, *11*(2), 8. http://dx.doi.org/10.5751/ES-01759-110208. DOI: 10.5751/ES-01759-110208

Cheddadi, R., Fady, B., François, L., Hajar, L., Suc, J.-P., Huang, K., Demarteau, M., Vendramin, G. G., & Ortu, E. (2009). Putative glacial refugia of *Cedrus atlantica* deduced from Quaternary pollen records and modern genetic diversity. *Journal of Biogeography*, *36*(7), 1361–1371. DOI: 10.1111/j.1365-2699.2008.02063.x

Cheddadi, R., Henrot, A.-J., François, L., Boyer, F., Bush, M., Carré, M., Coissac, E., De Oliveira, P. E., Ficetola, F., Hambuckers, A., Huang, K., Lézine, A.-M., Nourelbait, M., Rhoujjati, A., Taberlet, P., Sarmiento, F., Abel-Schaad, D., Alba-Sánchez, F., & Zheng, Z. (2017). Microrefugia, climate change, and conservation of *Cedrus atlantica* in the Rif Mountains, Morocco. *Frontiers in Ecology and Evolution, 5*, 114. DOI: 10.3389/fevo.2017.00114

Cosens, B. A., Craig, R. K., Hirsch, S. L., Arnold, C. A. T., Benson, M. H., DeCaro, D. A., Garmestani, A. S., Gosnell, H., Ruhl, J. B., & Schlager, E. (2017). The role of law in adaptive governance. *Ecology and Society*, *22*(1), 30. Advance online publication. DOI: 10.5751/ES-08731-220130

Craig, R. K., Garmestani, A. S., Allen, C. R., Arnold, C. A., Birgé, H., DeCaro, D. A., Fremier, A. K., Gosnell, H., & Schlager, E. (2017). Balancing stability and flexibility in adaptive governance: An analysis of tools available in U.S. *Environmental Law. Ecology and Society*, *22*(2), 3. http://dx.doi.org/10.5751/ES-08983-220203. DOI: 10.5751/ES-08983-220203

Cui, G., Kwak, H., Choi, S., Kim, M., Lim, C.-H., Lee, W.-K., Kim, J.-S., & Chae, Y. (2016). Assessing vulnerability of forests to climate change in South Korea. *Journal of Forestry Research*, *27*(3), 489–503. 10.1007/s11676-015-0201-2. DOI: 10.1007/s11676-015-0201-2

Cutter, S. L., Boruff, B. J., & Shirley, W. L. (2003). Social vulnerability to environmental hazards. *Social Science Quarterly*, *84*(2), 242–261. 10.1111/1540-6237.8402002. DOI: 10.1111/1540-6237.8402002

El Khaddari, M., & El Mousadik, A.. (2019). Seed germination and juvenile growth of *Tetraclinis articulata* from five Moroccan provenances. *Plant Cell Biotechnology and Molecular Biology*, *20*(11–12), 556–565.

El Mderssa, M., Malki, F., Ikraoun, H., Nassiri, L., & Ibijbijen, J. (2022). Détermination des paramètres d'évaluation du stock de carbone dans les écosystèmes forestiers (*Cedrus atlantica* Manetti, cèdre de l'Atlas au Maroc): Méthodes spécifiques et génériques. *Bois et Forêts des Tropiques*, *351*, 67–77. DOI: 10.19182/bft2022.351.a36330

Ellatifi, M. (2012). *L'économie de la forêt et des produits forestiers au Maroc: Bilan et perspectives* (Thèse de doctorat en économie). Université Montesquieu – Bordeaux IV.

Emberger, L. (1925). Le domaine naturel de l'arganier. *Bulletin de la Société Botanique de France*, *72*(4), 770–774. DOI: 10.1080/00378941.1925.10832788

Emberger, L. (1938). Contribution à la connaissance des cèdres et en particulier du déodar et du cèdre de l'Atlas. *Revue de Botanique Appliquée et d'Agriculture Coloniale, 18*(198), 77–92. DOI: 10.3406/jatba.1938.5828

Emberger, L. (1939). Aperçu général sur la végétation du Maroc. *Veröffentlichungen des Geobotanischen Instituts Rübel Zürich, 14*, 40–157.

Emberger, L. (1955). Une classification biogéographique des climats. *Recueil des Travaux du Laboratoire de Botanique et de Zoologie de la Faculté des Sciences de Montpellier, 7*, 3–43.

Fady, B., Farsakoglou, A. M., Caron, M., Abulaila, K., Aleksic, J., Alipour, S., Balian, D., Bedair, H., Bogunić, F., Cheikh Albassatneh, M., David-Schwartz, R., Delgado Clavero, C., Dönmez, A. A., Fennane, M., Gateble, G., Gauquelin, T., Hachi Illoul, M., Khaldi, A., Mahfoud-Saad, I., & Bou Dagher Karrat, M. (2025). Native trees of the Mediterranean region: Distribution, diversity and conservation challenges. *Current Forestry Reports, 11*(1), 20. DOI: 10.1007/s40725-025-00252-w

Fennane, M., & Ibn Tattou, M. (2012). Statistiques et commentaires sur l'inventaire actuel de la flore vasculaire du Maroc [Statistics and comments on the current inventory of the vascular flora of Morocco]. *Bulletin de l'Institut Scientifique, Rabat. Section Sciences de la Vie, 34*(1), 1–9.

Food and Agriculture Organization of the United Nations. (2000). FAO Forestry Paper: Vol. 140. *Thuriferous juniper stands in Morocco: Research and management.* Food and Agriculture Organization of the United Nations.

Food and Agriculture Organization of the United Nations. (2017). *Production d'outils d'aide à la décision et de gestion en matière de vulnérabilité des écosystèmes forestiers méditerranéens aux effets du changement climatique et valorisation des efforts d'optimisation des biens et services (Projet FFEM/Plan Bleu).* Food and Agriculture Organization of the United Nations.

Food and Agriculture Organization of the United Nations. (2017). *State of Mediterranean forests 2017.* Food and Agriculture Organization of the United Nations.

Food and Agriculture Organization of the United Nations, & Programme des Nations Unies pour l'Environnement. (2020). *La situation des forêts du monde 2020: Forêts, biodiversité et activité humaine.* Food and Agriculture Organization of the United Nations. DOI: 10.4060/ca8642fr

Food and Agriculture Organization of the United Nations. (2020). *Technical assistance for forest fire management in Algeria.* Food and Agriculture Organization of the United Nations. https://www.fao.org/in-action/nenfire/projects/algeria/fr/

Food and Agriculture Organization of the United Nations. (2023, November 10). *Innovative monitoring of forest and landscape restoration in the Mediterranean region.* Food and Agriculture Organization of the United Nations. https://www.fao.org/forest-resources-assessment/resources/news/detail/innovative-monitoring-of-forest-and-landscape-restoration-in-the-mediterranean-region/en

Food and Agriculture Organization of the United Nations. (2024). *La situation des forêts du monde 2024 – Innovations dans le secteur forestier pour un avenir plus durable*. Food and Agriculture Organization of the United Nations., DOI: 10.4060/cd1211fr

Food and Agriculture Organization of the United Nations. (2024). *Restoring the Mediterranean region: Status and challenges (Unasylva No. 255, Vol. 75, 2024/1).* Food and Agriculture Organization of the United Nations. DOI: 10.4060/cd1720en

Food and Agriculture Organization of the United Nations, & Silva Mediterranea. (2024, February). *Restoring Mediterranean forests — UN Decade Flagship.* Food and Agriculture Organization of the United Nations. https://www.fao.org/silva-mediterranea/un-decade-of-ecosystem-restoration/restoring-mediterranean-forests-flagship/en

Francos, M., Colino-Prieto, F., & Sánchez-García, C. (2024). How Mediterranean ecosystem deals with wildfire impact on soil ecosystem services and functions: A review. *Land (Basel), 13*(4), 407. DOI: 10.3390/land13040407

Füssel, H.-M. (2010). *Review and Quantitative Analysis of Indices of Climate Change Exposure, Adaptive Capacity, Sensitivity, and Impacts*. Development and Climate Change. DOI: 10.1596/9193

Gauquelin, T., Michon, G., Joffre, R., Duponnois, R., Genin, D., Fady, B., Bou Dagher-Kharrat, M., Derridj, A., Slimani, S., Badri, W., Alifriqui, M., Auclair, L., Simenel, R., Aderghal, M., Baudoin, E., Galiana, A., Prin, Y., Sanguin, H., Fernandez, C., & Baldy, V. (2018). Mediterranean forests, land use and climate change: A social-ecological perspective. *Regional Environmental Change, 18*(3), 623–636. DOI: 10.1007/s10113-016-0994-3

Ghanam, M. (2003, décembre). *La désertification au Maroc – Quelle stratégie de lutte?* 2nd FIG Regional Conference, Marrakech, Maroc.

Gharnit, Y., Outourakht, A., Boulli, A., & Hassib, A. (2023). Biodiversity, autecology and status of aromatic and medicinal plants in Geopark M'Goun (Morocco). *Annali di Botanica, 13*, 39–54.

Ghefar, M., Boutkhil, M., & Bouazzaoui, A. (2024). Assessing the impact of anthropogenic activities on land use and land cover changes in the semi-arid and arid regions of Algeria. *Environmental Monitoring and Assessment, 196*(4), 383. DOI: 10.1007/s10661-024-12524-2

Gimeno, T. E., Pías, B., Lemos-Filho, J. P., & Valladares, F. (2009). Plasticity and stress tolerance override local adaptation in the responses of Mediterranean holm oak seedlings to drought and cold. *Tree Physiology, 29*(1), 87–98. DOI: 10.1093/treephys/tpn007

Hakam, O., Ongoma, V., Beniaich, A., Meskour, B., El Kadi, M. A., Brouziyne, Y., & Baali, A. (2024). Assessment of the impact of climate change on argan tree in the Mediterranean GIAHS site, Morocco: Current and future distributions. *Modeling Earth Systems and Environment, 10*(4), 5529–5552. DOI: 10.1007/s40808-024-02077-1

Hamida, B., & Kouba, Y. (2025). Atlas cedar response to climate-driven disturbances at its southern range edge. *Environmental Monitoring and Assessment, 197*(6), 673. DOI: 10.1007/s10661-025-14140-0

Ilmen, R., & Benjelloun, H. (2013). Les écosystèmes forestiers marocains à l'épreuve des changements climatiques. *Forêt Méditerranéenne, 34*(3), 195–208.

Institut Royal des Études Stratégiques (IRES). (2024). *L'avenir de la forêt marocaine: Rapport de synthèse.* IRES.

IPCC. (2001). *Climate Change 2001: Impacts, Adaptation, and Vulnerability.* Cambridge University Press.

IPCC. (2007). [Synthesis Report. Contribution of Working Groups I, II and III to the Fourth Assessment Report of the Intergovernmental Panel on Climate Change.]. *Climatic Change*, •••, 2007.

Jha, S. K., Negi, A. K., Alatalo, J. M., Prabhu, V., Jha, M. B., & Kumar, H. (2022). Forest Degradation Index: A Tool for Forest Vulnerability Assessment in Indian Western Himalaya. *Sustainability (Basel), 14*(23), 15655. 10.3390/su142315655. DOI: 10.3390/su142315655

Karmoude, Y., Idbraim, S., Saidi, S., Masse, A., & Arbelo, M. (2025). Efficient argan tree deforestation detection using Sentinel-2 time series and machine learning. *Applied Sciences (Basel, Switzerland), 15*(6), 3231. DOI: 10.3390/app15063231

Kouachi, M. E.. (2024). Forty-year fire history reconstruction from Landsat data in Algeria. *Remote Sensing, 16*(13), 2500. DOI: 10.3390/rs16132500

Kumar, M., Kalra, N., Singh, H., Sharma, S., Singh Rawat, P., Kumar Singh, R., & Kumar Gupta, A. Kumar, P. Ravindranath, N. (2021). Indicator-based vulnerability assessment of forest ecosystem in the Indian Western Himalayas: an analytical hierarchy process integrated approach. *Ecol. Indic, 125*. 10.1016/j.ecolind.2021.107568

Laaribya, S. (2011). *Dynamique des espaces forestiers et plan stratégique de gestion et de développement socioéconomique de la forêt de la Maâmora* (Thèse de doctorat en sciences de la vie et de l'environnement). Faculté des Sciences de Kénitra.

Laaribya, S., Alaoui, A., Ayan, S., & Dindaroglu, T. (2024). Changes in the potential distribution of Atlas cedar in Morocco in the twenty-first century according to the emission scenarios of RCP 4.5 and RCP 8.5. *Forestist, 74*(1), 16–25. DOI: 10.5152/forestist.2023.0004

Laaribya, S., Alaoui, A., & Gmira, N. (2017). La forêt marocaine et le cas du développement durable de l'arganier *Argania spinosa* L. Skeels au Maroc. *Biodiversité Biologique et Conservation, 10*(2), 1–7.

Lahssini, S.. (2015). Modeling forest transition in the Maâmora landscape under climate change. *Annals of Forest Research, 58*(2), 279–296.

Lahssini, S., Lahlaoi, H., Alaoui, H., Hlal, E., Bagaram, M., & Ponette, Q. (2015). Predicting cork oak suitability in Maâmora forest using random forest algorithm. *Journal of Geographic Information System, 7*(2), 202–210. DOI: 10.4236/jgis.2015.72017

Lamhamedi, M. S.. (2015). Nursery technologies for drought-resistant reforestation in arid Morocco. *Revue Forestière Française, 67*(2), 175–188.

Lamhamedi, M. S.. (2015). *Techniques culturales et formation en pépinière forestière*. ENFI/FAO.

Lamhamedi, M. S., Bakry, S. H., & Hamrouni, L. (2015, décembre). Mise en application de nouvelles innovations techniques, technologiques et biotechnologiques pour la restauration, la domestication et l'intensification de la culture de l'arganier. In *Actes du 3e Congrès International de l'Arganier* (pp. 215–226). Agadir, Maroc.

Linares, J. C., Taïqui, L., & Camarero, J. (2011). Increasing drought sensitivity and decline of Atlas cedar (*Cedrus atlantica*) in the Moroccan Middle Atlas forests. *Forests, 2*(3), 777–796. DOI: 10.3390/f2030777

Locatelli, B., Herawati, H., Brockhaus, M., Idinoba, M., & Kanninen, M. (2008). *Methods and tools for assessing the vulnerability of forests and people to climate change* (Working Paper No. 43).

Lybbert, T. J., Aboudrare, A., Chaloud, D., Magnan, N., & Nash, M. (2011). L'essor du marché de l'huile d'argan marocaine semble bénéficier à certains ménages ruraux tout en menaçant la forêt d'arganiers endémique. *Proceedings of the National Academy of Sciences of the United States of America*, *108*, 13963–13968. DOI: 10.1073/pnas.1106382108

M'hirit, O., Benzyane, M., Benchekroun, F., El Yousfi, S. M., & Bendaoun, M. (1998). *L'arganier: Une espèce fruitière forestière à usage multiple.* Mardaga.

M'hirit, O., Samih, A., & Malagnoux, M. (Eds.). (1994). *Le cèdre de l'Atlas (Cedrus atlantica Manetti): Présentation générale et état des connaissances.* Réseau Silva Mediterranea; Annales de la Recherche Forestière au Maroc. https://www.fao.org/3/AH002F/AH002F00.pdf

Maghreb Arab Presse / Agence Nationale des Eaux et Forêts (MAP/ANEF). (2024, 1 août). L'ANEF renforce la digitalisation du secteur forestier. *Maghreb Arab Presse.* https://www.mapnews.ma/fr/actualites/social/l%E2%80%99anef-renforce-la-digitalisation-du-secteur-forestier

Maliha, S. N., Deborah, J. C., William, G. K., & Samuel, S. (2008). Regional assessment of landscape and land use change in the Mediterranean region: Morocco case study (1981–2003). In Liotta, P. H. (Eds.), *Environmental change and human security* (pp. 143–165). Springer.

MAP/ANEF. (2024, August 1). L'ANEF renforce la digitalisation du secteur forestier. Maghreb Arab Presse. https://www.mapnews.ma/fr/actualites/social/l%E2%80%99anef-renforce-la-digitalisation-du-secteur-forestier

MAPMDREF. (2020, February 27). *Présentation de la stratégie « Forêts du Maroc 2020–2030 ».* Haut-Commissariat aux Eaux et Forêts et à la Lutte Contre la Désertification. https://www.agriculture.gov.ma/fr/actualites/presentation-de-la-strategie-forets-du-maroc-2020-2030-skhirat

Martin-StPaul, N., Delzon, S., & Cochard, H. (2017). Plant resistance to drought depends on timely stomatal closure. *Ecology Letters*, *20*(11), 1437–1447. DOI: 10.1111/ele.12851

Médail, F., & Diadema, K. (2006). Biodiversité végétale méditerranéenne. *Annales de Géographie*, *115*(651), 618–640. DOI: 10.3406/geo.2006.21290

MedECC. (2019). *Report on climate and environmental changes in the Mediterranean Region.* Union for the Mediterranean.

MedECC. (2020). *Climate and Environmental Change in the Mediterranean Basin – Current Situation and Risks for the Future. First Mediterranean Assessment Report (MAR1).* Union for the Mediterranean, Plan Bleu, UNEP/MAP. DOI: 10.5281/zenodo.4768833

Moukrim, S., Lahssini, S., Naggar, M., Lahlaoi, H., Rifai, N., Arahou, M., & Rhazi, L. (2019b). Local community involvement in forest rangeland management: Case study of compensation on forest area closed to grazing in Morocco. *The Rangeland Journal, 41*(1), 43–53. DOI: 10.1071/RJ17119

Moukrim, S., Lahssini, S., Rhazi, M., Alaoui, H. M., Benabou, A., Wahby, I., El Madihi, M., Arahou, M., & Rhazi, L. (2019a). Climate change impacts on potential distribution of multipurpose agro-forestry species: *Argania spinosa* (L.) Skeels as case study. *Agroforestry Systems, 93*(4), 1209–1219. DOI: 10.1007/s10457-018-0232-8

Moukrim, S., Lahssini, S., Rhazi, M., Menzou, K., El Madihi, M., Rifai, N., Bouziani, Y., Azedou, A., Boukhris, I., & Rhazi, L. (2022). Climate change impact on potential distribution of an endemic species *Abies marocana* Trabut. *Ekologia (Bratislava), 41*(4), 329–339. DOI: 10.2478/eko-2022-0034

Moukrim, S., Lahssini, S., Rifai, N., Menzou, K., Mharzi-Alaoui, H., Labbaci, A., Rhazi, M., Wahby, I. W., El Madihi, M., & Rhazi, L. (2020). Modélisation de la distribution potentielle de *Cedrus atlantica* Manetti au Maroc et impacts du changement climatique. *Bois et Forêts des Tropiques, 344*, 3–16. DOI: 10.19182/bft2020.344.a31888

Naggar, M. (2020). La forêt et le nouveau modèle de développement au Maroc. *Forêt Méditerranéenne*, 2, 125–134. https://www.foret-mediterraneenne.org/upload/biblio/FORET_MED_2020_2_125-134.pdf

NFI. (2018). *National Agency for Water and Forests, Morocco. Map*. NFI.

Ozenda, P. (1975). Sur les étages de végétation dans les montagnes du bassin méditerranéen. *Documents de Cartographie Écologique, 16*, 1–32.

Pandey, R., Sharma, J., Singh, R., Rawat, M., Saklani, H., Tomar, P. K., Tiwari, L., Bhatt, I. D., Chand, T., Bala, N., Panwar, V. P., & Semeraro, T. (2022). Vegetation Characteristics Based Climate Change Vulnerability Assessment of Temperate Forests of Western Himalaya. *Forests, 13*(6), 848. 10.3390/f13060848. DOI: 10.3390/f13060848

Pausas, J. G., & Fernández-Muñoz, S. (2012). Fire regime changes in the Western Mediterranean Basin: From fuel-limited to drought-driven fire regime. *Climatic Change, 110*(1-2), 215–226. DOI: 10.1007/s10584-011-0060-6

Pausas, J. G., & Keeley, J. E. (2014). Abrupt climate-independent fire regime changes. *Ecosystems (New York, N.Y.), 17*(6), 1109–1120. DOI: 10.1007/s10021-014-9773-5

Peñuelas, J., & Sardans, J. (2021). Global change and forest disturbances in the Mediterranean Basin: Breakthroughs, knowledge gaps, and recommendations. *Forests, 12*(5), 603. DOI: 10.3390/f12050603

Programme d'action national de lutte contre la désertification (PAN-LCD). (2022). *Programme d'action national de lutte contre la désertification – Version actualisée*. Rabat, Maroc.

Quézel, P., & Médail, F. (2003). *Ecology and biogeography of Mediterranean forests*. Elsevier.

Ramírez-Valiente, J. A., Sánchez-Gómez, D., Aranda, I., & Valladares, F. (2010). Phenotypic plasticity versus local adaptation for leaf ecophysiological traits in thirteen contrasting cork oak populations under varying water availabilities. *Tree Physiology, 30*, 618–627. DOI: 10.1093/treephys/tpq013

Roshani, S. H., Rahaman, M. H., Masroor, M., Sharma, Y., Sharma, A., & Saha, T. K. (2024). Vulnerability assessment of forest ecosystem based on exposure, sensitivity and adaptive capacity in the Valmiki Tiger Reserve, India: A geospatial analysis. *Ecological Informatics, 80*, 17. 10.1016/j.ecoinf.2024.102494

Schultz, C. A., Timberlake, T. J., Wurtzebach, Z., McIntyre, K. B., Moseley, C., & Huber-Stearns, H. R. (2019). Policy tools to address scale mismatches: Insights from U.S. forest governance. *Ecology and Society, 24*(1), 13. https://www.jstor.org/stable/26796926. DOI: 10.5751/ES-10703-240121

Serbouti, S., Ettaqy, A., Boukcim, H., El Mderssa, M., El Ghachtouli, N., & Abbas, Y. (2023). Forests and woodlands in Morocco: Review of historical evolution, services, priorities for conservation measures and future research. *International Forestry Review, 25*(1), 121–145. DOI: 10.1505/146554823836838745

Sharma, J., Upgupta, S., Jayaraman, M., Chaturvedi, R. K., Bala, G., & Ravindranath, N. H. (2017). Vulnerability of forests in India: A national scale assessment. *Environmental Management, 60*(3), 544–553. DOI: 10.1007/s00267-017-0894-4

Sinsin, T. E., Mounir, F., & El Aboudi, A. (2020). Analyse comparative des dynamiques spatio-temporelles dans les écosystèmes d'arganiers de plaine et de montagne, Maroc. *The International Journal of Environmental Studies, 77*, 565–580. DOI: 10.1080/00207233.2020.1723960

Thorne, J. H., Choe, H., Stine, P. A., Chambers, J. C., Holguin, A., Kerr, A. C., & Schwartz, M. (2018). Climate Change Vulnerability Assessment of Forests in the Southwest USA. *Clim. Chang, 148*, 387–402. 10.1007/s10584-017-2010-4

Timzioura, R., Ezzine, S., Benomar, L., Lamhamedi, M. S., Ettaqy, A., Zine El Abidine, A., Zaher, H., Khasa, D. P., Pepin, S., & Abbas, Y. (2025). Bibliometric Analysis of Argan (*Argania spinosa* (L.) Skeels) Research: Scientific Trends and Strategic Directions for Climate-Resilient Ecosystem Management. *Forests*, *16*(6), 892. DOI: 10.3390/f16060892

Touhami, I., Rzigui, T., Zribi, L., Ennajah, A., Dhahri, S., Aouinti, H., Elaieb, M. T., Fkiri, S., Ghazghazi, H., Khorchani, A., Candelier, K., Khaldi, A., & Khouja, M. L. (2023). Climate change-induced ecosystem disturbance: A review on sclerophyllous and semi-deciduous forests in Tunisia. *Plant Biology*, *25*(4), 481–497. DOI: 10.1111/plb.13524

United Nations. (1997). *Morocco's commitment to sustainable development: Implementation of Agenda 21 and the Rio Conventions*. United Nations.

Valladares, F., Benavides, R., Rabasa, S., Díaz, M., Pausas, J. G., Paula, S., & Simonson, W. D. (2014). Global change and Mediterranean forests: Current impacts and potential responses. In Coomes, D. A., Burslem, D. F. R. P., & Simonson, W. D. (Eds.), *Forests and global change* (pp. 47–75). Cambridge University Press. DOI: 10.1017/CBO9781107323506.005

Weißhuhn, P., Müller, F., & Wiggering, H. (2018). Ecosystem Vulnerability Review: Proposal of an Interdisciplinary Ecosystem Assessment Approach. *Environmental Management*, *61*(6), 904–915. 10.1007/s00267-018-1023-8. DOI: 10.1007/s00267-018-1023-8

Zine El Abidine, A., Bouderrah, M., Lamhamedi, M., Mounir, F., & Ponette, Q. (2020). Choix des essences de reboisement pour la forêt de la Maâmora (Maroc) sur la base de la tolérance des plants juvéniles à la sécheresse. *Physio-Géo*, *15*, 133–160. DOI: 10.4000/physio-geo.11251

Chapter 3
Challenges and Threats to Forest Resiliency in Morocco: Environmental Pressures, Climate Change, and Human Impacts

Abderrahman Aafi
http://orcid.org/0009-0000-6438-0477
Directorate of Education, Training and Research, MAPMDREF, Rabat, Morocco

Said Moukrim
http://orcid.org/0000-0002-3172-3776
Research Center of Plant and Microbial Biotechnologies, Biodiversity and Environment, Faculty of Sciences. Mohammed V University in Rabat, Morocco

ABSTRACT

Morocco's forest ecosystems, spanning 5.8 million hectares, face many threats that compromise their ecological and socio-economic functions. This chapter employs the DPSIR framework combined with multi-criteria decision analysis to systematically evaluate and prioritize threats affecting Moroccan forests. Principal component analysis reveals water stress as the central node connecting all pressures through complex feedback loops and synergistic interactions. Critical threats are water stress, deforestation, timber overexploitation, and overgrazing. Major threats include forest fires, no-timber products overexploitation, and parasitic attacks. Minor threats include intersectoral coordination, scientific-management gaps and pollution. Those threats operate through self-reinforcing degradation spirals, causing massive dieback, regeneration failure, and land degradation. These pressures trigger

DOI: 10.4018/979-8-2600-0783-9.ch003

cascading impacts on biodiversity, ecosystem services, and rural livelihoods. To address root causes, break degradation and strengthen adaptive capacity, integrated strategies are needed.

INTRODUCTION

Morocco's forest ecosystems constitute a major natural heritage of considerable ecological and socio-economic importance. Covering 9 million hectares including esparto grass steppes (*Stipa tenacissima*), of which 5.8 million hectares consist of natural forests representing 12% of the national territory, these ecosystems encompass 7 ecoregions and 39 different ecosystems (Benabid, 2000). This forest mosaic, characterized by a 20% endemism rate and harboring more than 7,000 plant species including 4,500 vascular species belonging to 940 genera and 135 botanical families, provides essential ecosystem services related to water regulation, carbon storage, soil formation and stabilization, and biodiversity maintenance, while supporting rural livelihoods through the provision of fuelwood, livestock forage, mushrooms, aromatic and medicinal plants (AMP), and other non-timber forest products (NTFP) (Aafi, 2007; Benabou, 2022; Ghanmi et al., 2014). These forest systems are currently subjected to a growing and complex combination of climatic and anthropogenic constraints that weaken their resilience and profoundly modify their evolutionary trajectories. The continuous degradation of these ecosystems, estimated at 17,000 hectares per year, demonstrates the magnitude of the pressures exerted and the urgency of integrated action (CESE, 2022). The manifestations of this degradation are multiple and interconnected (Benzyane et al., 2010).

Faced with this complexity, it is insufficient to compile a descriptive inventory of threats or to treat each stressor in isolation. A truly integrative approach requires the adoption of an analytical framework capable of revealing underlying causal mechanisms, establishing a hierarchy among risk factors according to their relative importance and urgency, identifying synergistic interactions and feedback loops between threats, and guiding effective and prioritized responses. This work addresses this methodological requirement by combining a structured approach around a systematic analytical framework. The latter transcends any descriptive approach in order to reveal causal mechanisms with multi-criteria assessment and explicit analysis of non-linear interactions and feedbacks between stressors intended to prioritize management interventions (OECD, 1993).

To structure the analysis of threats weighing on Moroccan forests, the DPSIR framework (Driving forces, Pressures, State, Impact, Responses), as defined and progressively refined in the international environmental literature (OECD, 1993; Niemeijer & de Groot, 2008), was adopted. This conceptual framework, widely adopted by environmental and scientific institutions, allows moving beyond a fragmented vision of environmental problems by establishing a clear causal chain between observed phenomena and their underlying determinants. The DPSIR framework distinguishes underlying driving forces (e.g., socio-economic dynamics, demands for land and wood), pressures (deforestation, overgrazing, pollution, fires), changes in ecosystem state (degradation of forest structure and functions), and their impacts on ecosystem services and human well-being, before examining policy and operational responses. Threats were organized according to three hierarchical levels of interaction; driving forces, direct pressures, and state changes/impacts; to identify not only symptomatic manifestations but above all the root causes on which action maximizes response effectiveness (Niemeijer & de Groot, 2008).

Furthermore, a multi-criteria decision analysis (MCDA) approach was applied to establish priorities among competing threats. Each threat was evaluated according to five standardized criteria: spatial extent (proportion of national forest territory affected by the threat), temporal dynamics (speed and irreversibility: progression rate of the threat and degree of irreversibility of induced changes), degradation intensity (magnitude of changes in the structure, composition, and functioning of affected ecosystems), and propagation potential (cascading effects also integrating systemic interactions with other pressures) (Blagojević et al., 2019). This assessment was based on a documentary synthesis of available scientific and technical literature, complemented by an online survey of a panel of forestry experts and practitioners representing various institutions (forest administration, research, NGOs, civil society). The survey integrated field knowledge and scientific judgments to inform trade-offs between immediate priorities and long-term strategies.

Beyond a list of isolated threats, one of the methodological contributions of this work is the explicit examination of non-linear interactions and feedbacks between threats. Consistent with the principles of systems ecology, threats or stressors rarely act independently (Holling, 1973). Rather, they interact through feedback mechanisms that can amplify or attenuate threat effects, thereby generating multiplicative effects, non-linear dynamics, and sometimes abrupt regime shifts in ecosystems (Holling, 1973). Explicit mapping and discussion of the most important feedback loops and synergistic interactions among different threats were conducted to detect configurations likely to increase the vulnerability of ecological systems. These multiplicative interactions are particularly important for understanding forest degradation trajectories, enabling the identification of configurations where interventions can break degradation spirals.

By adopting this integrated approach, combining DPSIR with MCDA and systemic feedback analysis, this chapter aims to provide scientists and decision-makers with a clear framework for assessing the vulnerability of North African forests, prioritizing interventions, and designing adaptive management strategies that sustainably strengthen the resilience of forest ecosystems and the communities that depend on them.

2. PRIMARY DRIVERS

Within the adopted conceptual framework, driving forces are higher-level determinants that explain the emergence and intensification of pressures exerted on forests. They refer to socio-economic and biophysical dynamics that shape land uses, resource demand, and the vulnerability of ecological systems (OECD, 1993). The driving forces identified as most influential for Moroccan forests are: climate change (temperature and precipitation components) and major socio-economic drivers.

2.1. Climate Change

Climate observation series indicate a marked warming trend across the entire Moroccan territory. Various reports (IPCC, 2021; DMN, 2020) revealed an increase in average temperature in Morocco of approximately 1.5°C since the beginning of the 20th century. Multi-model climate projections anticipate an additional rise of 2 to 3°C by 2050 depending on greenhouse gas emission scenarios. Different bioclimatic zones, particularly in the interior and eastern regions of Moroccan territory, are especially affected by this warming. This temperature increase is accompanied by a progressive decrease in snow cover in the High and Middle Atlas as well as in southern regions. Relevant variables acting as drivers for forest state include: (i) increase in annual and seasonal average temperatures, (ii) increased frequency and intensity of heat waves with more days above critical thresholds, and (iii) lengthening of water deficit seasons expressed by increased potential evapotranspiration indices. National climate diagnostics indicate a trend toward greater precipitation irregularity and intensification of extremes, although average spatial trends remain heterogeneous across regions (DMN, 2020; CESE, 2022). Morocco has experienced substantial fluctuations in rainfall regimes in recent decades, affecting not only precipitation volume but also their temporal and spatial distribution. The country has alternated between episodes of prolonged drought and intense, brief precipitation events. Projections indicate that Morocco will experience more prolonged and intense drought episodes. Key variables to consider as drivers are: i) the downward trend in annual average precipitation in certain regions and increased spatial variability; ii) increased frequency and duration of rainfall deficit periods (droughts); and iii) occurrence of intense punctual rainfall events, measured by intensity and temporal dependence indices (IPCC, 2021; CESE, 2022; DMN, 2020).

Furthermore, extreme climate events, particularly heat waves, storms, and floods, are becoming more frequent in Morocco. The frequency of heat waves has increased, with summer temperatures regularly exceeding historical levels. Intense precipitation can trigger flash floods, particularly in urban areas and sectors lacking adequate drainage systems.

2.2 Socio-Economic Driving Forces

Demographic dynamics and spatial occupation patterns constitute major driving forces of pressure exerted on Moroccan forests. Population growth, combined with rural exodus and urban expansion processes, increases demand for land and forest resources, particularly in peri-urban areas where forest edges are converted for housing and peri-urban activities (CESE, 2022). These developments restructure land use trajectories and modify spatial configurations conducive to habitat fragmentation (Seto et al., 2012). Concurrently, economic determinants, notably relative fuel prices, energy supply policies, and market conditions, orient forest resource exploitation. Consumption of timber forest products (TFP) and non-timber forest products (NTFP) strongly depends on these economic variables, which determine the intensity and nature of harvesting (Ghanmi et al., 2014; Aafi, 2007). Market dynamics, including export demand for certain species (e.g., argan oil, essential oils), reinforce these pressures according to sectoral and territorial logics.

Agricultural expansion and intensification (conversion to crops or pastures, use of irrigation and inputs) sometimes constitute drivers of forest cover loss and degradation of forest-agriculture interfaces, influenced by agricultural policies and market access (Geist & Lambin, 2002). Infrastructure development, road networks, and tourism facilities increase the accessibility of forest territories, facilitating exploitation, conversion, and spatial fragmentation (Seto et al., 2012; CESE, 2022).

Furthermore, institutional and land tenure arrangement's structure stakeholder behaviors: insecurity of use rights, gaps in norm enforcement, and weakness of coordination and local governance mechanisms act as structural factors favoring overexploitation and degradation (Geist & Lambin, 2002). These driving forces operate simultaneously and interact at different scales, conditioning the distribution and intensity of direct pressures on forest ecosystems.

3. MAIN DIRECT PRESSURES ON FOREST ECOSYSTEMS

Analysis of the literature review and multi-criteria survey results aggregated scores on the selected operational criteria (spatial extent, intensity, progression speed, irreversibility, and cascade effect potential) to establish a threat ranking. The combined responses of these criteria and consideration of propagation potential via feedback loops distinguished threats of natural, anthropogenic, and governance origin and grouped the studied threats according to their relative importance into three classes, namely: critical threats (water stress; deforestation and conversion of forest areas; overexploitation of TFP; and overgrazing), major threats (fires; overexploitation of NTFP; and parasitic attacks), and minor threats (coordination/intersectoral deficits; gap between scientific research and management; pollution; and invasive species).

3.1. Multi-Criteria Threat Prioritization

Principal component analysis (PCA) was conducted after validating data suitability through appropriate statistical tests. The Kaiser-Meyer-Olkin test (KMO = 0.71) confirms sampling adequacy for factor analysis, with individual values per variable ranging from 0.57 to 0.76, attesting to sufficient correlations between descriptors. Bartlett's test of sphericity ($\chi^2 = 68.06$, df = 10, $p < 0.001$) rejects the hypothesis of no correlation between variables, validating the relevance of dimensional reduction. The first two principal components jointly explain 95.8% of total variance (PC_1: 81.4%; PC_2: 14.4%), demonstrating that a two-dimensional projection effectively captures the multidimensional structure of threats. This high proportion of explained variance validates the robustness of the analysis and justifies interpretation of the configurations observed in the biplot (Figure 1).

The first principal axis (PC_1, 81.4% variance) reveals a dominant gradient opposing threats with high spatial and temporal footprints to governance and diffuse pressures. The positive quadrant of axis 1 concentrates threats characterized by high values of extent, intensity, and progression speed (deforestation and conversion of forest areas, water stress, overgrazing, and overexploitation of TFP). These threats exhibit quasi-collinearity with extent, intensity, and speed vectors, indicating their structural interdependence: spatially extensive threats tend to progress rapidly and with high intensity. Conversely, the negative quadrant of axis 1 groups governance threats (intersectoral coordination deficits, research-management gap) and certain threats of natural or diffuse anthropogenic origin (invasive species, pollution, parasitic attacks), characterized by low values across all quantitative descriptors but playing a role as indirect aggravating factors. The second axis (PC_2, 14.4% variance) introduces an orthogonal dimension centered on the irreversibility of disturbances, discriminating threats according to their ecosystem recovery capacity. Irreversibility shows relative independence from other descriptors, confirming that it constitutes a distinct dimension from immediate severity and spatial magnitude.

Analysis of variable vectors reveals structuring correlations between descriptors. The cascade effect shows a marked positive correlation with extent, intensity, and progression speed, suggesting that spatial magnitude and rapid dynamics favor cascading disturbances. The quasi-collinearity between progression speed and cascade effect underlines the intrinsic link between the rapidity of threat propagation and its potential to trigger systemic disturbances in the forest ecosystem. This relationship is particularly evident for deforestation and water stress, positioned at the extremity of these vectors.

Figure 1. Biplot of principal component analysis of threats to forest ecosystems

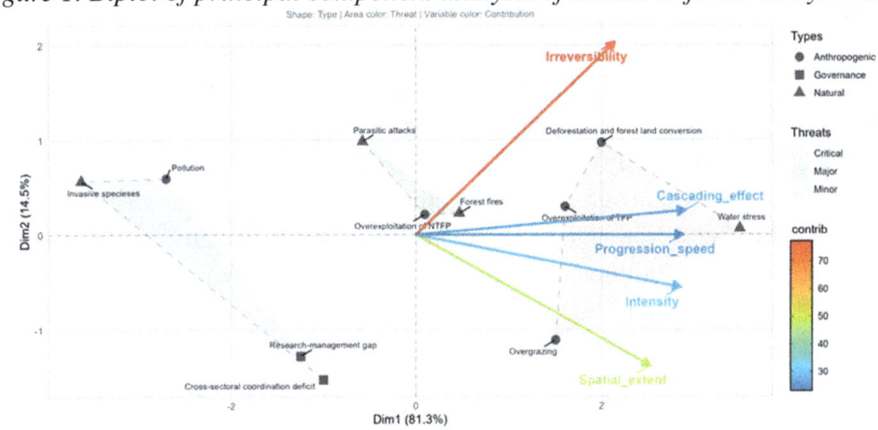

3.2. Threat Typology by Criticality Level

The combined responses of these criteria and consideration of propagation potential via feedback loops grouped the studied threats according to their relative importance into three classes exhibiting clear spatial segregation in the factorial plane.

3.2.1. Critical Threats

These threats occupy the zone characterized by maximum values of irreversibility, cascade effect, and progression speed (Figure 1, red shading). This category includes water stress, deforestation and forest lands conversion, overexploitation of TFP, and overgrazing.

Water stress in Morocco is characterized by a combination of climatic trends and water indicators revealing progressive degradation of water availability. National diagnostics indicate an increase in potential evapotranspiration and high rainfall variability with prolonged deficit periods (DMN, 2020; CESE, 2022). Renewable water resources per capita decreased from approximately 2,500 m^3 in 1960 to less than 650 m^3 currently, and certain projections estimate they could fall below 500 m^3 per capita/year by 2030; other scenarios predict a possible 20 to 50% reduction in mobilizable water volumes by 2030 (CESE, 2022). National diagnostics also indicate that, if trends continue, nearly 80% of currently available resources could experience significant scarcity situations over the next 25 years. Water services (rivers, aquifers) already show signs of stress related to overexploitation and local pollution, and climate series indicate an increase in the frequency and duration of drought episodes, mainly in certain territories experiencing several consecutive years of deficit (DMN, 2020). This temperature increase intensifies evapotranspiration and reduces water availability for native and introduced forest species (IPCC, 2021).

Permanent conversion of forest land to other uses remains a major driver of forest cover loss in Morocco. National diagnostics estimate significant historical losses and indicate growth in agricultural and peri-urban areas driven by food demand logics, urban expansion, and land valorization. Operational pressures include cultivation of peri-forest zones, conversion of rangelands to agricultural land, installation of infrastructure (roads) facilitating access to forest massifs, and densification of peri-urban and tourism occupations (Geist & Lambin, 2002). These dynamics are sustained by economic factors (land prices, market access), speculative logics (tourism, urbanization), and land tenure insecurities. Competition for these limited areas generates use conflicts between different land uses, particularly marked in overlap zones between collective lands and forest domain. Land tenure insecurity, overlapping use rights, and absence of effective arbitration mechanisms exacerbate these tensions. The weakness of integrated territorial planning and limited enforcement of regulations accelerate land use changes, making deforestation fronts spatially concentrated and recurrent, particularly in the Atlas, Rif, and coastal regions (Seto et al., 2012; CESE, 2022). Relevant indicators for measuring this pressure include annual cover loss rates, the share of land converted by final use (agriculture, urbanization, infrastructure), and land use changes at the scale of different watersheds (CESE, 2022).

Overexploitation of timber forest products in Morocco manifests through sustained legal and illegal harvesting, oriented toward providing fuelwood, timber, and materials for construction and handicrafts. Exploited volumes reach approximately 1.5 million m^3 per year, exceeding the sustainable regeneration capacities of stands (CESE, 2022). Structuring factors of this pressure include domestic energy demand, absence of accessible energy alternatives in many rural areas, particularly in mountain zones, and local and regional market demand for certain valued species. Harvesting practices vary between manual cutting, selective harvesting of sought-after species (e.g., holm oak, thuya, cedar), artisanal charcoal production, and collection of deadwoods. Increased accessibility through opening of tracks and roads facilitates large-scale commercial exploitation and log extraction to urban markets (FAO, 2020). Moreover, the sector includes significant informal components, sometimes making it difficult to control harvesting and trace extracted volumes. Pressures on TFP also operate via market dynamics (relative prices, seasonal cycles) and political decisions (energy substitutes, exploitation regulations), while institutional regulatory capacities remain uneven across territories.

Overgrazing, particularly by sheep and goats and in certain southern areas by camels, constitutes a major pressure on Moroccan forest ecosystems. Pastoral production in forests amounts to approximately 1.5 billion forage units in normal years, contributing 17% of the national fodder balance. Forests contribute to feeding 40% of the national livestock, a proportion that can double during drought years when forests become the only refuge for herds (Moukrim et al., 2019a). Repeated trampling by livestock compacts soils, reducing their porosity, infiltration capacity, and aeration. Intensive grazing systematically eliminates young plants and seedlings, preventing natural stand regeneration. This excessive pastoral pressure is observed across the entire forest domain, particularly in argan, holm oak, Atlas cedar, juniper, and pine stands.

3.2.2. Major Threats

These threats exhibit substantial correlations with intensity and extent and moderate correlations with irreversibility (Figure 1, green shading). They include forest fires, overexploitation of NTFP, and parasitic attacks and diseases.

Forest fires constitute a recurrent and growing threat to Moroccan forest ecosystems, exacerbated by climate change and land use modifications. Analysis of data for the period 1960-2019 reveals a total damaged area of nearly 184,400 hectares. Studies indicate that the number of forest fires in Morocco has doubled over the past two decades, with more intense fires occurring particularly in central and northern regions of the country. Climatic conditions like high temperatures, prolonged droughts, and strong winds, create longer and more intense fire seasons. Fires consume not only aboveground biomass but also release carbon stored in soils and trees, while degrading edaphic structure and compromising regeneration cycles.

NTFP products including AMP (*Origanum elongatum, O. compactum, Thymus satureoides, Rosmarinus officinalis, Artemisia herba-alba, Argania spinosa* oil), mushrooms (*Terfezia arenaria, T. arenaria, T. leptoderma, T. boudieri, Tuber spp:*

T. asa, T. oligospermum, T. gennadii, T. borchii var. phaerosperma, T. excavatum, T. brumale, T. rufum, T. unicinatum/aestivum, T. olbiensis, Delastria rosea, Tirmania spp., Picoa juniperi), and wild fruits are harvested at unsustainable levels (Ghanmi et al., 2014). Furthermore, widespread illegal hunting and poaching for meat and trophies result in overexploitation and decline of wild populations of various birds and mammals. Market dynamics, particularly growing export demand for certain products such as argan oil and AMP, reinforce these pressures according to sectoral and territorial logics. The economic valorization of these products, without regulation and sustainable management mechanisms, leads to intensive exploitation compromising resource sustainability (FAO, 2020).

Trees stressed by climate change, particularly by water stress and high temperatures, become more vulnerable to diseases and pest insect attacks. In the High and Middle Atlas regions as well as in the Rif, several notable parasitic threats affect forest stands, including bark beetles (*Ips sexdentatus, Orthotomicus erosus, Phaenops marmottani*), processionary moths (*Thaumetopoea bonjeani, T. pityocampa*), and pathogenic fungi such as root rot (*Armillaria mellea*) causing root decay (Linares et al., 2011).

3.2.3. Minor Threats

These threats exhibit weak correlations across all quantitative descriptors (Figure 1, blue shading). This category includes coordination and intersectoral deficits, gap between scientific research and management, pollution, and invasive species.

Weak intersectoral coordination among different actors and institutions involved in forest resource management and related sectors constitutes a major structural threat. Forest, agricultural, water, territorial planning, and tourism development policies are often developed and implemented in a compartmentalized manner, without effective mechanisms for integration and arbitration between competing objectives (Niemeijer & de Groot, 2008).

This institutional fragmentation translates into inconsistencies in strategic orientations, overlapping competencies, regulatory gaps, and limited capacity to address cross-cutting issues such as integrated watershed management or coordination of efforts to combat certain environmental issues. Local governance mechanisms remain weak, with limited participation of local communities in decision-making processes, despite their central role in daily forest resource management.

A significant gap persists between available scientific knowledge and its effective application in forest management practices. Research results on forest ecology, stand dynamics, species responses to climate change, and restoration techniques are often insufficiently transferred to practitioners and decision-makers. This gap results from several factors: inadequate communication channels between researchers and managers, lack of continuous training for technical staff, fragmented monitoring and evaluation systems, absence of institutionalized knowledge transfer mechanisms, and limited resources for adaptation and experimentation with new approaches in the field. Integration of local and traditional knowledge into management strategies also remains insufficient, despite its relevance for understanding local ecological dynamics and designing culturally appropriate interventions. This gap reduces intervention effectiveness and limits innovation capacity in the face of emerging challenges posed by climate change and growing anthropogenic pressures.

Pollution of agricultural origin constitutes a diffuse but growing threat to forest soils. Agricultural intensification, characterized by an increase of more than one million tons of fertilizers over the past 20 years, results in accumulation of nutrients (nitrogen, phosphorus) and chemical substances (pesticides, herbicides) in adjacent forest ecosystems through runoff and leaching. This chemical pollution alters the physicochemical and biological properties of soils, modifying their pH, disrupting natural biogeochemical processes, and affecting microbial communities essential to forest ecosystem functioning.

Furthermore, intensive irrigation without adequate drainage in arid zones causes progressive soil salinization, threatening approximately 500,000 hectares mainly located in major hydraulic zones. In the provinces of Ouarzazate and Errachidia alone, approximately 22,000 hectares of irrigated land and 5,000,000 hectares of rangelands are affected by salinization (Benzyane et al., 2010). Saline (4-6 g/L) and very saline (> 16 g/L) soils develop in palm groves, with annual deposits of 50,000 to 70,000 tons of salt in certain cultivated areas such as Tafilalt.

The introduction and proliferation of exotic invasive species modify the composition and structure of forest communities. These introductions result primarily from commercial activities, poorly designed reforestation programs, and land use modifications. Notable invasive species include *Acacia mollissima* and various *Eucalyptus* species (more than 250 species and clones introduced in 40 *arboreta* across different regions of Morocco, of which 30 are commonly used in plantations for timber production, industrial wood, service wood, fuelwood, pulpwood, soil protection and conservation, or ornamental purposes).

3.3. Feedback Loops and Synergies Between Threats

One of the most critical, yet often neglected, aspects in analyzing threats to forest ecosystems is recognizing that these threats do not operate in isolation. Rather, they interact through feedback loops and synergistic effects that create self-reinforcing degradation spirals far exceeding the simple addition of individual threats (Scheffer et al., 2001). These non-linear interactions can lead ecosystems to cross critical thresholds (tipping points) beyond which regime shifts occur, characterized by rapid and potentially irreversible modification of ecosystem structure and functioning (Folke et al., 2004).

Water stress constitutes a central node in this network of interactions, connecting directly or indirectly all observed pressures. Increasing temperatures and modified precipitation regimes intensify evapotranspiration, physiologically weakening trees and diminishing their natural defense mechanisms (Anderegg et al., 2015; Allen et al., 2010). This increased vulnerability amplifies susceptibility to parasitic attacks by bark beetles (*Ips sexdentatus, Orthotomicus erosus*), processionary moths (*Thaumetopoea bonjeani, T. pityocampa*), and pathogenic fungi (*Armillaria mellea*), causing massive dieback in Atlas cedar (Cedrus atlantica), cork oak (*Quercus suber*), Moroccan fir (*Abies maroccana*), Spanish juniper (*Juniperus thurifera*), and pine stands (Linares et al., 2011; Camarero et al., 2015). These mortalities reduce vegetation cover, exposing soils to accelerated erosion, decreasing their water retention capacity and fertility, thereby aggravating initial water stress in a positive feedback loop (Breshears et al., 2005). Simultaneously, water stress increases the flammability of desiccated vegetation, creating conditions favorable to fires whose release of stored carbon amplifies warming, prolonging droughts and favoring conditions even more conducive to new fires in a self-sustaining spiral (Mharzi Alaoui et al., 2017; Flannigan et al., 2009).

Deforestation and conversion of forest areas, driven by agricultural expansion and urbanization, fragment habitats by creating edge effects where altered microclimatic conditions increase stress on peripheral trees and facilitate establishment of invasive species that outcompete weakened native vegetation (Vilà et al., 2011). This fragmentation reduces forest productivity and intensifies pressure for overexploitation of TFP and NTFP, which accelerates degradation of remaining stands and closes a progressive overexploitation loop. Overgrazing compacts soils and eliminates natural regeneration, favoring erosion which, combined with cover loss due to fires and dieback, triggers in certain regions a desertification dynamic where wind erosion mobilizes loose substrates (Reynolds et al., 2007; D'Odorico et al., 2013).

These interactions and synergies are exacerbated by conversion of natural areas that intensify fragmentation, the gap between research and management that limits application of adapted solutions, and insufficient intersectoral coordination that prevents the integrated responses necessary to break these interconnected loops (Niemeijer & de Groot, 2008). The whole creates a system where disturbances amplify each other, with significant mortality rates and quasi-absent regeneration in vast areas, increasing the risk of crossing thresholds beyond which ecosystems shift toward degraded states difficult to restore, such as irreversible transformation of productive forests into open shrub formations or desertified zones already observed in several regions of the country.

4. IMPACTS

4.1. Impacts on Structural and Functional Integrity of Forest Ecosystems

Moroccan forest ecosystems undergo massive dieback resulting from synergistic interaction between climatic water stress, parasitic attacks, and anthropogenic pressures (Linares et al., 2011; Camarero et al., 2015). Field observations reveal mortality rates reaching up to 40% in certain localities, primarily affecting Atlas cedar (*Cedrus atlantica*) stands in the Middle and High Atlas, cork oak (*Quercus suber*) in Maamora forest, Moroccan fir (*Abies maroccana*) in the Rif, holm oak (*Quercus rotundifolia*) and Spanish juniper (Juniperus thurifera) in the High Atlas, as well as prickly juniper (*Juniperus oxycedrus*) in the Oriental region and acacia (*Acacia raddiana*) in the southern region (Benzyane et al., 2010) (Appendix: Figure 2, photos 1-10).

These diebacks are accompanied by quasi-generalized failure of natural regeneration in vast areas, resulting from the combination of water stress compromising seed germination and seedling establishment, overgrazing systematically eliminating young plants, and degradation of edaphic conditions. In certain ecosystems, overgrazing leads to total absence of natural regeneration accompanied by soil loss through severe erosion. This dynamic progressively transforms stand structure, with demographic aging, juvenile recruitment rarefaction, and evolution toward open and impoverished forest formations presenting considerably reduced resilience to future disturbances (Allen et al., 2010). Examples of successful natural regeneration of Atlas cedar and artificial regeneration of key species illustrate, however, the restoration potential when conditions are favorable (Appendix: Figure 2).

Conversion of forest land into agricultural areas, urban zones, and infrastructure resulted in historical loss of approximately 4 million hectares since the 1950s, with continuous degradation estimated at 17,000 hectares annually (CESE, 2022). This fragmentation creates a mosaic of isolated habitats exhibiting pronounced edge effects. In fragmented forests, forest cover reduction can reach 40%, modifying microclimatic conditions at forest-matrix interfaces: increased daytime temperatures, increased wind exposure, reduced atmospheric humidity, and increased light penetration (Fischer & Lindenmayer, 2007).

These edge conditions increase physiological stress of peripheral trees, rendering them more vulnerable to pathogens, pest insects, and droughts. Fragmentation also reduces ecological connectivity between populations, limiting gene flows and dispersal capacities, particularly critical in a climate change context requiring species migrations toward climatically more favorable zones. Spatial isolation of forest fragments compromises long-term population viability by increasing risks of genetic drift and local extinction, particularly for species with low dispersal capacity (Haddad et al., 2015).

Land degradation constitutes a multifaceted impact resulting from interaction between climate change, deforestation, overgrazing, and unsustainable agricultural practices, manifesting through three interconnected processes: erosion, salinization, and desertification (Figure 2, photos 12-16). Water and wind erosion intensifies with loss of protective vegetation cover, reaching rates of 6 to 30 tons per hectare per year depending on regions, particularly in mountainous and hilly areas. In southern regions, specific erosion exceeds 6 m^3/ha/year in the *Ziz* valley and reaches approximately 5.8 m^3/ha/year in the middle *Drâa* valley (Navas et al., 2009). Overgrazing aggravates this dynamic through soil compaction, reducing their porosity, infiltration capacity, and aeration, while increasing their vulnerability to removal by erosive forces. This erosion causes progressive fertility loss, reduced water retention capacity, and edaphic structure degradation compromising long-term ecosystem productivity (Breshears et al., 2005) (Figure 2, photo 11).

Desertification and sand encroachment represent an extreme form of land degradation, particularly manifest in southern and southeastern Morocco regions. Wind erosion mobilizes loose substrates and poorly developed soils, creating mobile dunes that progress at an average speed of 10 to 15 meters per year. Affected areas are considerable: 30,000 hectares in Ouarzazate and Zagora provinces, and 250,000 hectares in Errachidia province, with 60% of Ziz and Tafilalt oases area exposed to this phenomenon (Benzyane et al., 2010). Sandy deposits form dunes reaching 4 meters height and dune ridges that penetrate palm grove interiors, progressively covering present crops associated with date palm (Phoenix dactylifera, Lawsonia inermis, Medicago sativa), as well as spontaneous vegetation and trees such as tamarisks (Tamarix aphylla) and acacias (Acacia raddiana). This dynamic leads to progressive abandonment of lands that have become unproductive, extending bare surfaces that become in turn sources of wind sediment emission, amplifying sand encroachment of adjacent still-productive zones (Reynolds et al., 2007) (Figure 2, photos 12-15).

Salinization constitutes the third major land degradation process, threatening approximately 500,000 hectares mainly located in major hydraulic zones. In Ouarzazate and Errachidia provinces alone, approximately 22,000 hectares of irrigated land and 5,000,000 hectares of rangelands are affected by this phenomenon. A study conducted in Tafilalt reveals that 35% of palm grove soils are saline (4 to 6 g/L) and 18% very saline (> 16 g/L), with annual deposits of 50,000 to 70,000 tons of salt in the cultivated area (Benzyane et al., 2010). This phenomenon has extended over growing areas in recent decades due to reduced precipitation, increased evapotranspiration, and intensive irrigation without adequate drainage. Salinization modifies soil physicochemical properties, reduces their capacity to support native vegetation, and favors establishment of salt-tolerant bioindicator species such as Zygophyllum gaetulum, thus modifying plant community composition and structure and compromising ecosystem functions (D'Odorico et al., 2013) (Figure 2, photo 12).

4.2. Impacts on Biological Diversity

Combined pressures of water stress, habitat fragmentation, overgrazing, and overexploitation cause population declines particularly affecting endemic and restricted-range species. Several Moroccan Forest species now appear on the IUCN Red List as threatened due to habitat loss and degradation. The Barbary macaque (*Macaca sylvanus*), emblematic species of Rif and Atlas forests, experiences marked demographic decline resulting from increasing fragmentation of its forest habitat, limiting movements between populations and reducing genetic viability of isolated populations. Overexploitation of non-timber forest products, particularly AMP and mushrooms, causes population reductions affecting not only these species but also associated trophic networks. Rare and endemic species, often characterized by restricted ranges and specific ecological requirements, present particular vulnerability because they can neither migrate nor adapt sufficiently rapidly to rapid environmental changes, considerably increasing their risk of local or global extinction (Benzyane et al., 2010).

Climate change induces profound modifications in spatial and temporal distributions of forest species under current and future climatic conditions (Benabou et al. 2022; Moukrim et al., 2019; 2020; 2022). Studies conducted in Moroccan mountainous regions document altitudinal migrations of the order of 100 to 200 meters upward for several plant species, these movements being an adaptive response to temperature increase and search for more favorable thermal and water conditions (Benabid, 2000). Atlas cedar shows a tendency to migrate toward higher altitudes to maintain its optimal climatic envelope. Concurrently, longitudinal migrations northward are observed for certain species seeking to escape elevated temperatures and find appropriate water regimes. These spatial movements are accompanied by changes in forest community composition, with a tendency toward progression of thermophilic and drought-resistant species at the expense of mesophilic and montane species. Distribution modifications also generate phenological shifts affecting temporal synchronization between plants and their pollinators, seeds and their dispersers, or herbivores and their food resources. Early flowering induced by warmer spring temperatures can occur before pollinator emergence, creating temporal mismatches that disrupt mutualistic interactions and trophic networks, with cascading consequences on ecosystem structure and functioning (Peñuelas et al., 2009).

Anthropogenic and climatic disturbances create conditions favorable to establishment and proliferation of exotic invasive species. Habitat fragmentation generates disturbed habitats where invasive species find ecological niches conducive to their establishment. These invasive species, often characterized by rapid growth strategies, high seed production, and disturbance tolerance, enter into direct competition with native species weakened by water stress and other pressures (Vilà et al., 2011). Invasive species modify habitat structure by creating dense monospecific stands, alter biogeochemical cycles, and can change disturbance regimes, particularly by increasing fuel loads and fire risks. They progressively outcompete native vegetation, reducing local floristic diversity and modifying community composition. This biotic homogenization reduces ecosystem structural complexity and their capacity to provide diversified ecological niches for fauna, causing cascading effects across the entire trophic network (Gaertner et al., 2009).

4.3. Alteration of Ecosystem Service Provision

Forest cover degradation compromises ecosystem regulatory functions. Reduction of precipitation interception by forest canopies, decreased infiltration in compacted and degraded soils, and increased surface runoff profoundly modify water balances at watershed scales, resulting in reduced groundwater recharge, stream drying during low-flow periods, and aggravated flood risks during rainfall events (Andréassian, 2004). Forests also play a crucial role in climate regulation through atmospheric carbon sequestration. Fires massively release stored carbon, generating significant quantities of CO_2 annually in Mediterranean countries, while massive forest diebacks reduce sequestration capacity by decreasing living biomass and primary productivity, transforming forests from carbon sinks into net emission sources (IPCC, 2021). Forest cover loss directly exposes soils to erosive forces, primarily in mountainous regions, increasing vulnerability to landslides and mass movements. Fires also degrade air quality through emission of smoke and fine particles, presenting significant health risks for exposed populations, particularly individuals suffering from respiratory pathologies (Breshears et al., 2005; ANEF, 2019).

Forest degradation reduces availability of forest products essential to rural populations. Fuelwood production, primary energy source for many rural households, decreases with dieback and overexploitation, while availability of timber, construction materials, and NTFP, AMP, mushrooms, wild fruits, cork, honey, declines under combined effects of overexploitation and habitat degradation (Ghanmi et al., 2014). Degradation of forest and pastoral formations, combined with sand encroachment and desertification, progressively reduces pastoral carrying capacity of forest areas, threatening traditional livestock systems and creating a loop of overgrazing and accelerated degradation. Water availability constitutes a critical issue: renewable water resources per capita have fallen, with projection of possible reduction in mobilizable volumes, directly threatening forest ecosystem health, agricultural productivity, and socio-economic development (CESE, 2022).

Moroccan forest ecosystems hold considerable cultural and aesthetic value, constituting identity elements for local communities and attractions for tourism and ecotourism. Forest landscape degradation through dieback, recurrent fires, sand encroachment, and desertification erodes these aesthetic values and reduces natural site attractiveness. Other traditional systems and practices, representing exceptional cultural heritage illustrating population ingenuity in natural resource management and adaptation to arid environments, are threatened. Loss of these systems represents not only environmental degradation but also erosion of intangible cultural heritage, traditional knowledge, and territorial identities linked to these ecosystems.

4.4. Socio-Economic Impacts on Communities and Territories

Moroccan rural communities maintain close dependence on forest resources for their livelihoods. Forest degradation directly threatens these populations by reducing availability and accessibility of essential resources: fuelwood for cooking and heating, construction materials for housing, NTFP for food, supplementary income and medicinal uses, and forage for livestock feeding. Income-generating activities related to forests such as beekeeping, AMP collection and commercialization, mushroom gathering, and ecotourism experience marked decline, compromising household economic stability and increasing their vulnerability to poverty (FAO, 2020).

Food and pastoral security are particularly threatened. Reduced forest forage production directly affects national livestock that depends on forest rangelands. Sand encroachment and salinization lead to land abandonment and compromise local agricultural production systems. Irrigation infrastructure suffers significant damage caused by sand encroachment, requiring costly maintenance and rehabilitation investments. This productive base degradation threatens traditional production systems and associated lifestyles, accumulated over centuries of ingenious adaptation to environmental constraints.

Forest degradation increases territorial vulnerability to natural hazards. Floods represent a recurrent threat causing human losses and considerable economic damages, with 94% of flood-prone areas affected by urbanization on floodplains. Desertification and sand encroachment cause definitive territory loss. These processes generate considerable economic costs related to productivity loss, restoration necessity, and investments required for infrastructure protection. Health risks associated with forest fires, particularly exposure to smoke and fine particles, affect population respiratory health, particularly vulnerable groups (ANEF, 2019).

Forest ecosystem degradation and land productivity loss generate significant migratory pressures. Populations confronted with impossibility of maintaining traditional activities in areas that have become uninhabitable or unproductive are constrained to migration, generally toward urban zones. This migratory dynamic contributes to urbanization acceleration (3.5% annual urban growth rate), exerting in turn new pressures on peri-urban ecosystems and aggravating problems related to unplanned urban expansion. These movements cause erosion of traditional knowledge and sustainable natural resource management practices accumulated over generations.

5. PATHWAYS TO RESILIENCE AND SUSTAINABLE FOREST MANAGEMENT (SFM)

Analysis of driving forces, pressures, their systemic interactions, and their impacts reveals that Moroccan forest ecosystems are engaged in degradation trajectories characterized by self-reinforcing feedback loops where disturbances amplify each other, increasing the risk of crossing critical thresholds leading to potentially irreversible regime shifts (Scheffer et al., 2001; Folke et al., 2004). Faced with this complexity, a fragmented approach targeting each threat in isolation proves insufficient. Forest socio-ecosystem resilience requires an integrated sustainable forest management (SFM) strategy intervening simultaneously on root causes, proximate pressures, and adaptive capacities, according to multi-level logic (Messier et al., 2015). This multi-level architecture reflects an adaptive management philosophy where continuous evaluation, learning by experience, and iterative strategy adjustment allow navigating the uncertainty inherent in complex socio-ecological systems (Figure 2, photos 17-24) (Berkes, 2009).

5.1. Strategic and Institutional Framework

Establishing a robust strategic framework constitutes the foundation on which all operational interventions rest. Integrated national policies, notably the Morocco Forests 2020-2030 strategy, Green Generation 2020-2030 strategy, and national program for drinking water supply and irrigation 2020-2027, provide an overall vision articulating sustainable forest management objectives with water security, agricultural development, and climate adaptation issues. These strategic frameworks primarily target critical threats of structural water stress, deforestation, and overexploitation by establishing quantified objectives, financing mechanisms, and monitoring-evaluation systems. Climate-forest-water integration in these policies explicitly recognizes intersectoral interdependencies and the need to avoid counterproductive trade-offs where gains in one sector generate costs in another.

Establishing multi-stakeholder platforms bringing together forest, agricultural, water, territorial planning, and tourism development departments, as well as civil society and private sector representatives, allows developing arbitration mechanisms between competing objectives and designing coherent strategies maximizing intersectoral coordination and synergies (Sayer et al., 2013). Integrated territorial planning, operationalized at watershed and landscape scales, constitutes the operational tool of this coordination, allowing managing land use conflicts, which represents an aggravating factor of fragmentation, through participatory and transparent zoning and resource allocation processes.

Participatory and decentralized governance transferring part of responsibilities and decision-making powers to local communities, addressing both efficiency issues, given fine knowledge of local dynamics, and equity issues, as they bear direct conservation costs. Formalizing partnerships through signing agreements between the department responsible for forest heritage management and populations via inclusive formal structures such as associations and cooperatives creates a legal framework for resource co-management. This approach particularly targets critical threats of overexploitation and overgrazing by transforming incentives: when communities benefit directly and sustainably from sustainable resource management, they become active conservation agents rather than pressure sources (Agrawal et al., 2008). Securing use rights on one side and land tenure on the other constitutes an essential prerequisite to this long-term commitment.

Strengthening legal frameworks and their effective enforcement targeting illegal deforestation and unsustainable overexploitation of TFP and NTFP. Increased repression for illegal activities, combined with developing surveillance systems using both advanced technologies (satellite imagery, drones, geographic information systems, sensors, and others) and traditional patrol methodologies, improves deterrence and detection (Kissinger et al., 2012). Strengthened collaboration with local authorities ensures rapid response adapted to local context, avoiding often counterproductive uniform repressive approaches.

5.2. Technical and Operational Interventions

Adaptive water stress management, critical threat identified as central node in the threat interaction network, requires multi-scale interventions. At the water infrastructure level, modernizing irrigation systems favoring water-efficient irrigation methods, seawater desalination projects, and interconnection between hydraulic basins aim to improve utilization efficiency and diversify supply sources. At the soil-water conservation practices level, developing rainwater harvesting systems, improving infiltration through anti-erosive techniques, and promoting methods increasing soil moisture retention reduce losses and maximize ecosystem availability. At the forest ecology level, selecting native forest species resistant to climate stress, notably cork oak, holm oak, argan tree, Barbary thuja, Atlas cypress, Atlas pistachio, wild olive, and carob tree. Practices improving resilience include planting in mixed-species systems, particularly agroforestry integrating aromatic and medicinal plants, maximizing functional diversity and ecological complementarities (Messier et al., 2015).

Controlling deforestation and fragmentation combines strict protection of priority areas, regulating land conversions through zoning and authorization systems, and active restoration of ecological connectivity. Protecting forest massifs with high conservation value, endemic species refuges, ecological corridors, watershed headwaters, through their classification as protected areas with effective management plans constitutes a first line of defense. Regulating conversions in buffer and transition zones requires rigorous application of environmental impact assessments and anticipatory territorial planning avoiding opportunistic deforestation fronts. Restoring connectivity through rehabilitating ecological corridors linking isolated fragments facilitates gene flows and species migrations, critical in a climate change context requiring movements toward climatically more favorable zones (Haddad et al., 2015).

Regulating TFP and NTFP overexploitation requires establishing sustainable quotas based on scientific assessments of stand productivity and regeneration capacity. For timber products, promoting certified forest management systems guaranteeing legality and harvesting sustainability, combined with developing energy alternatives (butane gas, renewable energies) reducing pressure on fuelwood, decreases exploitation beyond sustainability thresholds (FAO, 2020). For non-timber products, AMP, mushrooms, wild fruits, regulating harvest through permit systems and calendars respecting biological cycles, associated with valorization and certification initiatives favoring quality over quantity, transforms economic incentives toward sustainability.

Managing overgrazing, critical threat affecting the entire forest domain, requires defining carrying capacities based on forage productivity and regeneration needs, implementing grazing rotation systems avoiding continuous pressure on the same areas, and improving forage productivity outside forests through forage intensification in agricultural areas. This approach recognizes that forests contribute to feeding 40% of national livestock, a proportion that can double during drought years, and that total prohibition of forest grazing would be socially and economically unacceptable for pastoral communities. The challenge is therefore to maintain a pastoral function compatible with regeneration and proper forest functioning through integrated livestock-forest management in a concerted and negotiated manner with communities.

Integrated fire management, major threat whose frequency and intensity have doubled over the past two decades, combines prevention, early detection, efficient suppression, and post-fire restoration. Early detection systems using surveillance technologies allow rapid intervention limiting propagation. However, fuel management through controlled burns reducing loads in high-risk areas and establishing strategic firebreaks creates less flammable landscape mosaics (Pausas & Keeley,

2019). Community engagement through awareness campaigns on ignition causes and risky behaviors reduces anthropogenic fire starts. Post-fire restoration through planting adapted species and erosion protection accelerates recovery and improves ecosystem resilience.

Phytosanitary surveillance and parasitic attack management of bark beetles (*Ips sexdentatus, Orthotomicus erosus*), processionary moths (*Thaumetopoea* spp.), fungal pathogens (*Armillaria mellea*), primarily target vulnerable stands weakened by water stress. Regular monitoring allows early detection of infestation outbreaks, while targeted interventions favor biological and ecological approaches preserving natural balances. Reducing tree stress through improving water availability and avoiding unnecessary disturbances constitutes the most effective preventive strategy (Jactel et al., 2005). Preventing and controlling biological invasions requires early detection protocols, rapid risk assessment, and targeted intervention before populations definitively establish. Strict regulation of introductions and awareness of forest stakeholders constitute effective and less costly preventive strategies (Vilà et al., 2011).

Reforestation programs primarily based on native species maximize ecological compatibility and resilience. Degraded land rehabilitation through mechanical and biological sand fixation, soil health improvement, and hydrological function restoration addresses areas affected by sand encroachment and desertification. Agroforestry systems combining production and conservation offer integrated solutions particularly adapted to arid and semi-arid zones (Reynolds et al., 2007). Community engagement through creating income-generating activities associated with reforestation programs transforms restoration from technical action into locally appropriate socio-ecological process.

5.3. Transversal Capacities and Support Mechanisms

Applied research, innovation, and knowledge transfer address the threat of gap between scientific research and forest management, whose effects limit all intervention effectiveness. Developing studies on climate change impacts on forest ecosystems, species physiological and demographic responses to stresses, and comparative effectiveness of restoration techniques orients management choices toward evidence-based approaches. Establishing institutionalized knowledge transfer mechanisms such as demonstration networks, continuous training and publication accessibility, and integrating local and traditional knowledge into management strategies enrich good practices and the repertoire of available actions (Berkes, 2009). Furthermore, establishing open and interoperable data infrastructure addresses information fragmentation and institutional compartmentalization that limit collective response capacity to forest threats. Although Morocco has adopted prospective legislative frameworks in the data domain, their implementation remains uneven across public entities. For the forest sector, systematic opening of geographic data and their metadata constitutes a prerequisite for improving transparency, facilitating knowledge transfer, and generating additional scientific value.

Integrated monitoring of climate, biodiversity, and forest health provides data necessary for adaptive management. Developing monitoring frameworks collecting standardized indicators on climatic variables, stand composition and structure, regeneration rates, parasitic infestations, and ecosystem service provision allows tracking change trajectories, evaluating intervention effectiveness, and early detecting warning signals of regime transitions (Lindenmayer & Likens, 2010). Public and stakeholder access to monitoring information promotes transparency and facilitates involvement in collective management. Capacity building through training forest managers, technicians, and community members on sustainable management approaches, restoration techniques, and monitoring tools empowers actors at all levels. Environmental education campaigns particularly targeting risky behaviors: ignitions, overexploitation, illegal conversions, progressively modify practices. Managing use conflicts through participatory planning processes, mediation mechanisms, and recognition of customary rights reduces tensions and favors emergence of negotiated solutions acceptable to different stakeholders.

Aligning national sustainable forest management strategies with United Nations Sustainable Development Goals, particularly SDG 15 (Life on land), SDG 6 (Clean water and sanitation), and SDG 13 (Climate action), and Morocco's commitments under the Paris Agreement on climate change ensures coherence with international frameworks and facilitates access to climate finance (UNFCCC, 2015). Well-managed forests improve their carbon sequestration capacity, transforming the forest sector from net emission source, through fires and dieback, into carbon sink contributing to Nationally Determined Contributions. Mobilizing international financing through Green Climate Fund mechanisms and Mediterranean and African regional technical cooperation partnerships provides financial and technical resources necessary for large-scale deployment of identified interventions. Establishing public-private partnerships for restoration and sustainable valorization of forest products diversifies financing sources and promotes innovation.

CONCLUSION

Morocco's rich and diverse forests face multiple and interdependent threats that seriously compromise their resilience. Integrated and multi-criteria analysis identified threats according to their criticality level and the central role of water stress as threat multiplier. This analysis also revealed that these threats operate through complex feedback loops and synergistic effects that trigger degradation cascades well beyond the sum of individual impacts and can lead to irreversible ecological regime shifts.

Faced with these non-linear dynamics, isolated interventions prove ineffective. Sustainable rehabilitation imperatively requires integrated strategies that simultaneously break several links in feedback loops, target critical leverage points, maintain ecosystems far from identified thresholds, and strengthen research and intersectoral coordination. In this sense, the challenge consists of translating existing political commitments into concrete actions that recognize the systemic complexity of threats while exploiting positive synergies between conservation, climate adaptation, and sustainable development. Morocco's forest resilience will fundamentally depend on our collective capacity to manage these complex interactions in an integrated long-term perspective.

REFERENCES

Aafi, A. (2007). *Etude de la diversité floristique de l'écosystème de Chêne-liège de la forêt de la Mamora.* Thèse Doct. Es-Sc. Agronomiques, IAV Hassan II, Rabat, 190 p.

Agrawal, A., Chhatre, A., & Hardin, R. (2008). Changing governance of the world's forests. *Science, 320*(5882), 1460–1462. DOI: 10.1126/science.1155369 PMID: 18556552

Allen, C. D., Macalady, A. K., Chenchouni, H., Bachelet, D., McDowell, N., Vennetier, M., Kitzberger, T., Rigling, A., Breshears, D. D., Hogg, E. H. T., Gonzalez, P., Fensham, R., Zhang, Z., Castro, J., Demidova, N., Lim, J.-H., Allard, G., Running, S. W., Semerci, A., & Cobb, N. (2010). A global overview of drought and heat-induced tree mortality reveals emerging climate change risks for forests. *Forest Ecology and Management, 259*(4), 660–684. DOI: 10.1016/j.foreco.2009.09.001

Anderegg, W. R., Hicke, J. A., Fisher, R. A., Allen, C. D., Aukema, J., Bentz, B., Hood, S., Lichstein, J. W., Macalady, A. K., McDowell, N., Pan, Y., Raffa, K., Sala, A., Shaw, J. D., Stephenson, N. L., Tague, C., & Zeppel, M. (2015). Tree mortality from drought, insects, and their interactions in a changing climate. *The New Phytologist, 208*(3), 674–683. DOI: 10.1111/nph.13477 PMID: 26058406

Andréassian, V. (2004). Waters and forests: From historical controversy to scientific debate. *Journal of Hydrology (Amsterdam), 291*(1-2), 1–27. DOI: 10.1016/j.jhydrol.2003.12.015

ANEF. (2019). *Bilan des activités 2018.* Ministry of Agriculture.

Benabid, A. (2000). *Flore et écosystème du Maroc « Evaluation et préservation de la biodiversité.* Ibis Press.

Benabou, A., Moukrim, S., Laaribya, S., Aafi, A., Chkhichekh, A., El Maadidi, T., & El Aboudi, A. (2022a). Mapping ecosystem services of forest stands: Case study of Maamora, Morocco. *Geography, Environment, Sustainability, 15*(1), 141–149. DOI: 10.24057/2071-9388-2021-047

Benabou, A., Moukrim, S., Lahssini, S., El Aboudi, A., Menzou, K., Elmalki, M., & Rhazi, L. (2022b). Impact of climate change on potential distribution of Quercus suber in the conditions of North Africa. *Biosystems Diversity, 30*(3), 289–294. DOI: 10.15421/012231

Benzyane, M., Aafi, A., Sbay, H., El Antry, S., Yassine, M., Ilmen, R., & Ghailoule, D. (2010). *Les écosystèmes naturels marocains et les changements climatiques: la résilience écologique à l'épreuve.* Centre de Recherche Forestière.

Berkes, F. (2009). Evolution of co-management: Role of knowledge generation, bridging organizations and social learning. *Journal of Environmental Management, 90*(5), 1692–1702. DOI: 10.1016/j.jenvman.2008.12.001 PMID: 19110363

Blagojević, B., Jonsson, R., Björheden, R., Nordfjell, T., & Lindroos, O. (2019). Multi-criteria decision analysis (MCDA) in forest operations – an introductional review. *Croatian Journal of Forest Engineering, 40*(1), 191–205.

Breshears, D. D., Cobb, N. S., Rich, P. M., Price, K. P., Allen, C. D., Balice, R. G., Romme, W. H., Kastens, J. H., Floyd, M. L., Belnap, J., Anderson, J. J., Myers, O. B., & Meyer, C. W. (2005). Regional vegetation die-off in response to global-change-type drought. *Proceedings of the National Academy of Sciences of the United States of America, 102*(42), 15144–15148. DOI: 10.1073/pnas.0505734102 PMID: 16217022

Camarero, J. J., Gazol, A., Sangüesa-Barreda, G., Oliva, J., & Vicente-Serrano, S. M. (2015). To die or not to die: Early warnings of tree dieback in response to a severe drought. *Journal of Ecology, 103*(1), 44–57. DOI: 10.1111/1365-2745.12295

D'Odorico, P., Bhattachan, A., Davis, K. F., Ravi, S., & Runyan, C. W. (2013). Global desertification: Drivers and feedbacks. *Advances in Water Resources, 51*, 326–344. DOI: 10.1016/j.advwatres.2012.01.013

Direction de la Météorologie Nationale (DMN). (2020). *Synthèse du climat au Maroc en 2024*. Rabat. Disponible sur: https://www.marocmeteo.ma/fr/climat-2024

Economique, C. Social et Environnemental (CESE). (2022). Les écosystèmes forestiers du Maroc: Risques, défis et opportunités. Rabat, Maroc.

Fischer, J., & Lindenmayer, D. B. (2007). Landscape modification and habitat fragmentation: A synthesis. *Global Ecology and Biogeography, 16*(3), 265–280. DOI: 10.1111/j.1466-8238.2007.00287.x

Flannigan, M. D., Krawchuk, M. A., de Groot, W. J., Wotton, B. M., & Gowman, L. M. (2009). Implications of changing climate for global wildland fire. *International Journal of Wildland Fire, 18*(5), 483–507. DOI: 10.1071/WF08187

Folke, C., Carpenter, S., Walker, B., Scheffer, M., Elmqvist, T., Gunderson, L., & Holling, C. S. (2004). Regime shifts, resilience, and biodiversity in ecosystem management. *Annual Review of Ecology, Evolution, and Systematics, 35*(1), 557–581. DOI: 10.1146/annurev.ecolsys.35.021103.105711

Food and Agriculture Organization (FAO). (2020). *State of the World's Forests 2020: Forests, Biodiversity and People*. FAO.

Gaertner, M., Den Breyen, A., Hui, C., & Richardson, D. M. (2009). Impacts of alien plant invasions on species richness in Mediterranean-type ecosystems: A meta-analysis. *Progress in Physical Geography, 33*(3), 319–338. DOI: 10.1177/0309133309341607

Geist, H. J., & Lambin, E. F. (2002). Proximate causes and underlying driving forces of tropical deforestation: Tropical forests are disappearing as the result of many pressures, both local and regional, acting in various combinations in different geographical locations. *Bioscience, 52*(2), 143–150. DOI: 10.1641/0006-3568(2002)052[0143:PCAUDF]2.0.CO;2

Ghanmi, M., Aafi, A., Satrani, B., Aberchane, M., Khia, A., & Yakhlef, S. B. (2014). Aromatic and medicinal plants in North Africa: Opportunities, constraints and prospects. In Gurib-Fakim, A. (Ed.), *Novel Plant Bioresources: Applications in Food, Medicine and Cosmetics* (pp. 411–421). Wiley-Blackwell. DOI: 10.1002/9781118460566.ch29

Haddad, N. M., Brudvig, L. A., Clobert, J., Davies, K. F., Gonzalez, A., Holt, R. D., Lovejoy, T. E., Sexton, J. O., Austin, M. P., Collins, C. D., Cook, W. M., Damschen, E. I., Ewers, R. M., Foster, B. L., Jenkins, C. N., King, A. J., Laurance, W. F., Levey, D. J., Margules, C. R., & Townshend, J. R. (2015). Habitat fragmentation and its lasting impact on Earth's ecosystems. *Science Advances, 1*(2), e1500052. DOI: 10.1126/sciadv.1500052 PMID: 26601154

Holling, C. S. (1973). Resilience and stability of ecological systems. *Annual Review of Ecology and Systematics, 4*(1), 1–23. DOI: 10.1146/annurev.es.04.110173.000245

Intergovernmental Panel on Climate Change (IPCC). (2021). *Climate Change 2021: The Physical Science Basis*. Contribution of Working Group I to the Sixth Assessment Report. Cambridge University Press.

Jactel, H., Brockerhoff, E. G., & Duelli, P. (2005). A test of the biodiversity-stability theory: meta-analysis of tree species diversity effects on insect pest infestations, and re-examination of responsible factors. In Scherer-Lorenzen, M., Körner, C., & Schulze, E. D. (Eds.), *Forest Diversity and Function* (pp. 235–262). Springer. DOI: 10.1007/3-540-26599-6_12

Kissinger, G., Herold, M., & De Sy, V. (2012). *Drivers of Deforestation and Forest Degradation: A Synthesis Report for REDD+ Policymakers*. Lexeme Consulting.

Linares, J. C., Camarero, J. J., & Carreira, J. A. (2011). Competition modulates the adaptation capacity of forests to climatic stress: Insights from recent growth decline and death in relict stands of the Mediterranean fir Abies pinsapo. *Journal of Ecology, 99*(2), 592–603.

Lindenmayer, D. B., & Likens, G. E. (2010). The science and application of ecological monitoring. *Biological Conservation, 143*(6), 1317–1328. DOI: 10.1016/j.biocon.2010.02.013

Messier, C., Puettmann, K., Chazdon, R., Andersson, K. P., Angers, V. A., Brotons, L., Filotas, E., Tittler, R., Parrott, L., & Levin, S. A. (2015). From management to stewardship: Viewing forests as complex adaptive systems in an uncertain world. *Conservation Letters, 8*(5), 368–377. DOI: 10.1111/conl.12156

Mharzi Alaoui, H., Assali, F., Rouchdi, M., Tahiri, D., Lahssini, S., Aafi, A., & Moukrim, S. (2017). Cartographie de l'intensité du feu et de la sensibilité aux feux de cimes en forêts Méditerranéennes. Cas de la province de Chefchaouen au nord du Maroc. [La Terre et La Vie]. *Revue d'Écologie, 72*(4), 387–409. DOI: 10.3406/revec.2017.1900

Moukrim, S., Lahssini, S., Naggar, M., Lahlaoi, H., Rifai, N., Arahou, M., & Rhazi, L. (2019). Local community involvement in forest rangeland management: Case study of compensation on forest area closed to grazing in Morocco. *The Rangeland Journal, 41*(1), 43–53. DOI: 10.1071/RJ17119

Moukrim, S., Lahssini, S., Rhazi, M., Menzou, K., El Madihi, M., Rifai, N., Bouziani, Y., Azedou, A., Boukhris, I., & Rhazi, L. (2022). Climate change impact on potential distribution of an endemic species Abies marocana Trabut. *Ekologia (Bratislava), 41*(4), 329–339. DOI: 10.2478/eko-2022-0034

Moukrim, S., Lahssini, S., Rifai, N., Menzou, K., Mharzi-Alaoui, H., Labbaci, A., Rhazi, M., Wahby, I. W., El Madihi, M., & Rhazi, L. (2020). Modélisation de la distribution potentielle de Cedrus atlantica Manetti au Maroc et impacts du changement climatique. *Bois et Forêts des Tropiques, 344*, 3–16. DOI: 10.19182/bft2020.344.a31888

Navas, A., Machin, J., & Gaspar, L. (2009). Les sols dans le pays du Ziz (Sud-est marocain), caractéristiques et aspects de dégradation. *Applied Geography (Sevenoaks, England), 20*, 369–394.

Niemeijer, D., & de Groot, R. S. (2008). A conceptual framework for selecting environmental indicator sets. *Ecological Indicators, 8*(1), 14–25. DOI: 10.1016/j.ecolind.2006.11.012

Organisation de Coopération et de Développement Économiques (OCDE). (1993). *OECD core set of indicators for environmental performance reviews*. Environment Monographs No. 83. Paris: OECD Publications.

Pausas, J. G., & Keeley, J. E. (2019). Wildfires as an ecosystem service. *Frontiers in Ecology and the Environment*, *17*(5), 289–295. DOI: 10.1002/fee.2044

Peñuelas, J., Rutishauser, T., & Filella, I. (2009). Phenology feedbacks on climate change. *Science*, *324*(5929), 887–888. DOI: 10.1126/science.1173004 PMID: 19443770

Reynolds, J. F., Stafford Smith, D. M., & Lambin, E. F.. (2007). Global desertification: Building a science for dryland development. *Science*, *316*(5826), 847–851. DOI: 10.1126/science.1131634 PMID: 17495163

Sayer, J., Sunderland, T., Ghazoul, J., Pfund, J. L., Sheil, D., Meijaard, E., Venter, M., Boedhihartono, A. K., Day, M., Garcia, C., van Oosten, C., & Buck, L. E. (2013). Ten principles for a landscape approach to reconciling agriculture, conservation, and other competing land uses. *Proceedings of the National Academy of Sciences of the United States of America*, *110*(21), 8349–8356. DOI: 10.1073/pnas.1210595110 PMID: 23686581

Scheffer, M., Carpenter, S., Foley, J. A., Folke, C., & Walker, B. (2001). Catastrophic shifts in ecosystems. *Nature*, *413*(6856), 591–596. DOI: 10.1038/35098000 PMID: 11595939

Seto, K. C., Güneralp, B., & Hutyra, L. R. (2012). Global forecasts of urban expansion to 2030 and direct impacts on biodiversity and carbon pools. *Proceedings of the National Academy of Sciences of the United States of America*, *109*(40), 16083–16088. DOI: 10.1073/pnas.1211658109 PMID: 22988086

United Nations Framework Convention on Climate Change (UNFCCC). (2015). *Paris Agreement*. UNFCCC.

Vilà, M., Espinar, J. L., Hejda, M., Hulme, P. E., Jarošík, V., Maron, J. L., Pergl, J., Schaffner, U., Sun, Y., & Pyšek, P. (2011). Ecological impacts of invasive alien plants: A meta-analysis of their effects on species, communities and ecosystems. *Ecology Letters*, *14*(7), 702–708. DOI: 10.1111/j.1461-0248.2011.01628.x PMID: 21592274

APPENDIX

Figure 2. Manifestations of the main anthropogenic and natural threats to Moroccan forest ecosystems: Forest stand dieback (photos 1-11) and land degradation (photos 12-16); and examples of resilience pathways and sustainable management (photos 17-20) (© Aafi Abderrahman)

01- *Cedrus atlantica* (Middle Atlas)

02 -*Juniperus oxycedrus* (Oriental region)

03- *Abies maroccana* (Rif-Chefchaouen)

4- *Pinus maghrebiana* (Rif-Chefchaouen)

5- *Juniperus thurifera* (Central High Atlas)

06- *Quercus suber* (Central region)

07- *Acacia raddiana* (Eastern region, Tata)

8- *Argania spinosa* (Southwest -Agadir)

9- *Cedrus atlantica* & *Quercus rotundifolia* (High Atlas)

10- *Acacia raddiana* (Eastern region, Zagora)

11- water and wind erosion phenomena (Middle Atlas)

12- Salinization (Southern region, Smara) Development of *Zygophyllum gaetulum*

13- Sand cover and buried palm tree (Eastern region, Zagora)

14- Individuals trees of *Tamarix aphylla* buried by sand (Eastern region, Errachidia)

15- Sand encroachment and desertification water stress (Southern region, Smara)

16- Spontaneous vegetation buried by maritime dunes (Central region, Essaouira)

17- Natural regeneration of the *Cedrus atlantica* ecosystem (Middle Atlas)

18- Artificial regeneration of *Cedrus atlantica* (Middle Atlas)

19- Artificial regeneration of *Argania spinosa* (Central region, Essaouira)

20- Artificial regeneration of *Acacia raddiana* in (Southern region, Smara)

21 Natural regeneration (*A. gummifera*) after planting *Opuntia ficus-indica* (Central region, Marrakech)

22- Effective agroforestry system for conservation of biodiversity & combating desertification (Eastern region, Zagora)

23- Mechanical and biological (*Tamarix aphylla*) approach for sand dunes revegetation (Southern region, Laayoune)

24- Restoration of vegetation using mechanical and biological sand fixation (Southern region, Laayoune)

Chapter 4
Climate Change Resilience in Mediterranean Mixed-Oak Forest Landscapes

Amel Ennajah
http://orcid.org/0000-0003-0740-6508
National Research Institute of Rural Engineering, Water, and Forests, Carthage University, Ariana, Tunisia

Salma Sai-Kachout
Research National Institute of Agronomy, Carthage University, Tunisia

Rania Mechergui
http://orcid.org/0000-0001-6690-2224
National Research Institute of Rural Engineering, Water, and Forests, Carthage University, Ariana, Tunisia

Sondess Fkiri
National Research Institute of Rural Engineering, Water, and Forests, Carthage University, Ariana, Tunisia

Bouthaina Stiti
National Research Institute of Rural Engineering, Water, and Forests, Carthage University, Ariana, Tunisia

Mokhtar Baraket
National Research Institute of Rural Engineering, Water, and Forests, Carthage University, Ariana, Tunisia

Zouhaier Nasr
National Research Institute of Rural Engineering, Water, and Forests, Carthage University, Ariana, Tunisia

Sihem Jebari
National Research Institute of Rural Engineering, Water, and Forests, Carthage University, Ariana, Tunisia

ABSTRACT

In the Kroumirie region of north-western Tunisia, populations of cork oak (Quercus suber) are suffering from drought. This decline in cork oak forests requires a re-

DOI: 10.4018/979-8-2600-0783-9.ch004

evaluation of sustainable forest management techniques. A plant association is defined as a group of plants with a specific floristic composition, uniform physiognomy and growth under uniform site conditions. In this context, the floristic composition would include Quercus suber and one or more companion plants intended to support its persistence. As part of this initiative, the present study, aims to test the impact of such plant associations on the growth and hydraulic conductance PLC of cork oak. This chapter details the role and functioning of symbiotic associations in Mediterranean ecosystems. The study was conducted in a mixed forest of cork oak and zeen oak in the Kroumirie region of north-western Tunisia (Jendouba Governorate).

INTRODUCTION

Climate change presents a major interdisciplinary challenge, uniting the efforts of ecologists, biologists, and modelers to assess its potential consequences on forest-derived ecosystem services. Climate change exerts worldwide pressures on forest ecosystems (Dale et al., 2001; Parmesan & Yohe, 2003; Trumbore et al., 2015). A rise in the mean temperature and variation in rainfall patterns could modify the current distribution of plant species (Mechergui el al., 2022; Lopez-Tirado et al., 2018). Over the last few decades, the scientific community has become increasingly concerned with climate change. More extreme climatic conditions, such as rising mean temperatures and variations in rainfall patterns, are expected within the present century (IPCC, 2014). Indeed, recent climate projections for the Mediterranean basin anticipate a rise in air temperature and a reduction in precipitation, particularly during the summer months, leading to more prolonged periods of soil water deficit (Gratani et al., 2016; IPCC, 2014). Extended dry spells lead to a soil water deficit, where evapotranspiration losses exceed the water inputs from rainfall and atmospheric moisture (Brunetti et al., 2002; Moretti et al., 2015). Such conditions heighten the risk of local species extinction and subsequent habitat fragmentation, particularly in vulnerable regions like North Africa (Benito Garzon et al., 2008; Klausmeyer & Shaw, 2009). The resilience of Mediterranean forests in the face of climate change will be heavily influenced by the ability of tree species to adapt to shifting temperatures and rainfall patterns. Across the Mediterranean Basin, rising temperatures and greater aridity are expected to significantly alter species composition (Dukes et al., 2005; Petit et al., 2005). These changes are likely to drive shifts in species ranges, both in latitude and elevation (Benito et al., 2014). Ultimately, the capacity for local adaptation will shape the future of these forests, offering critical insights into the scale and geography of climate change impacts. This understanding is essential for developing strategies to mitigate its future effects. The anticipated rise in the frequency and severity of extreme climatic events poses significant threats to

Mediterranean tree species. These species, however, possess diverse mechanisms to respond to environmental shifts (Limousin et al., 2012; Soheili et al., 2023). Among these, enhanced phenotypic plasticity and increased intra-specific genetic variation in adaptive traits (i.e., local adaptation) are considered key survival strategies (Matesanz & Valladares, 2014; Pérez-Ramos et al., 2019). Phenotypic plasticity, defined as the capacity of a single genotype to produce different phenotypes in response to specific environmental conditions (Bongers et al., 2017; Bradshaw, 1965), allows for immediate acclimation to changing conditions, a process vital for plant persistence (Bussotti et al., 2015). Local adaptation, conversely, arises from natural selection favoring genotypes with traits conferring higher fitness in their native environment, thus providing a competitive advantage over other genotypes (Peguero-Pina et al., 2014; Savolainen et al., 2007).

As suitable habitats contract or become less viable, ecological niches may be colonized by other species (Pons & Pausas, 2006; Lopez-Tirado & Hidalgo, 2016). Tree species, especially those at the edges of their climatic envelopes, can experience rapid range shifts due to global change (Lopez-Tirado & Fidalgo, 2016). Climate change will inevitably affect species geographical distribution, change species composition and richness, impact biodiversity, and influence the structure and function of ecosystems (Thuiller et al., 2011; Scheffers et al., 2016).

Oak is an evergreen sclerophyllous tree species native to western Mediterranean region, mainly distributed along southwest Europe and northwest Africa. It is well adapted to nonsevere droughts and high temperatures, being regarded as a transitional species between the high drought tolerant evergreen oaks (such as *Quercus ilex* or *Q. coccifera*, David et al., 2007) and the mesic deciduous oaks (such as *Q. faginea* or *Q. pyrenaica*, Mediavilla and Escudero, 2004). oak forests play an important ecological role in key ecosystem functions such as net primary production, carbon sequestration, soil formation, hydrological flow regulation and reservoir for biodiversity (Bugalho et al., 2011; Torralba et al., 2016; Reis et al., 2021), while they are also of critical economic importance due to their large contribution to the global cork market. Even though oaks can adapt to environmental conditions and their genetic variability allow them to cope with climatic variation (Leite et al., 2019), the ongoing climate change scenarios of higher water deficit in combination with other natural and human-induced factors such as diseases caused by exotic pathogens are alarmingly threatening their natural populations (Aguilera et al., 2020; González et al., 2020; Serrano et al., 2021).

However, mixed forests, characterized by the presence of diverse tree species, possess exceptional water retention and soil nourishment capabilities (Zhou, et al., 2018), thereby playing a pivotal role in managing watershed dynamics (Andersan, et al., 1976; Zhou, et al., 2002; Goeking, et al., 2020). Research indicates that mixed forest stands, such as those comprising *Q. suber* and *Q. canariensis*, exhibit higher levels of soil organic matter compared to monoculture stands of *Q. suber* alone. This elevated organic content enhances the soil's water retention capacity, effectively reducing runoff. Additionally, the varied structure of mixed forests facilitates greater canopy interception of rainfall, leading to increased water capture and subsequent evaporation, thereby minimizing runoff volumes.Moreover, mixed forests create a more stable microclimate, characterized by reduced temperature fluctuations, which can influence evapotranspiration rates and subsequently impact runoff dynamics. Furthermore, their inherent resilience to pests, diseases, and climate variations positions mixed forests as robust ecosystems capable of maintaining their runoff-regulating function amidst changing environmental conditions. While both mixed and single-species forests play indispensable roles in runoff control, the diversity inherent in mixed forests provides additional benefits, making them vital components in sustainable management strategies.

The mountains of Kroumirie and Mogods in Tunisia constitute an important potential in terms of biodiversity and offer a remarkable opportunity to enhance sustainable development. The landscape diversity characterizing this region includes a significant number of different types of habitats, such as coastal sand dunes covered by a mosaic of vegetation, freshwater, peatland habitats and numerous forest and shrub formations, distinguished by a significant preserved forest of dominant oaks; *Q. suber* and *Q. canariensis* (Frouja et al., 2022). The forests of Kroumirie and Mogods have significant diversity in terms of species and habitats, which are nowadays threatened by anthropogenic pressure and changes in land use. These lands gave way to large forest plots following reforestation policies resulting from the forest ecosystem restoration strategy (Frouja et al., 2022). Irrespective of the degree of land use intensification, many researches have demonstrated that diversifying species on the same land at the same time enhances biodiversity, resilience to pests and diseases, and ecosystem service provisions compared to agriculture or forest monocrops (Kahane et al., 2013; Torralba et al., 2016; Veldkamp et al., 2023). However, the moderate shading and evapotranspiration from trees can improve the microclimate by reducing air temperatures and atmospheric evaporative demand (Kanzler et al., 2019). To realize the multiple productive and environmental benefits of mixed forests, appropriate planning must be implemented to maximize synergies among species, particularly in regions subjected to climate change and drought stress.

ECOLOGICAL PROFILES OF OAK FOREST MANAGEMENT IN MEDITERRANEAN REGIONS

Zeen oak (*Quercus canariensis* Wild.)

Quercus canariensis is a deciduous oak species adapted to Mediterranean-type climates (Peguero et al., 2016). Its classification presents challenges due to leaf polymorphism and geographical distribution (Ennajah, 2010). Its taxonomic classification is as follows:

- Kingdom: Plantae
- Clade / Division: Tracheophyta
- Clade / Class: Angiosperms
- Clade / Order: Eudicots
- Order: Fagales
- Family: Fagaceae
- Genus: *Quercus*
- Species: *Quercus canariensis* Wild.

The global range of canariensis oak is restricted to the southwestern Mediterranean basin, with its distribution encompassing Spain, Portugal, Morocco, Tunisia, Algeria, and occasionally southeastern France (Aissi et al., 2020) (Figure 1).

Figure 1. Mondial distribution range of the Q. canariensis (Aissi et al., 2020)

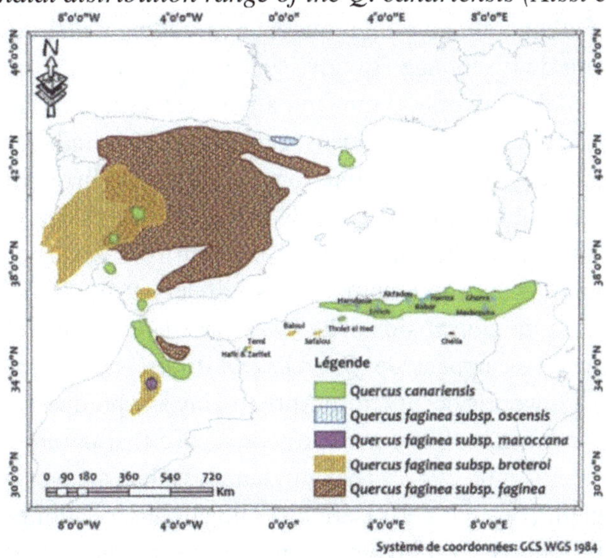

In Tunisia, the Zeen oak covers an area of 100.000 hectares (Nouri et al., 2022). It found mainly in the Northwest Kroumirie: El-Feidja, Djbel Tegma, Ain-Draham, Oued Zeena, Chihia (Figure 2).

Figure 2. Tunisian distribution range of the Q. canariensis (Nouri et al., 2022)

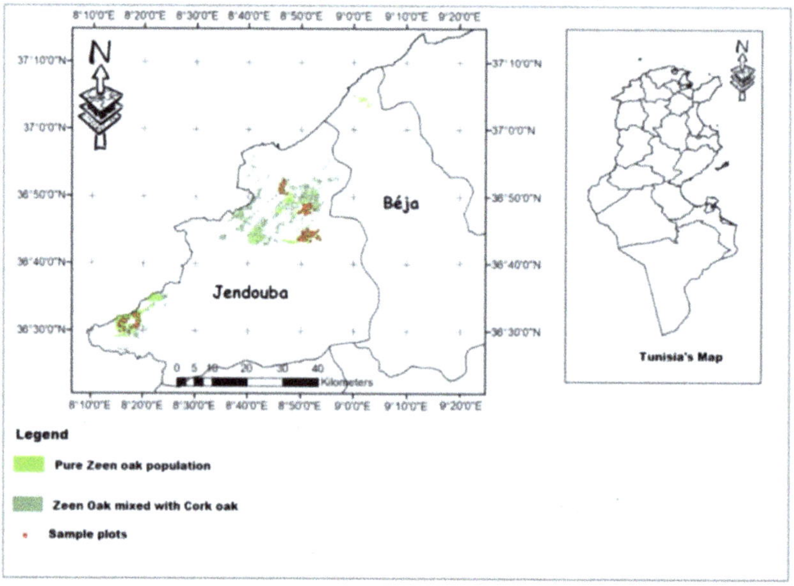

The Zeen oak (*Q. canariensis*) is a large tree that can reach 30 metres in height. It has a clear bole of 20 metres and a circumference exceeding 6 metres, and develops a broad crown with dense foliage (Boudy, 1950). This species thrives in regions with an annual rainfall of over 800 mm and a northern aspect. It has dark brown, fissured bark and, in the largest specimens, can exceed 30 m in height at maturity. Its crown has a spreading form. The trunk can reach 2 metres in diameter at 1 metre above ground level (Rabhi, 2011) (Figure 3). Zeen oak leaves can reach 20 cm in length and 12 cm in width. They are oval-shaped with 10–12 pairs of lobes, which typically correspond to the same number of secondary veins (Fig. 3). The main vein is prominent on the dorsal side. The leaves are dark green on the underside and glaucous green on the upper side. *Quercus canariensis* is a monoecious species (Chaabani, 2013). Flowering occurs from April to May and produces annual acorns, which become abundant after 30 years. The acorns have shallow cups covered with lanceolate, flat scales (Taibi, 2012). They measure less than 4 cm in length and less than 1.5 cm in width (Fig. 3). Fertilisation is allogamous, resulting in significant genetic variability and marked polymorphism (Ksontini, 1996). Zeen oak wood is of considerable importance in North Africa. It is used for constructing bridge beams and also serves as firewood, producing high-quality charcoal (Chaabani, 2013). The bark contains high concentrations of tannin, which is used in cosmetics and for leather tanning (Rabhi, 2011).

This species thrives in acid soils, typically in the most humid parts of valleys and coastal Mediterranean areas, often associated with streams and valley bottoms (Borreguero et al., 2024). It is sensitive to drought and prefers regions with stable rainfall and moderate temperatures (0–24°C) (Frouja et al., 2022; Borreguero et al., 2024). Its distribution is limited by cold and drought, with populations declining in drier or more extreme climates (shuxia Sun et al., 2024). *Quercus canariensis* is considered a relict species, with fragmented and isolated populations due to historical climate changes and human pressures (Vila-Viçosa et al., 2015; Nouri et al., 2022; Bouandas et al., 2024). Many stands show a predominance of either old or young individuals, reflecting regeneration challenges and the impact of proximity to human settlements.

Figure 3. Morphological characteristics of Zeen oak: a, trunk; b, leaf; c, bud; d, acorn (Thorand, 2018)

In Kroumiria, around two-thirds of the 23,000 hectares occupied by *Q. canariensis* are found alongside *Q. suber* (Frouja et al., 2022). These mixed stands are primarily found above 700 metres, but can also be found at lower altitudes when there is sufficient soil and atmospheric moisture, particularly in valley bottoms and along wadi banks. Notable examples include the sites of Hamdia, Khroufa, Aïn Zana, El Feija and Hammam Bourguiba (Frouja et al., 2022). Outside of this association with cork oak, pure stands of *Q. canariensis* are generally found above an altitude of 1000 metres (Frouja et al., 2022).

Due to climate change, it is anticipated that cork oak forests will gradually replace all *Q. canariensis* stands below 1,000 metres, while open formations will be colonised by Aleppo pine (*Pinus halepensis*) and the warmest areas will evolve into low maquis formations (Vessella et al., 2017; Touhami et al., 2020; Frouja et al., 2022).

Cork oak (*Quercus suber* L.)

Quercus suber L., commonly known as cork oak, is an evergreen tree that belongs to the Quercus genus and the Fagaceae family.

Globally, cork oak woodlands span approximately 2.2 million hectares, nearly all of which is located in the Mediterranean nations of Algeria, France, Italy, Morocco, Portugal, Spain, and Tunisia (De Riguo & Caudullo, 2016). The majority vast woodlands are located along the Atlantic shoreline of the Iberian Peninsula. Forests of *Q. suber* span nearly 1.5 million hectares in Europe and 700,000 hectares in North Africa. Its distribution is dispersed, implying that it is primarily a remnant. The species has additionally been presented to nations beyond the mediterranean area in the twentieth century, either for cork production or merely as a decorative tree, and a restricted quantity can be located in Bulgaria, California, Chile, New Zealand, southern Australia, and Turkey (Figure 4).

Figure 4. Frequency of Q. suber occurrences within the field observations as reported by the National Forest Inventories. The chorology of the native spatial range for Q. suber is derived after EUFORGEN (EUFORGEN, 2009)

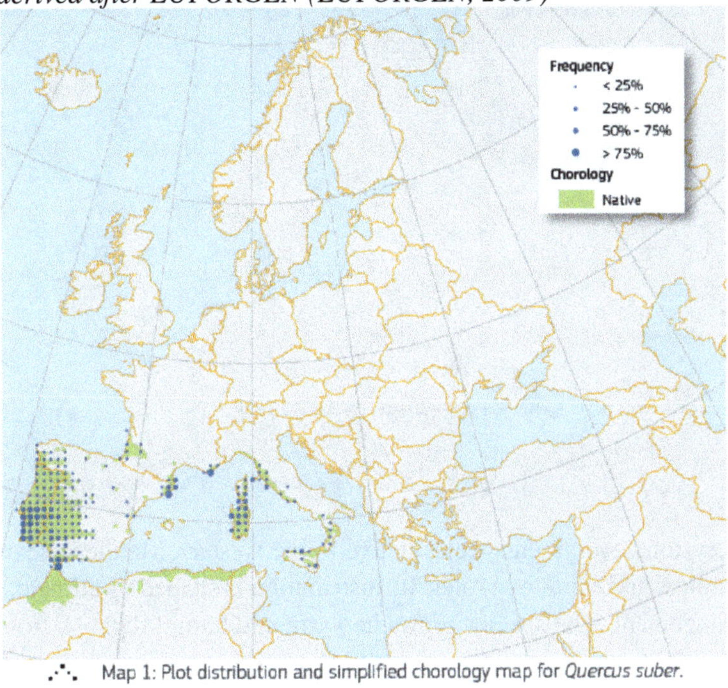

In Tunisia, Cork oak woodlands occupy a northern strip between the coastal mountains and the Mediterranean Sea, with the core area in the Northwest, especially the Kroumirie and Mogods regions, extending from the Algerian border to the city of Sejnane (Figure 5). Additional stands are found near Bizerte, in Cap Bon (Nabeul governorate), and minor populations in Zaghouan, about 50 km from Tunis. The total area covered by cork oak in Tunisia is estimated at around 70,000 hectares (Ennajah, 2010; Bellahirech et al., 2019).

Figure 5. Tunisian distribution range of the Q. suber (Stiti et al., 2014)

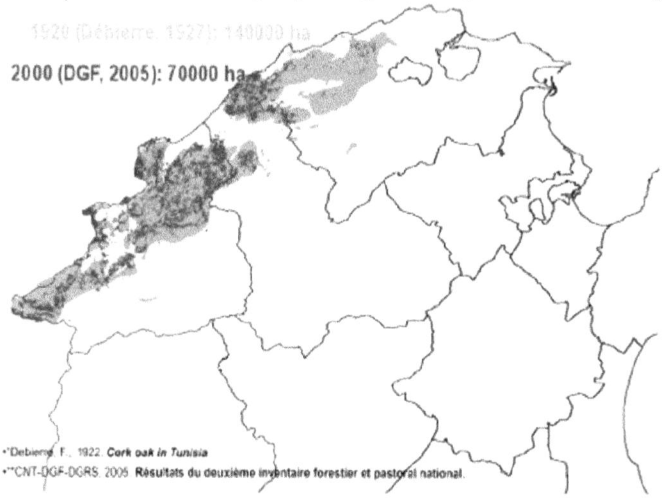

It is a keystone evergreen species native to the western Mediterranean, valued for its ecological and economic roles. Its distribution is shaped by climate, soil, and human management. The species grows in warm and humid regions, from the sea elevation to 2,000 m above sea elevation and favor siliceous- or carbonate-free soils and have the ability to establish in highly acidic soils. Its soil pH tolerance ranges from 3.4 to 7.8, though optimal development occurs between pH 4.7 and 6.5. The species prefers well-aerated loamy and sandy soils, avoiding waterlogged or compacted substrates. Occasionally, it may also establish it self in decarbonated soils.

Q. suber L. displays several xeromorphic adaptations to Mediterranean climatic constraints. Mature trees reach up to 25 m in height with trunk diameters over 2 m (Fig. 6), developing a dense, rounded crown (Silva et al., 2023). The sclerophyllous, oval leaves possess a dark green adaxial surface and a lighter abaxial side covered with dense stellate trichomes and abundant stomata. This functional morphology contributes to reduced transpirational water loss (Simões et al., 2020). *Q. suber* L. is defined by its thick, fissured, greyish-brown bark—a renewable tissue that, when harvested, exposes a distinctive red trunk layer. The crown develops from an oval juvenile form to a broadly spreading, irregular structure, often fracturing into separate rounded crowns in aged trees (Vicente et al., 2025). Mature dimensions generally range from 10 to 15 meters in both height and diameter (Figure 6). *Q. suber* L. is a monoecious, wind-pollinated species bearing both male and female reproductive structures on the same individual. Flowering occurs in spring, with successful pollination leading to acorn maturation by autumn. The fruit of *Q. suber* L., a glabrous acorn, is characterized by a smooth, brownish-red testa, typically marked with a distinctive green apex. The species exhibits marked interannual variation in acorn production (masting habit), with fruit maturation occurring predominantly in the autumn season (Simões et al., 2020) (Fig. 6). The cork oak woodlands support a abundant diversity of flora and fauna and rank among the most significant social, economic, cultural, and ecological worth (Benabou et al., 2022).

Figure 6. Morphological characteristics of cork oak: a, trunk; b, leaf; c, bud; d, acorn. (Image Owen Johnson)

In terms of climate, Cork oak thrives in regions with annual precipitation typically between 400–2400 mm, but it is most productive above 600–800 mm. It tolerates maximum temperatures up to 31°C, with photosynthetic activity optimal around 25°C and declining above 35°C. Some genotypes can withstand critical temperatures up to 42–56°C, especially with acclimatization and adequate soil moisture (Vizinho et al., 2023). The species is highly drought-tolerant but sensitive to prolonged or extreme drought, which can limit growth and regeneration. Climate change, particularly the increasing frequency of droughts and extreme temperatures, jeopardizes the survival, growth, and cork production of cork oaks. Genetic studies, however, indicate that the species possesses a relatively homogeneous genetic pool and sufficient adaptive capacity to withstand moderate climatic shifts. Nevertheless, populations at the continental or range limits are more vulnerable (Pina-Martins et al., 2018; Vanhove et al., 2021). Management strategies, such as fostering beneficial microclimates through north-facing slope plantings or the maintenance of shrub cover, can enhance seedling survival under increasingly arid conditions.

Cork oak (*Q. suber*) shows notable resilience to climate change, but its long-term survival is closely linked to its interactions with other species and the overall biodiversity of its ecosystem. Cork oak populations exhibit high genetic diversity and phenotypic plasticity, allowing adaptation to varying temperature and precipitation regimes across the Mediterranean. Most populations are not expected to require large genetic shifts to survive predicted climate changes, though southern and continental populations are more vulnerable to rapid or extreme shifts (Vanhove et al., 2021; Pina-Martins et al., 2018).

The resilience of cork oak (*Q. suber*) forests is positively correlated with high species richness and the diversity of associated plant and microbial communities. Mixed stands, where cork oak grows alongside other native hardwoods, demonstrate greater resilience to climate change impacts compared to monocultures. This enhanced stability is attributed to the buffering capacity of biodiversity against environmental stressors and pest outbreaks (Mechergui et al., 2021). Furthermore, the presence of a shrub layer and companion trees can generate favorable microclimates, which have been shown to improve seedling establishment and growth during drought periods (Vizinho et al., 2023). Conversely, competitive interactions pose a significant threat; the climate-driven expansion of species such as Aleppo pine (Pinus halepensis) can reduce cork oak habitat availability and productivity (Mechergui et al., 2021).

Conservation strategies should prioritize maintaining or increasing species diversity within cork oak woodlands, supporting natural regeneration, and managing competitive or invasive species. Promoting mixed-species stands and protecting microhabitats are key to enhancing ecosystem resilience in the face of climate change (Vizinho et al., 2023). Sustaining cork oak forests in the face of climate change will depend on management practices that bolster the species' inherent resilience factors—namely, its genetic diversity and capacity for local adaptation—as well as the stability provided by diverse species associations. Therefore, conservation efforts must prioritize the enhancement of biodiversity and the active management of competitive interactions.

Mixed cork oak - zeen oak forest

Mixed forests of cork oak (*Q. suber*) and Zeen oak (*Q. canariensis*) represent key ecological components of Mediterranean woodlands, particularly across southern Spain and North Africa. While these stands exhibit unique ecological interactions, their contemporary structure and dynamics are increasingly shaped by the interplay of human management and escalating environmental pressures.

In high-precipitation areas (>800 mm), *Q. suber* and *Q. canariensis* often co-dominate, yet exhibit distinct edaphic preferences: *Q. suber* dominates on drier, nutrient-poor ridges, and *Q. canariensis* in moister valleys. Their coexistence, most prevalent on midslopes, creates a mosaic of soil nutrient, moisture, and microbial conditions that supports high understory biodiversity (Aponte et al., 2013). Over the past century, forest management has increasingly prioritized *Q. suber* due to its high economic value for cork production. This has driven a successional shift towards cork oak monocultures, particularly in more productive, mesic areas, at the expense of *Q. canariensis*. Management practices such as selective planting, directed acorn sowing, and the exploitation of *Q. suber*'s vigorous resprouting capacity have accelerated this transition (Urbieta et al., 2008). In contrast, *Q. canariensis*, once valued for timber and charcoal, is now largely marginalized in contemporary management frameworks (Urbieta et al., 2013).

Q. canariensis and *Q. suber* exhibit distinct resource-use strategies: *Q. canariensis* enhances soil fertility through higher nutrient inputs and rapid litter decomposition, whereas *Q. suber*'s slower nutrient cycling promotes drought resilience. This functional complementarity can create positive plant-soil feedbacks that stabilize their coexistence and boost ecosystem productivity (Aponte et al., 2013). However, such mixed stands may also present an ecological trade-off by facilitating the spread of generalist root pathogens like *Phytophthora cinnamomi*, which exhibits higher infectivity in diverse oak assemblages (Gómez et al., 2020). Maintaining both species is crucial for biodiversity and ecosystem function. Conservation strategies should balance economic interests with ecological integrity, promote natural regeneration, and monitor disease risks. Management should avoid excessive favoring of *Q. suber* to preserve the unique structure and resilience of these mixed forests. Mixed *Q. suber–Q. canariensis* forests are ecologically rich but face challenges from management-driven species shifts and disease. Sustaining their diversity and function requires integrated conservation and adaptive management approaches.

Operationalizing the multi-dimensional concept of resilience requires consistent quantification of forest attributes that empirically relate to both resistance to, and recovery from, disturbances (Nikinmaa et al., 2020). Below, we outline key resilience dimensions and corresponding indicators, along with their ecological relevance. These dimensions—forest diversity, individual tree quality, and biological legacies—are interconnected and can be actively shaped through ecosystem management (Lafond et al., 2014; Mina et al., 2022).

At the stand scale, forest diversity encompasses structural heterogeneity and compositional variety (Keren et al., 2020). Greater diversity in these attributes is generally associated with enhanced resistance to disturbances, as well as with more effective recovery, due to a broader range of species- and trait-specific responses post-disturbance (Silva Pedro et al., 2015). Similarly, individual tree quality contributes to both dimensions: tree health and vigor improve stand resistance, while the fecundity and growth of surviving individuals facilitate post-disturbance recovery.

However, in mixed-species stands, variations in crown architecture and apical dominance can lead to distinct growth forms, including increased stem curvature and leaning. This occurs when a species is compelled to extend into unoccupied or better-lit growing spaces to compete for light—particularly among light-demanding species that are overtopped by neighbors with low canopy light transmittance (Grotta et al., 2004) or by those with greater resistance to mechanical branch abrasion (Putz et al., 1984; Hajek et al., 2015). Such adaptive growth responses can also result in irregular crown shapes (Seidel et al., 2011; Brisson 2011) and corresponding variability in branch dimensions, where branches receiving more sunlight tend to be longer and thicker (Kint et al., 2010). Furthermore, crown asymmetry may induce asymmetric stem cross-sections and the formation of reaction wood.

CASE STUDY RESULTS: MIXED-SPECIES STAND DYNAMICS IN OUED ZENA FOREST

The Oued Zeda forest is located in northwestern Tunisia, within the Kroumiria region. It spans a high-altitude area between 600 and 800 meters, which confers a humid to sub-humid Mediterranean climate. Annual precipitation is abundant, frequently exceeding 1000 mm and reaching up to 1500 mm in the highest sectors. The area is characterized by mild, dry summers and cold, humid winters, marked by frequent frosts and annual snowfall (Fig. 7). The region is also notable for its persistent fog, which significantly contributes to the ambient humidity. The predominant soils are Mediterranean red-brown soils (fersiallitic), typical of the humid forest regions of the Maghreb. They have often developed from sandstone and quartzite substrates, which are acidic and siliceous, making them particularly suitable for Cork Oak (*Q.*

suber). The depth of these soils is variable; they are generally well-developed on gentle slopes but can become shallower and skeletal on steep slopes. The humus associated with the cork oak forest is primarily of the "mull" or "moder" type, indicating intense biological activity.

Figure 7. Study area: Administrative map of Jendouba Governorate with the location of the Oued Zena forest and its bioclimatic zone

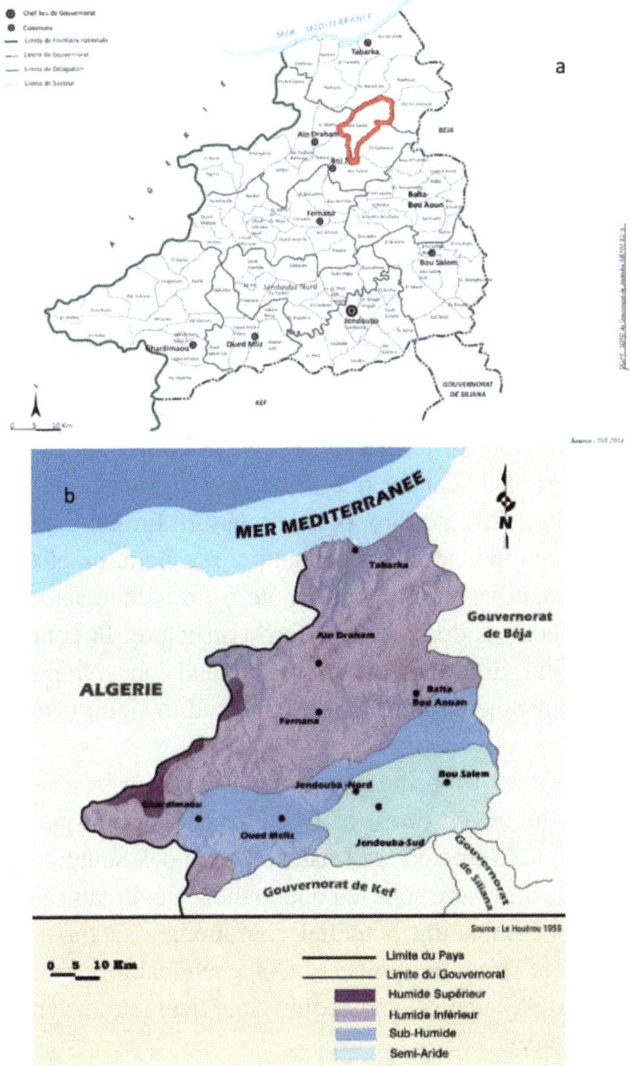

INM (National Institute of Meteorology), 2024

The Kroumirie region, including Oued Zena, is among Tunisia's wettest zones and supports diverse habitats with significant stands of *Q. suber* and *Q. canariensis*. These forests contribute to soil conservation, water regulation, and biodiversity, providing habitat for a rich array of flora and fauna (Frouja et al., 2022). The area's landscape diversity is a key asset for sustainable development and ecosystem services (Frouja et al., 2022).

The forests of Oued Zena are recognized for their high ecological value, yet their integrity is increasingly compromised by a combination of climatic and anthropogenic threats. Key among these are prolonged drought, intensified fire regimes, and pervasive human disturbance. Ensuring the long-term resilience of these Mediterranean woodlands requires conservation strategies that explicitly integrate climate adaptation, proactive fire management, and the stringent protection of natural regeneration zones. The vulnerability of these ecosystems is exemplified by the widespread decline of cork oak stands across the Kroumirie region in recent decades. A period of severe drought and rising temperatures between 1988 and 1995 resulted in the mortality of an estimated 63,622 cork oak trees, a trend exacerbated by ongoing climate change (Touhami et al., 2020). Such droughts not only induce direct tree mortality but also reduce growth vigour and predispose forests to secondary infestations by pests and pathogens (Touhami et al., 2020; Frouja et al., 2022). Concurrently, forest fires have increased in both frequency and severity, with 2021 marking a record year for tree cover loss in Tunisia due to major fire events (Touhami et al., 2023). Anthropogenic pressure further critically impedes forest recovery. Research indicates that the natural regeneration of *Q. canariensis* in Kroumirie is most successful in areas remote from human settlements, where seedling and sapling cohorts dominate the forest structure. In contrast, activities such as livestock grazing, firewood collection, and land conversion near populated areas severely disrupt regeneration processes and lead to significant alterations in forest architecture.

To assess the influence of forest composition, a study was carried out in three adjacent sites within the mixed forest of Oued Zeda (Kroumiria, Tunisia): CZ2 (pure zeen oak stand), CZ3 (mixed cork oak and zeen oak stand), and CZ4 (pure cork oak stand). These sites were selected due to their bioclimatic and pedological homogeneity, thereby eliminating potential confounding factors. The measurements encompassed leaf structural traits (LMA), chemical properties (N), as well as biometric (height and diameter) and ecophysiological parameters ('PLC' Loss of hydraulic conductivity).

This forest is characterized by a temperate, lower humid climate and features a heterogeneous stand composed of a cluster of Zeen Oak (*Q. canariensis*) and Cork Oak (*Q. suber*) (Table. 1), accompanied by the following humicolous species: *Erica arborea*, *Myrtus communis*, *Arbutus unedo*, and *Viburnum tinus* (Figure 8).

Table 1. Study site characteristics

Site	Dominant Species	Longitude	Latitude	Altitude (m)
CZ2	Zeen Oak (*Q. canariensis*)	36°44.323' N	008°52.058' E	317
CZ3	Zeen Oak & Cork Oak (*Q. canariensis* & *Q. suber*)	36°45.214' N	008°52.405' E	316
CZ4	Cork Oak (*Q. suber*)	36°46.264' N	008°53.251' E	315

Figure 8. Overview of the study stands (CZ2, CZ3, and CZ4)

Results indicated that Cork Oak exhibits superior height growth when associated with Zeen Oak (Figure 9). Similarly, Cork Oak trees in the mixed stand (CZ3) displayed the lowest Leaf Mass per Area (LMA) coupled with the highest nitrogen concentrations (Figure 10). Soil nitrogen content in the monospecific stands (CZ2 and CZ4) was lower than in the mixed stand (CZ3) (Figure 10). In the associated stand, Cork Oak trees presented the lowest, and statistically significant, LMA values (Figure 10).

Figure 9. Growth parameters: (a) height, (b) circumference and (c) density of cork oak and zeen oak trees at three different sites: CZ2 site with zeen oak, CZ3 site with cork oak and zeen oak, CZ4 site with cork oak alone

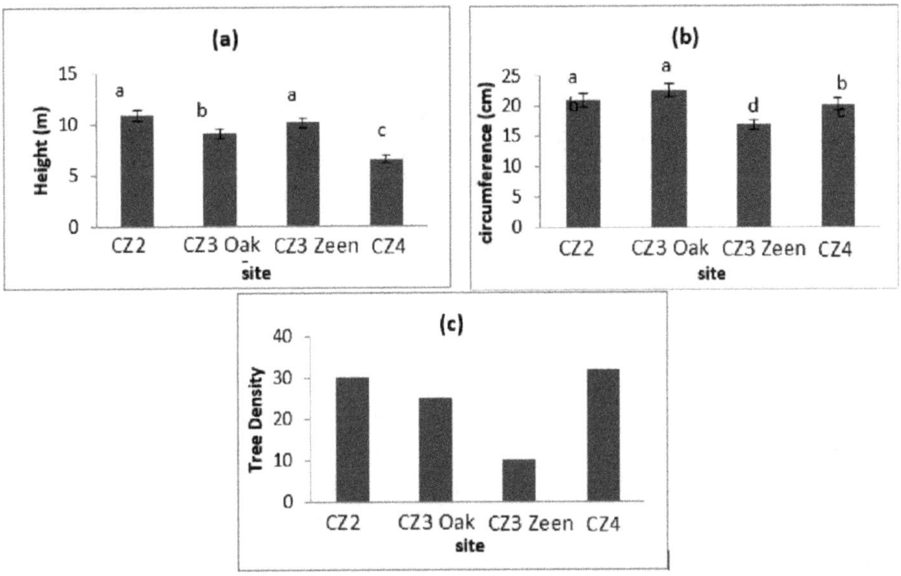

Figure 10. (a) Leaf structural parameters, (b) leaf chemical parameters, and (c) soil chemical parameters at the three sites: CZ2 site with zeen oak, CZ3 site with cork oak + zeen oak, CZ4 site with cork oak alone

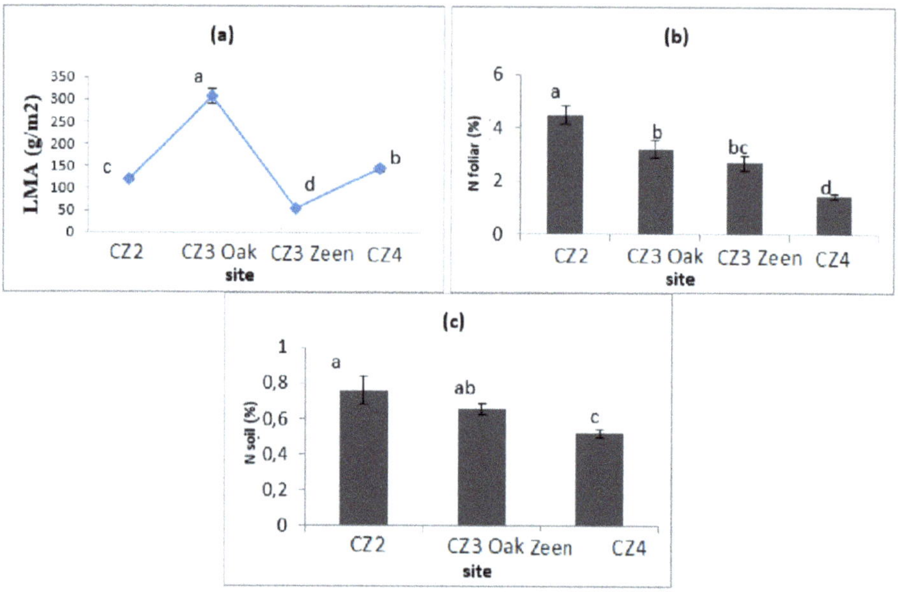

Physiological data were analyzed by constructing vulnerability to cavitation curves (PLC). These curves describe the relationship between the mean percentage loss of hydraulic conductivity (PLC) in the samples and the water potential (Ψ) measured in the originating branch. The water potential inducing a 50% loss of conductivity (PLC50) is a standard parameter for quantifying vulnerability to xylem cavitation (Cochard et al., 1999). This threshold value provides an objective basis for evaluating a species' drought tolerance and enables direct interspecific comparisons using a single, unified metric.

The vulnerability curves yielded P50 values of approximately -2.2 MPa and -3.6 MPa for CZ3 and CZ4, respectively (Figure 11). The P50 value for CZ4 was less negative than that for CZ3. Consequently, the cork oak in CZ4 was more vulnerable to xylem cavitation than in the mixed stand (CZ3).

Figure 11. Vulnerability curves for cork oak trees (Excel). CZ3 site with cork oak + zeen oak, CZ4 site with cork oak alone

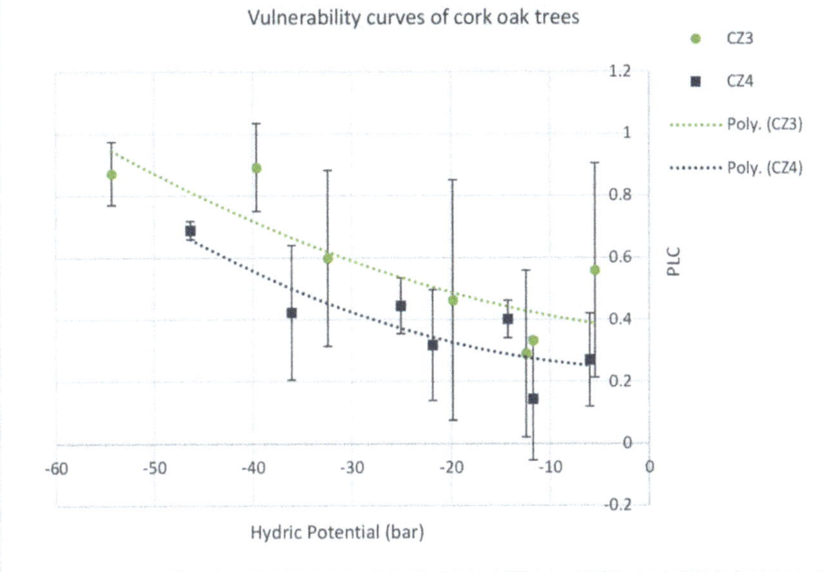

For Zeen oak, the water potential inducing a 50% loss of hydraulic conductivity (P50) was -2.4 MPa in the monospecific stand (CZ2), compared to -2.8 MPa in the mixed stand (CZ3) (Figure 12). The less negative P50 value in CZ3 indicates that Zeen oak was more vulnerable to drought-induced xylem cavitation in the mixed stand than in the monospecific stand.

Figure 12. Vulnerability curves for zeen oak trees (Excel). CZ2 site with zeen oak, CZ3 site with cork oak + zeen oak

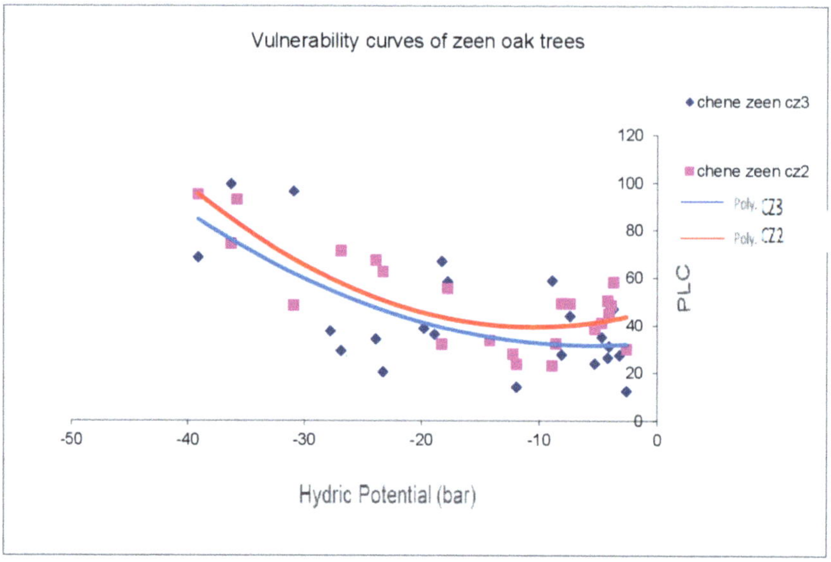

Analysis of the vulnerability curves revealed a distinct hierarchy of resistance to xylem cavitation. Cork oak in the monospecific stand (CZ4) was the most vulnerable, followed by Zeen oak in the mixed stand (CZ3). The most resistant was Zeen oak in the monospecific stand (CZ2). Notably, Cork oak trees within the mixed stand (CZ3) exhibited greater resistance (a more negative P50) than their counterparts in the pure Cork oak stand (CZ4).

This study was conducted under conditions of high environmental homogeneity, where the bioclimatic conditions and soil type are constant across the sites. Consequently, the primary variable explaining the observed differences lies in the specific stand composition, namely the presence of monospecific versus mixed stands of Cork oak (*Q. suber*) and Zeen oak (*Q. canariensis*). The Leaf Mass per Area (LMA) is a recognized functional trait that serves as a robust indicator of site fertility and species productivity (Herzi, 2016). In accordance with this principle, our results demonstrate that the mixed stand CZ3 (Cork oak + Zeen oak), which has the highest soil nitrogen content, is considered the most fertile. This enhanced fertility is crucial, as nitrogen is an essential element for the growth and development of Cork oak (Ennajah, 2010). The relationship between LMA and tree vigor is well-established: higher LMA values are generally interpreted as a response to environmental stress and are correlated with slower leaf growth rates. LMA is one of the key traits of the so-called "leaf economics spectrum" (LES), which globally describes the coordinated variations in leaf structural, chemical, and photosynthetic traits (Wright et al., 2004). Thus, according to LES, it could be expected that populations from more xeric habitats with higher LMA showed lower photosynthesis on a mass basis due to a higher investment in non-photosynthetic structural tissues (Alonson-Forn et al., 2020; Wright et al., 2004). Our data corroborate this relationship. Indeed, Cork oak trees in the monospecific stand (CZ4) exhibited the highest foliar LMA, coupled with the lowest foliar nitrogen concentrations. This profile, which contrasts sharply with that of individuals of the same species in the mixed stand (CZ3), suggests that Cork oaks in pure conditions experience greater nutritional and/or environmental stress, limiting their growth performance.

These results highlight an asymmetrical interspecific interaction between Cork oak (*Q. suber*) and Zeen oak (*Q. canariensis*). The data confirm that Cork oak growth is significantly enhanced in the mixed stand, suggesting a net benefit from this association for this species. This configuration therefore presents a serious option for forest management, likely to optimize the productivity of cork oak forests.

However, this study reveals a complex dynamic of interspecific competition. The analysis of physiological and nutritional traits indicates that Zeen oak derives no apparent benefit from this cohabitation. On the contrary, data on vulnerability to cavitation (P50) and foliar nitrogen content suggest that Zeen oak in the mixed stand experiences greater water and nutritional stress than in pure stands. This indicates direct competition, primarily for access to soil water resources and, to a lesser extent, for available nitrogen, where Cork oak appears to be an effective, if not dominant, competitor. Thus, the improved performance of Cork oak in the association may partially result from resource acquisition at the expense of Zeen oak, outlining an interaction akin to commensalism rather than a symmetrical mutualism.

CONCLUSION

These findings enable the proposal of a targeted and adaptive silvicultural strategy. Given its asymmetric benefit favoring Cork oak, the mixed Cork oak/ Zeen oak association should not be applied uniformly. Its ecological and economic relevance is highest in marginal or transitional zones where Zeen oak stands are already degraded, vulnerable, or at the limit of their optimal distribution range. In these ecologically sensitive contexts, the declining productivity of Zeen oak limits its silvicultural prospects. Therefore, proactive management could involve enrichment trials through the gradual introduction of Cork oak saplings into these degraded Zeen oak stands. This approach aims to leverage the relative competitiveness of Cork oak, observed in our study, to initiate a gentle stand transition. The long-term goal would be to encourage a partial or complete ecological replacement, allowing Cork oak—better adapted to the new site conditions (particularly hydrological) —

to recolonize the space, thereby maintaining forest cover, ecosystem services, and economic value in landscapes otherwise prone to degradation. The comprehensive dataset gathered in this study-including growth parameters, leaf functional traits (LMA), hydraulic vulnerability curves, and soil nutrient dynamics-represents a significant contribution to developing climate-adaptive management strategies for Mediterranean cork oak ecosystems.

Our findings yield several critical insights for forest managers and ecological modellers. For management practice, this research provides quantitative evidence of cork oak's competitive advantage in mixed stands with Zeen oak under environmental stress, establishing mixed plantations as a viable climate adaptation strategy. The physiological thresholds identified, particularly P50 values, offer practical early-warning indicators for monitoring drought stress. Furthermore, the demonstrated relationship between soil nitrogen availability and leaf traits suggests potential for targeted nutrient management to enhance stand resilience.

For modelling applications, this study supplies essential ecophysiological parameters for two keystone Mediterranean species. The quantified relationships between stand composition, resource availability, and physiological responses significantly improve our capacity to predict forest productivity and vulnerability under future climate scenarios. Specifically, these data enable more accurate parameterization of vegetation models simulating forest dynamics in Northwest Tunisia.

Ultimately, this work establishes an empirical foundation for evidence-based silvicultural interventions. By integrating these findings into management frameworks, we can develop robust strategies to maintain cork oak forest functionality, secure cork production, and ensure ecosystem service provision under changing climatic conditions. The asymmetric species responses documented here particularly highlight the importance of species-specific approaches in climate-smart forestry.

REFERENCES

Aguilera, E., Díaz-Gaona, C., García-Laureano, R., Reyes-Palomo, C., Guzmán, G. I., Ortolani, L., Sánchez-Rodríguez, M., & Rodríguez-Estévez, V. (2020). Agroecology for adaptation to climate change and resource depletion in the Mediterranean region. A review. *Agricultural Systems*, *181*, 102809. Advance online publication. DOI: 10.1016/j.agsy.2020.102809

Aissi, A., Beghami, Y., Lepais, O., & Vela, E. (2020). Analyse morphologique et taxonomique du complexe Quercus faginea (Fagaceae) en Algérie. *Botany*, *99*(2), 99–113. DOI: 10.1139/cjb-2020-0075

Alonso-Forn, D., Sancho-Knapik, D., Ferrio, J. P., Peguero-Pina, J. J., Bueno, A., Onoda, Y., Cavender-Bares, J., Niinemets, Ü., Jansen, S., Riederer, M., Cornelissen, J. H. C., Chai, Y., & Gil-Pelegrín, E. (2020). Revisiting the Functional Basis of Sclerophylly within the Leaf Economics Spectrum of Oaks: Dierent Roads to Rome. *Current Forestry Reports*, *6*(4), 260–281. DOI: 10.1007/s40725-020-00122-7

Andersan, H. W., Hoover, Marvin D., Reinhart, K. G. (1976). Forests and water effects of forest management on floods, sedimentation and water supply [Report]. - US: USDA Forest service general technical report. *Pacific Southwest Forest and Range Exp*. Stn., Berkeley, Calif. PSW-18, 115 p., illus.

Aponte, C., García, L., & Marañón, T. (2013). Tree species effects on nutrient cycling and soil biota: A feedback mechanism favouring species coexistence. *Forest Ecology and Management*, *309*, 36–46. DOI: 10.1016/j.foreco.2013.05.035

Bellahirech, A., Branco, M., Catry, F., Bonifácio, L., Sousa, E., & Jamâa, M. (2019). Site- and tree-related factors affecting colonization of cork oaks Quercus suber L. by ambrosia beetles in Tunisia. *Annals of Forest Science*, *76*(2), 45. DOI: 10.1007/s13595-019-0815-1

Benabou, A., Moukrim, S., Laaribya, S., Aafi, A., Chkhichekh, A., Maadidi, T. E., & El Aboudi, A. (2022). Mapping ecosystem services of forest stands: Case study of Maamora, Morocco. *Geography, Environment, Sustainability*, *15*(1), 141–149. DOI: 10.24057/2071-9388-2021-047

Benito, B. M., Lorite, J., Pérez-Pérez, R., Gómez-Aparicio, L., & Peñas, J. (2014). Forecasting plant range collapse in a mediterranean hotspot: When dispersal uncertainties matter. *Diversity & Distributions*, *20*(1), 72–83. DOI: 10.1111/ddi.12148

Benito Garzon, M., Sanchez de Dios, R., & Sainz Ollero, H. (2008). Effects of climate change on the distribution of Iberian tree species. *Applied Vegetation Science*, *11*(2), 169–178. DOI: 10.3170/2008-7-18348

Bongers, F. J., Olmo, M., Lopez-Iglesias, B., Anten, N. P. R., & Villar, R. (2017). Drought responses, phenotypic plasticity and survival of Mediterranean species in two different microclimatic sites. *Plant Biology*, *19*(3), 386–395. DOI: 10.1111/plb.12544 PMID: 28054449

Borreguero, I., Sánchez-Miranda, Á., Martínez-Sancho, E., Rubio-Casal, A., Sánchez-Salguero, R., & Matías, L. (2024). Negative resilience trends to more frequent extreme droughts in Mediterranean oak species. *bioRxiv*. https://doi.org/ DOI: 10.1101/2024.11.19.624310

Bouandas, A., Belhoucine-Guezouli, L., Oliva, F., Semir, B., Bendjebbar, K., Pardo, F., & Pujade-Villar, J. (2024). Morphological and Taxonomic Analysis of the Quercus faginea and Quercus canariensis (Fagaceae) Complexes in Algeria. *International Journal of Plant Biology*, *15*(4), 927–943. Advance online publication. DOI: 10.3390/ijpb15040066

Boudy, P. (1950). Economie forestière Nord-Africaine, TOME 2: Monographie et traitement des essences forestières. Ed. Larosa, Paris, 525p.

Bradshaw, A. D. (1965). Evolutionary significance of phenotypic plasticity in plants. *Advances in Genetics*, *13*, 115–155. Advance online publication. DOI: 10.1016/S0065-2660(08)60048-6

Brisson, J. (2011). Neighborhood competition and crown asymmetry in Acer saccharum. *Canadian Journal of Forest Research*, *31*(12), 2151–2159. DOI: 10.1139/x01-161

Brunetti, M., Maugeri, M., Nanni, T., & Navarra, A. (2002). Droughts and extreme events in regional daily Italian precipitation series. *International Journal of Climatology*, *22*(5), 543–558. DOI: 10.1002/joc.751

Bugalho, M. N., Caldeira, M. C., Pereira, J. S., Aronson, J., & Pausas, J. G. (2011). Mediterranean cork oak savannas require human use to sustain biodiversity and ecosystem services. *Frontiers in Ecology and the Environment*, *9*(5), 278–286. DOI: 10.1890/100084

Bussotti, F., Pollastrini, M., Holland, V., & Brüggemann, W. (2015). Functional traits and adaptive capacity of European forests to climate change. *Environmental and Experimental Botany*, *111*, 91–113. DOI: 10.1016/j.envexpbot.2014.11.006

Chaabani, S. (2013). Influence du climat sur la croissance radicale de chene zeen en Kroumirie, projet de fin d'étude, Ecole supérieur d'Agriculture de Mograne, 57p.

Cochard, H. & Granier A. (1999). Fonctionnement hydraulique des arbres forestiers. Rev. For. Fr. Li, 121-133 pp.

Dale, V. H., Joyce, L. A., Mcnulty, S., Neilson, R. P., Ayres, M. P., Flannigan, M. D., & Wotton, M. B. (2001). Climate change and forest disturbances. *Bioscience*, *51*(9), 723–734. DOI: 10.1641/0006-3568(2001)051[0723:CCAFD]2.0.CO;2

David, T. S., Henriques, M. O., Kurz-Besson, C., Nunes, J., Valente, F., Vaz, M., Pereira, J. S., Siegwolf, R., Chaves, M. M., Gazarini, L. C., Gazarini, L. C., & David, J. S. (2007). Water-use strategies in two co-occurring Mediterranean evergreen oaks: Surviving the summer drought. *Tree Physiology*, *27*(6), 793–803. DOI: 10.1093/treephys/27.6.793 PMID: 17331898

De Riguo, D., & Caudullo, G. (2016). Quercus suber in Europe: distribution, habitat, usage and threats. Eds.), European Atlas of Forest, Tree Species. Publ. Off. EU, Luxembourg, p 165.

Dukes, J. S., Chiariello, N. R., Cleland, E. E., Moore, L. A., Rebecca Shaw, M., Thayer, S., Tobeck, T., Mooney, H. A., & Field, C. B. (2005). Responses of grassland production to single and multiple global environmental changes. *PLoS Biology*, *3*(10), e319. DOI: 10.1371/journal.pbio.0030319 PMID: 16076244

Ennajah, A. (2010). Croissance et productivité des forêts de chênes lièges « Quercus-suber L.» en Tunisie. Vulnérabilité aux changements climatiques. Thèse de Doctorat en Sciences Biologiques – faculté des Sciences de Tunis - 192 p.

EUFORGEN. Distribution map of cork oak (Quercus suber) (2009). www.euforgen.org

Frouja, O., Hammami, M., Dakhlaoui, S., Wannes, W., Hessini, K., Msaada, K., & Ahmed, B. (2022). Intra and interspecific variability of Quercus suber and Quercus canariensis, an intrinsic water-use efficiency approach for differentiation. *Environmental Research Communications*, *4*(10), 105002. Advance online publication. DOI: 10.1088/2515-7620/ac9729

Goeking, S. A., & Tarboton, D. G. (2020). Forests and water yield: A synthesis of disturbance effects on streamflow and snowpack in western coniferous forests. *Journal of Forestry*, *118*(2), 172–192. DOI: 10.1093/jofore/fvz069

Gómez, M., González, M., Gómez-Aparicio, L., & Serrano, M. (2020). Coexistent Mediterranean woody species as a driving factor of Phytophthora cinnamomi infectivity and survival. *Annals of Applied Biology*, *177*(1), 41–50. DOI: 10.1111/aab.12599

González, M., Romero, M.-Á., García, L.-V., Gómez-Aparicio, L., & Serrano, M.-S. (2020). Unravelling the role of drought as predisposing factor for Quercus suber decline caused by Phytophthora cinnamomi. *European Journal of Plant Pathology*, *156*(4), 1015–1021. DOI: 10.1007/s10658-020-01951-9

Gratani, L., Catoni, R., & Varone, L. (2016). Evergreen species response to Mediterranean climate stress factors. *IForest (Viterbo)*, *9*(6), 946–953. DOI: 10.3832/ifor1848-009

Grotta, A. T., Gartner, B. L., & Radosevich, S. R. (2004). Influence of species proportion and timing of establishment on stem quality in mixed red alder – Douglas-fir plantations. *Canadian Journal of Forest Research*, *34*(4), 863–873. DOI: 10.1139/x03-259

Hajek, P., Seidel, D., & Leuschner, C. (2015). Mechanical abrasion, and not competition for light, is the dominant canopy interaction in a temperate mixed forest. *Forest Ecology and Management*, *348*, 108–116. DOI: 10.1016/j.foreco.2015.03.019

Herzi, M. (2016). Etude de l'impact des légumineuses associées au chêne liège sur sa croissance, cas de cytise triflore Cytisus triflorus. L'Hérit., Mémoire de Mastère de Recherche, INAT, 36-37 p.

IPCC. (2014). Climate Change 2014: Synthesis Report. Contribution of Working Groups I, II and III to the Fifth Assessment Report of the Intergovernmental Panel on Climate Change. IPCC, Geneva. AR5 Synthesis Report [— IPCC]. *Climatic Change*, •••, 2014.

Kahane, R., Hodgkin, T., Jaenicke, H., Hoogendoorn, C., Hermann, M., Hughes, J. D. A., Padulosi, S., & Looney, N. (2013). Agrobiodiversity for food security, health and income. *Agronomy for Sustainable Development*, *33*(4), 671–693. DOI: 10.1007/s13593-013-0147-8

Kanzler, M., Böhm, C., Mirck, J., Schmitt, D., & Veste, M. (2019). Microclimate effects on evaporation and winter wheat (*Triticum aestivum* L.) yield within a temperate agroforestry system. *Agroforestry Systems*, *93*(5), 1821–1841. DOI: 10.1007/s10457-018-0289-4

Keren, S., Svoboda, M., Janda, P., & Nagel, T. A. (2020). Relationships between structural indices and conventional stand attributes in an old-growth forest in Southeast Europe. *Forests*, *11*(1), 4. DOI: 10.3390/f11010004

Kint, V., Hein, S., Campioli, M., & Muys, B. (2010). Modelling self-pruning and branch attributes for young Quercus robur L. and Fagus sylvatica L. trees. *Forest Ecology and Management*, *260*(11), 2023–2034. DOI: 10.1016/j.foreco.2010.09.008

Klausmeyer, K. R., & Shaw, M. R. (2009). Climate change, habitat loss, protected areas and the climate adaptation potential of species in Mediterranean ecosystems worldwide. *PLoS One*, *4*(7), 1–9. DOI: 10.1371/journal.pone.0006392 PMID: 19641600

Ksontini, F. (1996). *Contribution à l'étude de la diversité génétique des peuplements de chêne-liège en Tunisie: implications pour la conservation*. Mémoire de Mastère de Recherche, Institut National Agronomique de Tunisie.

Lafond, V., Lagarrigues, G., Cordonnier, T., & Courbaud, B. (2014). Uneven-aged management options to promote forest resilience for climate change adaptation: Effects of group selection and harvesting intensity. *Annals of Forest Science*, *71*(2), 173–186. DOI: 10.1007/s13595-013-0291-y

Leite, C., Oliveira, V., Lauw, A., & Pereira, H. (2019). Cork rings suggest how to manage Quercus suber to mitigate the effects of climate changes. *Agricultural and Forest Meteorology*, *266–267*, 12–19. DOI: 10.1016/j.agrformet.2018.11.032

Léo. Thorand. (2018). Résistance à l'embolie chez deux espèces en association: chêne liège/chêne zeen. Ecole Nationale Supérieure Agronomique de Toulouse. Stage d'été. 65p.

Limousin, J.-M., Rambal, S., Ourcival, J.-M., Rodríguez-Calcerrada, J., Pérez-Ramos, I. M., Rodríguez-Cortina, R., Misson, L., & Joffre, R. (2012). Morphological and phenological shoot plasticity in a Mediterranean evergreen oak facing long-term increased drought. *Oecologia*, *169*(2), 565–577. DOI: 10.1007/s00442-011-2221-8 PMID: 22159896

Lopez-Tirado, J., & Hidalgo, P. J. (2016). Predictive modelling of climax oak trees in southern Spain: Insights in a scenario of global change. *Plant Ecology*, *217*(4), 451–463. DOI: 10.1007/s11258-016-0589-6

Lopez-Tirado, J., Vessella, F., Schirone, B., & Hidalgo, P. J. (2018). Trends in evergreen oak suitability from assembled species distribution models: Assessing climate change in south-western Europe. *New Forests*, *49*(4), 471–487. DOI: 10.1007/s11056-018-9629-5

Matesanz, S., & Valladares, F. (2014). Ecological and evolutionary responses of Mediterranean plants to global change. *Environmental and Experimental Botany*, *103*, 53–67. DOI: 10.1016/j.envexpbot.2013.09.004

Mechergui, K., Jaouadi, W., Altamimi, A., Naghmouchi, S., & Ammari, Y. (2021). Effect of climate change on the spatial distribution and cork production of Quercus suber L., the risk of exclusion by the Aleppo pine expansion, and management practices to protect Q. suber habitat: A review. *Notulae Botanicae Horti Agrobotanici Cluj-Napoca, 49*(1), 12218. DOI: 10.15835/nbha49112218

Mechergui, K., Jaouadi, W., & Ammari, Y. (2022). Le chêne zéen (Quercus canariensis Willd.) en Tunisie: Écologie, production, état de dégradation et les pratiques d'aménagement. *Revue Marocaine des Sciences Agronomiques et Vétérinaires, 10*(1). https://www.agrimaroc.org/index.php/Actes_IAVH2/article/view/1075

Mediavilla, S., & Escudero, A. (2004). Stomatal responses to drought of mature trees and seedlings of two co-ccurring Mediterranean oaks. *Forest Ecology and Management, 187*(2-3), 281–294. DOI: 10.1016/j.foreco.2003.07.006

Mina, M., Messier, C., Duveneck, M. J., Fortin, M. J., & Aquilu'e, N. (2022). Managing for the unexpected: Building resilient forest landscapes to cope with global change. *Global Change Biology, 28*(14), 4323–4341. DOI: 10.1111/gcb.16197 PMID: 35429213

Moretti, V., Di Bartolomei, R., Sorgi, T., Aromolo, R., & Salvati, L. (2015). Soil water deficit and climate conditions during the dry season along the coastal-inland gradient in Castelporziano forest, central Italy. *Rendiconti Lincei. Scienze Fisiche e Naturali, 26*(S3), 283–288. DOI: 10.1007/s12210-014-0336-7

Nikinmaa, L., Lindner, M., Cantarello, E., Jump, A. S., Seidl, R., Winkel, G., & Muys, B. (2020). Reviewing the Use of Resilience Concepts in Forest Sciences. *Current Forestry Reports, 6*(2), 61–80. DOI: 10.1007/s40725-020-00110-x PMID: 35747899

Nouri, J., Houcine, S., Foued, A., & Chââbane, A. (2022). Dynamics of the natural regeneration of a natural zeen oak (Quercus canariensis Willd.) forests in Kroumirie, north-western Tunisia. *Arabian Journal of Geosciences, 15*(22), 1661. DOI: 10.1007/s12517-022-10783-2

Parmesan, C., & Yohe, G. (2003). A globally coherent fingerprint of climate change impacts across natural systems. *Nature, 421*(6918), 37–42. DOI: 10.1038/nature01286 PMID: 12511946

Peguero-Pina, J. J., Sancho-Knapik, D., Barrón, E., Camarero, J. J., Vilagrosa, A., & Gil-Pelegrín, E. (2014). Morphological and physiological divergences within Quercus ilex support the existence of different ecotypes depending on climatic dryness. *Annali di Botanica, 114*(2), 301–313. DOI: 10.1093/aob/mcu108 PMID: 24941998

Peguero-Pina, J. J., Sisó, S., Sancho-Knapik, D., Diaz-Espejo, A., Flexas, J., Galmés, J., & Gil-Pelegrín, E. (2016). Leaf morphological and physiological adaptations of a deciduous oak (Quercus faginea Lam.) to the Mediterranean climate: A comparison with a closely related temperate species (Quercus robur L.). *Tree Physiology*, *36*(3), 287–299. DOI: 10.1093/treephys/tpv107 PMID: 26496958

Pérez-Ramos, I. M., Matías, L., Gómez-Aparicio, L., & Godoy, Ó. (2019). Functional traits and phenotypic plasticity modulate species coexistence across contrasting climatic conditions. *Nature Communications*, *10*(1), 2555. Advance online publication. DOI: 10.1038/s41467-019-10453-0 PMID: 31186418

Petit, R. J., Hampe, A., & Cheddadi, R. (2005). Climate changes and tree phylogeography in the Mediterranean. *Taxon*, *54*(4), 877–885. DOI: 10.2307/25065568

Pina-Martins, F., Baptista, J., Pappas, G.Jr, & Paulo, O. (2018). New insights into adaptation and population structure of cork oak using genotyping by sequencing. *Global Change Biology*, *25*(1), 337–350. DOI: 10.1111/gcb.14497 PMID: 30358018

Pons, J., & Pausas, J. G. (2006). Oak regeneration in heterogeneous landscape: The case of fragmented Quercus suber forests in the eastern Iberian Peninsula. *Forest Ecology and Management*, *231*(1-3), 196–204. DOI: 10.1016/j.foreco.2006.05.049

Putz, F. E., Parker, G. G., & Archibald, R. M. (1984). Mechanical abrasion and intercrown spacing. *American Midland Naturalist*, *112*(1), 24–28. DOI: 10.2307/2425452

Rabhik., (2011). Ajustement de modèles hauteur – Circonférence-age pour le chêne zéen (Quercus canariensis Willd) dans la forêt d'Akfadou (TiziOuzou); effet de la densité et de la station, Thèse de magister en foresterie, Gestion et consèrvation des écosystèmes . Univ Tlemcen. 67p.

Ramírez-Valiente, J., Valladares, F., Gil, L., & Aranda, I. (2009). Population differences in juvenile survival under increasing drought are mediated by seed size in cork oak (Quercus suber L.). *Forest Ecology and Management*, *257*(8), 1676–1683. DOI: 10.1016/j.foreco.2009.01.024

Reis, F., Pereira, A. J., Tavares, R. M., Baptista, P., & Lino-Neto, T. (2021). Cork oak forests soil bacteria: Potential for sustainable agroforest production. *Microorganisms*, *9*(9), 1973. DOI: 10.3390/microorganisms9091973 PMID: 34576868

Savolainen, O., Pyhäjärvi, T., & Knürr, T. (2007). Gene flow and local adaptation in trees. *Annual Review of Ecology, Evolution, and Systematics*, *38*(1), 595–619. DOI: 10.1146/annurev.ecolsys.38.091206.095646

Scheffers, B. R., Meester, L. D., Bridge, T. C. L., Hoffmann, A. A., Pandolfi, J. M., Corlett, R. T., & Watson, J. E. M. (2016). The broad footprint of climate change from genes to biomes to people. *Science*, *354*(6313), aaf7671. Advance online publication. DOI: 10.1126/science.aaf7671 PMID: 27846577

Seidel, D., Leuschner, C., Müller, A., & Krause, B. (2011). Crown plasticity in mixed forests—Quantifying asymmetry as a measure of competition using terrestrial laser scanning. *Forest Ecology and Management*, *261*(11), 2123–2132. DOI: 10.1016/j.foreco.2011.03.008

Serrano, M. S., Pérez, F. J., & Gómez-Aparicio, L. (2021). Disentangling the interactive effects of climate change and Phytophthora cinnamomi on coexisting Mediterranean tree species. *Agricultural and Forest Meteorology*, *298–299*, 108295. Advance online publication. DOI: 10.1016/j.agrformet.2020.108295

Shibani, C., Pradeep, L., & Indra, P. (2025). Enhancing flood resilience through mixed forest cover: A case study of the Upper Sabarmati River Basin, Gujarat, India. *Journal of Information Systems Engineering and Management, 10* (33s) e-ISSN: 2468-4376. https://www.jisem-journal.com/. https://doi.org/DOI: 10.52783/jisem.v10i33s.5457

Silva, J., Araújo, S., Sales, H., Pontes, R., & Nunes, J. (2023). Quercus suber L. Genetic Resources: Variability and Strategies for Its Conservation. *Forests*, *14*(9), 1925. DOI: 10.3390/f14091925

Silva Pedro, M., Rammer, W., & Seidl, R. (2015). Tree species diversity mitigates disturbance impacts on the forest carbon cycle. *Oecologia*, *177*(3), 619–630. DOI: 10.1007/s00442-014-3150-0 PMID: 25526843

Simões, R., Rodrigues, A., Ferreira-Dias, S., Miranda, I., & Pereira, H. (2020). Chemical Composition of Cuticular Waxes and Pigments and Morphology of Leaves of Quercus suber Trees of Different Provenance. *Plants*, *9*(9), 1165. Advance online publication. DOI: 10.3390/plants9091165 PMID: 32916803

Soheili, F., Heydari, M., Woodward, S., & Naji, H. R. (2023). Adaptive mechanism in Quercus brantii Lindl. leaves under climatic differentiation: Morphological and anatomical traits. *Scientific Reports*, *13*(1), 3580. DOI: 10.1038/s41598-023-30762-1 PMID: 36869142

Stiti, B., Piazzetta, R., Khaldi, A. (2014). Régénération de la suberaie tunisienne: état des lieux, contraintes et avancées techniques. *Forêt Méditerranéenne, XXXV* (2), pp.151-160. hal-03556655.

Sun, S., Zhang, Y., Wang, N., Yang, W., Zhai, Y., Wang, H., Fan, P., You, C., Zheng, P., & Wang, R. (2024). Changing effects of energy and water on the richness distribution pattern of the Quercus genus in China. *Frontiers in Plant Science*, *15*, 1301395. Advance online publication. DOI: 10.3389/fpls.2024.1301395 PMID: 38298826

Taibi., (2012). Contribution à une étude descriptive des peuplements à chêne zéen (Quercus faginea Lamk) dans les forêts de Zariffet et Hafir (Wilaya de Tlemcen), 70p.

Thuiller, W., Lavergne, S., Roquet, C., Boulangeat, I., Lafourcade, B., & Araujo, M. B. (2011). Consequences of climate change on the tree of life in Europe. *Nature*, *470*(7335), 531–534. DOI: 10.1038/nature09705 PMID: 21326204

Torralba, M., Fagerholm, N., Burgess, P. J., Moreno, G., & Plieninger, T. (2016). Do European agroforestry systems enhance biodiversity and ecosystem services? A meta-analysis. *Agriculture, Ecosystems & Environment*, *230*, 150–161. DOI: 10.1016/j.agee.2016.06.002

Touhami, I., Chirino, E., Aouinti, H., Khorchani, E., Elaieb, M., Khaldi, A., & Nasr, Z. (2020). Decline and dieback of cork oak (Quercus suber L.) forests in the Mediterranean basin: A case study of Kroumirie, Northwest Tunisia. *Journal of Forestry Research*, *31*(5), 1–17. DOI: 10.1007/s11676-019-00974-1

Touhami, I., Rzigui, T., Zribi, L., Ennajah, A., Dhahri, S., Aouinti, H., Elaieb, M., Fkiri, S., Ghazghazi, H., Khorchani, A., Candelier, K., Khaldi, A., & Khouja, M. (2023). Climate change-induced ecosystem disturbance: A review on sclerophyllous and semi-deciduous forests in Tunisia. *Plant Biology*, *25*(4), 481–497. DOI: 10.1111/plb.13524 PMID: 37014233

Trumbore, S., Brando, P., & Hartmann, H. (2015). Forest health and global change. *Science*, *349*(6250), 814–818. https://www.science.org/doi/10.1126/science.aac6759. DOI: 10.1126/science.aac6759 PMID: 26293952

Urbieta, I., Zavala, M., & Marañón, T. (2008). Human and non-human determinants of forest composition in southern Spain: Evidence of shifts towards cork oak dominance as a result of management over the past century. *Journal of Biogeography*, *35*(9), 1688–1700. Advance online publication. DOI: 10.1111/j.1365-2699.2008.01914.x

Urbieta, M. S., Toril, E. G., Aguilera, Á., Giaveno, M. A., & Donatile, E. (2013). Cyanobacteria and photosynthetic species as part of the microbial community structure of biofilms in copahue geothermal springs (Neuquén, Argentina). *Advanced Materials Research*, *825*, 11–14. . DOI: 10.4028/www.scientific.net/AMR.825.11

Vanhove, M., Pina-Martins, F., Coelho, A., Branquinho, C., Costa, A., Batista, D., Silva, A., Sousa, P., Henriques, A., Marques, I., Belkadi, B., Knowles, L., & Paulo, O. (2021). Using gradient Forest to predict climate response and adaptation in Cork oak. *Journal of Evolutionary Biology*, *34*(6), 910–923. DOI: 10.1111/jeb.13765 PMID: 33484040

Veldkamp, E., Schmidt, M., Markwitz, C., Beule, L., Beuschel, R., Biertümpfel, A., Bischel, X., Duan, X., Gerjets, R., Göbel, L., Grab, R., Guerra, V., Heinlein, F., Komainda, M., Langhof, M., Luo, J., Potthoff, M., van Ramshorst, J. G. V., Rudolf, C., & Wachendorf, M. (2023). Multifunctionality of temperate alley-cropping agroforestry outperforms open cropland and grassland. *Communications Earth & Environment*, *4*(1), 20. Advance online publication. DOI: 10.1038/s43247-023-00680-1

Vessella, F., López-Tirado, J., Simeone, M., Schirone, B., & Hidalgo, P. (2017). A tree species range in the face of climate change: Cork oak as a study case for the Mediterranean biome. *European Journal of Forest Research*, *136*(3), 555–569. DOI: 10.1007/s10342-017-1055-2

Vicente, E., Carme, M., Silva, F., Stiti, B., Vizcaíno-Palomar, N., & Garzón, M. (2025). Phenotypic integration of post-germination traits in Quercus suber: Morphology is mediated by acorn mass and leaf physiology by populations' climate. *Annals of Botany*, *136*(1), 153–166. DOI: 10.1093/aob/mcaf068 PMID: 40220265

Vila-Viçosa, C., Vázquez, F., Mendes, P., Río, S., Musarella, C., Cano-Ortiz, A., & Meireles, C. (2015). Syntaxonomic update on the relict groves of Mirbeck's oak (Quercus canariensis Willd. and Q. marianica C. Vicioso) in southern Iberia. *Plant Biosystems*, *149*(3), 512–526. DOI: 10.1080/11263504.2015.1040484

Vizinho, A., Príncipe, A., Vasconcelos, A., Rebelo, R., Branquinho, C., & Penha-Lopes, G. (2023). Using and Creating Microclimates for Cork Oak Adaptation to Climate Change. *Land (Basel)*, *12*(3), 531. Advance online publication. DOI: 10.3390/land12030531

Wright, I. J., Reich, P. B., Westoby, M., Ackerly, D. D., Baruch, Z., Bongers, F., Cavender-Bares, J., Chapin, T., Cornelissen, J. H. C., Diemer, M., Flexas, J., Garnier, E., Groom, P. K., Gulias, J., Hikosaka, K., Lamont, B. B., Lee, T., Lee, W., Lusk, C., & Villar, R. (2004). The world-wide leaf economics spectrum. *Nature*, *428*(6985), 821–827. DOI: 10.1038/nature02403 PMID: 15103368

Zhou, G. Y., Morris, J. D., Yan, J. H., Yu, Z. Y., & Peng, S. L. (2002). Hydrological impacts of reafforestation with eucalyptus and indigenous specious: A case study in Southern China [Journal] // Elsevier Journal, *Forest ecology and management*. 209-222, ISSN 0378-1127, https://doi.org/DOI: 10.1016/S0378-1127(01)00694-6

Zhou, Q., Keith, D. M., Zhou, X., Cai, M., Cui, X., Wei, X., & Luo, Y. (2018). Comparing the water holding characteristics of broad leaved, coniferous and mixed forest litter layers in a Karst Region. *Mountain Research and Development*, *38*(3), 220–229. DOI: 10.1659/MRD-JOURNAL-D-17-00002.1

Section 2
Disturbances and Risk Dynamics

Chapter 5
An Operational Spatio-Temporal and Explainable Machine Learning Framework for Daily Wildfire Risk Prediction in Morocco

Hicham Hajji
IAV Hassan II Rabat, Morocco

Hicham Mharzi Alaoui
http://orcid.org/0000-0003-2226-5133
National Center For Climate and Environmental Risks, Morocco

Fouad Assali
National Agency for Water and Forest, Morocco

Taoufik Aadel
National Agency for Water and Forest, Morocco

Samir Taberkant
National Agency for Water and Forest, Morocco

Hajar Benaissa
National Agency for Water and Forest, Morocco

ABSTRACT

This chapter introduces AI4Fire, an operational and explainable spatio-temporal machine learning framework for daily wildfire risk prediction in Northern Morocco. The system integrates large-scale geospatial data, advanced spatio-temporal feature engineering, high-performance computing, and explainable artificial intelligence to generate high-resolution risk and interpretability maps over more than one million grid points. AI4Fire combines meteorological dynamics, heatwave indicators,

DOI: 10.4018/979-8-2600-0783-9.ch005

vegetation status, topography, socio-environmental factors, and historical wildfire pressure within a scalable Big Data architecture enabling near real-time inference. Validation over the 2021-2023 fire seasons demonstrates strong and stable predictive performance, with more than 75% of wildfire events occurring in areas classified as very high risk. By coupling accuracy, scalability, and transparency, AI4Fire provides an effective decision-support tool for proactive wildfire prevention and sustainable forest management in Mediterranean ecosystems.

1. INTRODUCTION

Wildfires represent one of the most critical natural hazards affecting forest ecosystems worldwide, causing severe ecological degradation, economic losses, and human casualties. In recent decades, the Mediterranean basin has emerged as a global hotspot for wildfire risk, driven by rising temperatures, prolonged droughts, and increasing anthropogenic pressures. According to the Intergovernmental Panel on Climate Change, climate projections indicate a significant intensification of fire-prone weather conditions across the region by mid-century, particularly in Southern Europe and North Africa (IPCC, 2021). Empirical studies further confirm a rising likelihood of large, heat-induced wildfires in Mediterranean ecosystems (Ruffault et al., 2020).

Morocco is among the Mediterranean countries most exposed to this escalating threat. Its complex topography, heterogeneous vegetation cover, expanding wildland–urban interfaces, and strong interannual climate variability create favorable conditions for the ignition and rapid spread of forest fires. Previous studies conducted in Northern Morocco have demonstrated the strong influence of topography, land cover, climate variability, and anthropogenic activities on wildfire ignition and propagation (Assali et al., 2016; Mharzi Alaoui et al., 2017; Mharzi Alaoui et al., 2019). Consequently, early warning systems and predictive modeling tools have become essential components of national wildfire prevention and mitigation strategies.

Recent advances in artificial intelligence (AI), Earth observation technologies, and geospatial analytics have opened new perspectives for wildfire risk assessment. The increasing availability of high-resolution meteorological data, satellite imagery, sensor networks, and spatial databases has enabled the development of data-driven models capable of capturing complex nonlinear relationships between environmental drivers and wildfire occurrence. Numerous studies have demonstrated the potential of machine learning methods, such as artificial neural networks and gradient boosting algorithms, for estimating wildfire occurrence probability in Mediterranean landscapes (Elia et al., 2020; Chen & Guestrin, 2016; Kuter et al., 2011).

More recently, deep learning techniques have shown remarkable capabilities for wildfire-related tasks, leveraging advanced architectures including Transformers (Yu et al., 2023; Qiao et al., 2023), attention-based segmentation models (Niknejad & Bernardino, 2021), self-supervised learning frameworks (Fernandes et al., 2023; Chen et al., 2020), and multimodal deep learning approaches combining optical, thermal, and meteorological data (Alipour et al., 2023; Bhamra et al., 2023). These approaches significantly improve detection accuracy and spatio-temporal generalization.

Despite these methodological advances, the transition from experimental research models to fully operational wildfire prediction systems remain limited. Only a small number of platforms currently deliver daily wildfire risk maps using machine learning methods (Cardil et al., 2023; McCormick, 2020). Most operational systems continue to rely primarily on physically based fire danger indices derived from meteorological and topographic variables, often neglecting historical wildfire occurrence, vegetation stress dynamics, and complex spatio-temporal dependencies (Mharzi Alaoui et al., 2017; Mharzi Alaoui et al., 2019). Furthermore, the lack of model interpretability remains a critical barrier to operational deployment, as decision-makers require transparent and explainable outputs to support risk communication, resource allocation, and tactical planning.

In parallel, the growing scale and complexity of geospatial datasets raise significant computational challenges. Efficient processing of massive spatio-temporal data streams requires scalable spatial data abstractions and high-performance computing solutions capable of parallel spatial joins, buffering operations, and feature engineering pipelines (Zhou et al., 1998; Rocklin et al., 2015; Dugré et al., 2023; Dask-GeoPandas, 2024).

In this context, this chapter introduces AI4Fire, the first operational wildfire risk prediction system in Morocco, and among the first in the Mediterranean region, to combine spatio-temporal machine learning, spatial big data processing, and explainable artificial intelligence (XAI) within a unified architecture. Operational since 2022, AI4Fire generates daily wildfire risk and interpretability maps over more than one million grid points, enabling near real-time monitoring and decision support.

The main objectives of this work are summarized as follows:

1. Development of an operational spatio-temporal machine learning framework integrating meteorological dynamics, heatwave indicators (Hulley et al., 2020; Guion et al., 2022), vegetation condition, topography, and historical wildfire pressure to deliver daily wildfire risk maps.
2. Implementation of a scalable spatial big data computing architecture, deployed on an HPC infrastructure using distributed task scheduling and spatial data

partitioning techniques (Rocklin et al., 2015; Dugré et al., 2023; Zhou et al., 1998).
3. Integration of an explainable AI module, providing local interpretability maps that identify the dominant drivers of wildfire risk at each spatial location (Chen & Guestrin, 2016; ELI5, 2024).
4. Deployment of a fully automated operational pipeline, ensuring seamless integration between meteorological data centers, HPC facilities, and analytical platforms without human intervention.

Through this integrated framework, AI4Fire aims to strengthen wildfire early warning capabilities in Morocco and to support evidence-based forest management and disaster risk reduction policies across Mediterranean environments.

1.1 Climate Change, Mediterranean Vulnerability, and the Escalation of Wildfire Risk

Wildfires have become one of the most visible and destructive manifestations of climate change impacts on terrestrial ecosystems. Among global regions, the Mediterranean basin is increasingly recognized as a climate change hotspot, characterized by rising mean temperatures, increasing frequency and duration of heatwaves, declining precipitation, and enhanced interannual climate variability (IPCC, 2021). These trends are projected to intensify throughout the twenty-first century, leading to longer fire seasons, higher fire intensity, and increased difficulty in wildfire suppression.

Recent empirical and modeling studies indicate a strong coupling between heat extremes, prolonged drought conditions, and large wildfire occurrence across Mediterranean landscapes (Ruffault et al., 2020). The combined effect of fuel desiccation, reduced live fuel moisture, and atmospheric instability significantly amplifies ignition probability and fire spread potential. In addition, climate-driven shifts in vegetation composition and structure may further exacerbate fire susceptibility by increasing fuel continuity and altering fire regimes.

Morocco is particularly vulnerable to these evolving dynamics due to its bioclimatic diversity and transitional position between temperate Mediterranean and arid Saharan systems. Northern Morocco, in particular, concentrates a large proportion of the country's forest resources, including cork oak, holm oak, pine, and mixed forest ecosystems, which are highly sensitive to heat stress and drought. These ecological vulnerabilities are compounded by complex topography, which influences local wind regimes and fire behavior, as well as by expanding wildland–urban interfaces, increasing both ignition sources and potential impacts.

Beyond climatic drivers, anthropogenic pressures play a critical role in shaping wildfire risk patterns. Agricultural expansion, infrastructure development, population growth, and increased mobility contribute to higher ignition likelihood, particularly during extreme weather conditions. Previous studies in Morocco have demonstrated that wildfire occurrence results from the interaction of environmental susceptibility and human activity, rather than from meteorological factors alone (Assali et al., 2016; Mharzi Alaoui et al., 2019).

In this context, wildfire risk can no longer be considered a static phenomenon. Instead, it emerges as a dynamic spatio-temporal process, shaped by the cumulative effects of past fire activity, evolving climate conditions, vegetation stress, and human influence. This evolving risk landscape calls for adaptive, data-driven, and operationally responsive wildfire management strategies, capable of integrating heterogeneous information sources and updating risk assessments on a daily basis.

1.2 Artificial Intelligence and the Digital Transformation of Wildfire Management

The rapid expansion of digital technologies over the last two decades has profoundly transformed environmental monitoring and natural hazard management. In the wildfire domain, the increasing availability of Earth observation data, high-resolution meteorological forecasts, in situ measurements, and geospatial information systems has laid the foundation for a new generation of data-driven wildfire intelligence systems.

Artificial intelligence (AI), and machine learning in particular, has emerged as a key enabling technology for extracting actionable knowledge from these large and complex datasets. Unlike traditional physically based fire danger indices, which rely on simplified representations of fire–weather relationships, AI-based approaches can capture nonlinear interactions, multiscale dependencies, and latent patterns embedded within historical data. As a result, machine learning models have demonstrated growing potential for wildfire ignition prediction, fire spread modeling, and impact assessment.

However, despite significant progress in research-oriented applications, the operational adoption of AI in wildfire management remains limited. Many existing studies focus on algorithmic performance under experimental conditions, often neglecting key operational constraints such as computational scalability, data latency, system robustness, and user interpretability. Moreover, the "black-box" nature of many advanced AI models poses a major challenge for their acceptance by decision-makers, who require transparent and explainable outputs to support risk communication and operational planning.

In response to these limitations, a paradigm shift is currently underway, moving from isolated predictive models toward integrated wildfire intelligence platforms. Such platforms aim to combine Big Data infrastructures, scalable computing environments, spatio-temporal machine learning, and explainable artificial intelligence (XAI) within unified operational systems. This integration enables continuous data ingestion, near real-time risk assessment, and interpretable decision support tailored to the needs of forest managers and emergency services.

Explainability plays a particularly critical role in this transformation. By providing insights into the relative contribution of meteorological conditions, vegetation status, historical fire activity, and human factors, XAI approaches enhance trust, accountability, and situational awareness. They also facilitate learning and adaptation, allowing practitioners to better understand evolving risk drivers under changing climatic conditions.

Within this digital transformation framework, AI is not intended to replace expert knowledge, but rather to augment human decision-making by synthesizing large volumes of information into coherent, timely, and interpretable risk indicators. When embedded within operational pipelines and supported by high-performance computing infrastructures, AI-based wildfire risk prediction systems can significantly improve prevention planning, surveillance prioritization, and resource allocation, ultimately contributing to more resilient forest ecosystems and reduced wildfire impacts.

2. WILDFIRES AND BIG DATA: TOWARD DATA-DRIVEN FIRE RISK INTELLIGENCE

2.1 Big Data Foundations in Environmental and Wildfire Applications

The concept of Big Data has been widely formalized by Gartner, which defines Big Data as data assets characterized by high volume, velocity, and variety, requiring innovative forms of information processing to enhance insight generation, decision-making, and process automation. This definition has since been extended by the inclusion of additional dimensions such as veracity and value, particularly relevant in environmental and risk-related applications where data quality and interpretability are critical (Gandomi & Haider, 2015).

The emergence of Big Data was initially driven by major web-based companies such as Google, Facebook, Yahoo, and Twitter, which faced unprecedented challenges related to the storage, processing, and analysis of massive and continuously growing datasets. These challenges led to a paradigm shift away from traditional centralized relational database systems toward distributed data storage and parallel computing architectures, enabling horizontal scalability and fault tolerance. Technologies such as distributed file systems, task scheduling frameworks, and data-parallel programming models have since become foundational components of modern Big Data ecosystems.

The geospatial domain has been profoundly affected by this transformation. Initiatives such as OpenStreetMap illustrate the democratization of geographic information production, enabling large communities of contributors to collaboratively generate and update geospatial knowledge at global scale. Beyond volunteered geographic information, the proliferation of satellite imagery, geolocation-enabled devices, Internet of Things (IoT) sensors, unmanned aerial vehicles (UAVs), connected vehicles, and mobile applications has resulted in an exponential growth of geolocated and time-stamped data streams.

Recent estimates indicate that the number of connected devices worldwide continues to increase rapidly, generating billions of geospatially referenced observations on a daily basis. These data streams vary widely in spatial resolution, temporal frequency, semantic content, and uncertainty, thereby challenging conventional Geographic Information Systems (GIS). While traditional GIS platforms excel at storing, visualizing, and querying structured spatial datasets, they exhibit significant limitations when confronted with large-scale, high-velocity, and heterogeneous spatio-temporal data, particularly in near real-time operational contexts.

In wildfire management, these challenges are especially pronounced. Fire-related data originate from diverse sources, including meteorological forecasts, satellite-based fire detections, vegetation indices, historical fire records, socio-economic datasets, and real-time sensor observations. The integration and analysis of such heterogeneous data at national or regional scale necessitate Big Data infrastructures capable of distributed spatial processing, scalable analytics, and continuous data ingestion, forming the technological backbone of modern wildfire intelligence systems.

2.2 Artificial Intelligence as a Core Enabler of Wildfire Intelligence

Artificial intelligence (AI) represents one of the most transformative technological developments of recent decades, with profound implications for scientific research, industrial applications, and public decision-making. Contrary to common perception, AI does not constitute a sudden technological rupture, but rather the culmination of decades of advances in computing power, algorithmic research, and data availability.

The recent acceleration of AI adoption is largely attributable to two converging factors: (i) the exponential growth in computational resources, including parallel processors and graphics processing units (GPUs), and (ii) the unprecedented availability of large-scale datasets, particularly from the digital and geospatial domains.

At the core of modern AI lies machine learning, a family of methods that enable computational systems to learn patterns and relationships directly from data, without explicit rule-based programming. Machine learning algorithms infer models by optimizing objective functions that minimize prediction error or maximize explanatory power, allowing them to adapt to complex, nonlinear, and noisy environments.

In environmental and wildfire applications, AI offers a powerful alternative to traditional deterministic and physically based models. While physics-based approaches rely on simplified assumptions and parameterizations, AI-based models can integrate multiple interacting drivers, capture nonlinear dependencies, and adapt dynamically as new data become available. This flexibility is particularly valuable in wildfire contexts, where ignition and propagation processes emerge from the interaction of climate, vegetation, topography, and human activity.

2.2.1 Machine Learning Paradigms

Machine learning encompasses a broad family of algorithms designed to learn patterns from data and generalize to unseen situations. These methods are typically categorized into three principal paradigms depending on the availability of labeled information. Supervised learning relies on labeled datasets where explicit input–output relationships are known and used to train predictive models. This paradigm is widely applied in wildfire research for tasks such as ignition prediction, wildfire risk classification, and burned area estimation, where historical fire occurrences are used as target variables (Jain et al., 2020). In contrast, unsupervised learning focuses on discovering hidden structures or patterns within unlabeled datasets. In wildfire science, these techniques are frequently used for fire regime analysis, environmental pattern discovery, and landscape classification, enabling the identification of spatial and temporal structures in large environmental datasets (Hastie, Tibshirani, & Friedman, 2009). A third paradigm, semi-supervised learning, combines both labeled and unlabeled data in order to improve learning performance when labeled observations are scarce or incomplete—an issue frequently encountered in environmental and disaster-related datasets (Bishop, 2006).

Classical machine learning techniques, including decision trees, random forests, and gradient boosting algorithms, have demonstrated strong performance in wildfire risk modeling due to their ability to handle heterogeneous predictor variables, nonlinear relationships, and missing data, which are typical characteristics of geospatial and meteorological datasets (Breiman, 2001; Friedman, 2001). Among these approaches, tree-based ensemble models have become particularly popular in environmental applications because they provide both high predictive accuracy and robustness to complex data structures. Furthermore, these models offer an important advantage in operational contexts: intrinsic interpretability, allowing the estimation of feature importance and the identification of key environmental drivers influencing wildfire occurrence. This interpretability facilitates model validation, transparency, and user trust, which are critical factors for the adoption of predictive analytics in wildfire management and decision-support systems (Jain et al., 2020).

2.2.2 Deep Learning and Advanced Neural Architectures

Deep learning represents a specialized subfield of machine learning characterized by multi-layer artificial neural networks capable of learning hierarchical feature representations from raw or minimally processed data. These architectures are particularly effective for unstructured data, including images, videos, and sequential time series.

Deep neural networks are inspired by simplified models of biological neurons, with layers of interconnected units that progressively transform input data into higher-level abstractions. Learning is achieved by iteratively adjusting network parameters to minimize a loss function, typically using gradient-based optimization techniques. Recent advances have enabled the training of networks comprising tens to hundreds of millions of parameters, with even larger-scale architectures expected in the coming decade.

In wildfire applications, deep learning has demonstrated significant potential for tasks such as fire detection from satellite and UAV imagery, smoke identification, fuel mapping, and spatio-temporal forecasting. However, the increasing complexity of deep learning models introduces challenges related to computational cost, data requirements, interpretability, and operational robustness.

Consequently, the selection of an appropriate AI architecture involves a critical trade-off between model expressiveness, training complexity, computational efficiency, and explainability. For operational wildfire risk prediction, this trade-off motivates hybrid approaches that combine the predictive strength of advanced machine learning models with scalable computing infrastructures and explainable AI techniques, ensuring both performance and usability.

2.3 Big Data Challenges in Wildfire Applications: From Data Explosion to Actionable Intelligence

Despite the unprecedented availability of wildfire-related data, transforming Big Data into actionable and reliable wildfire intelligence remains a major scientific and operational challenge. These challenges are commonly structured around the five dimensions of Big Data, namely Volume, Velocity, Variety, Veracity, and Value, each of which poses specific constraints in the context of wildfire risk management.

2.3.1 Volume

Wildfire-related data are generated at increasingly large spatial and temporal scales. High-resolution satellite imagery, continuous meteorological forecasts, historical fire archives, and dense spatial grids used for operational prediction can rapidly reach terabyte-scale volumes. For national or regional wildfire monitoring systems, such as those deployed in Mediterranean countries, daily processing may involve millions of spatial units and hundreds of predictive variables.

This data volume exceeds the capacity of traditional centralized data storage and sequential processing workflows. As a result, distributed storage systems and parallel computing architectures are required to ensure scalability and acceptable processing times. In operational contexts, the inability to process large volumes efficiently may directly compromise the timeliness and usefulness of early-warning systems.

2.3.2 Velocity

Velocity refers to the speed at which data are generated, transmitted, and must be processed. In wildfire management, data velocity is particularly critical due to the dynamic nature of fire risk. Meteorological forecasts are updated multiple times per day, satellite-based fire detections are produced in near real time, and fuel conditions can change rapidly during heatwave episodes.

Operational wildfire risk prediction systems must therefore support continuous data ingestion and rapid model inference, often under strict time constraints. Delayed processing may result in outdated risk assessments, reducing the effectiveness of preventive measures. This requirement motivates the adoption of streaming data pipelines, automated workflows, and near real-time analytics, integrated within high-performance computing environments.

2.3.3 Variety

Wildfire intelligence systems must integrate data from highly heterogeneous sources, including numerical weather prediction outputs, remote sensing imagery, vector-based geographic data, socio-economic indicators, and historical event records. These datasets differ not only in format and spatial resolution, but also in temporal granularity, uncertainty levels, and semantic meaning.

Managing such heterogeneity poses significant challenges for data harmonization, spatial alignment, and feature engineering. In particular, combining raster and vector data, synchronizing time series with different temporal resolutions, and ensuring spatial consistency across datasets require advanced geospatial data abstraction layers.

Without such abstractions, data integration becomes error-prone, computationally inefficient, and difficult to maintain in operational settings.

2.3.4 Veracity

Data veracity refers to data quality, reliability, and uncertainty, which are critical concerns in wildfire applications. Fire occurrence records may contain positional inaccuracies, missing attributes, or reporting biases. Satellite-based products may be affected by cloud cover, sensor limitations, or classification errors, while socio-economic datasets often suffer from temporal lag or spatial aggregation issues.

In wildfire risk modeling, poor data quality can propagate through machine learning pipelines, leading to biased predictions and reduced trust among end users. Addressing veracity requires robust data preprocessing, quality control, uncertainty handling, and validation protocols, as well as transparent documentation of data limitations. Explainable AI approaches further contribute to veracity by allowing users to inspect model behavior and identify potential inconsistencies.

2.3.5 Value

Ultimately, the goal of Big Data analytics in wildfire management is not data accumulation, but value creation. Value emerges when data-driven insights translate into improved decision-making, reduced wildfire impacts, and more sustainable forest management practices. However, extracting value from Big Data is often the most challenging dimension, as it requires aligning technical outputs with operational needs, institutional workflows, and policy objectives.

In this context, predictive accuracy alone is insufficient. Wildfire intelligence systems must deliver interpretable, timely, and actionable information, tailored to the constraints of forest managers, emergency services, and policymakers. The integration of explainable artificial intelligence, operational dashboards, and spatial decision-support tools is therefore essential to bridge the gap between complex analytics and real-world application.

2.3.6 Toward Integrated and Explainable Wildfire Big Data Systems

Addressing the five Big Data challenges in a unified manner requires integrated system architectures that combine scalable computing, advanced analytics, and user-centered design. In wildfire applications, this integration enables the transition from static hazard indicators toward dynamic, explainable, and adaptive wildfire intelligence systems.

Such systems not only support short-term operational decision-making, but also facilitate long-term strategic planning, climate adaptation strategies, and policy development. By embedding Big Data and AI within operational wildfire management frameworks, it becomes possible to move from reactive fire suppression toward anticipatory risk management and prevention-oriented forest governance.

This conceptual framework directly motivates the design choices adopted in the AI4Fire system, which integrates scalable Big Data infrastructures, spatio-temporal machine learning, and explainable AI to address the full spectrum of wildfire intelligence challenges. The following sections detail how these principles are operationalized within the proposed methodology.

3. DATA AND METHODS

3.1 Study Area and Data Sources

As a first operational deployment, the proposed framework focuses on Northern Morocco (Figure 1), a region historically characterized by high wildfire frequency and severity, accounting for the majority of large fire events nationwide (Assali et al., 2016; Mharzi Alaoui et al., 2017; Mharzi Alaoui et al., 2019). This area exhibits a complex interplay between Mediterranean climatic influences, mountainous terrain, heterogeneous vegetation cover, and increasing anthropogenic pressure, making it particularly prone to wildfire ignition and propagation.

To comprehensively capture the multifactorial nature of wildfire occurrence, the AI4Fire platform integrates a wide range of heterogeneous geospatial and temporal datasets, which are structured into six main categories:

1. Historical wildfire events: spatial point and polygon datasets describing ignition locations, burned areas, occurrence dates, and fire duration, compiled from national forest service records.
2. Topo-climatic variables: elevation, slope, aspect, and high-resolution meteorological variables, including temperature, humidity, wind speed, and precipitation.
3. Forest fuel characteristics: land cover type, vegetation density, forest structure, and fuel load proxies derived from remote sensing products.
4. Vegetation status indicators: dynamic indices capturing vegetation vigor, dryness, and seasonal stress, derived from satellite-based spectral metrics.
5. Neighborhood and spatial configuration metrics: spatial descriptors characterizing landscape fragmentation, fuel continuity, and proximity-based interactions.

6. Socio-economic and anthropogenic factors: proximity to infrastructure, settlements, road networks, agricultural activities, and population density, extracted through advanced spatial queries.

Figure 1. Localization of the study area

3.2 Overview of the Methodological Framework

Given the inherently spatio-temporal nature of wildfire processes, the proposed framework is designed to explicitly integrate both spatial interactions and temporal dynamics. The methodology relies on a multi-stage pipeline involving spatio-temporal feature engineering, large-scale parallel computing, machine learning-based prediction, and explainable AI-driven interpretation.

More than 100 spatio-temporal features are automatically generated and streamed into a gradient boosting model based on XGBoost (Chen & Guestrin, 2016). Daily wildfire risk scores are computed for over one million regular grid points covering Northern Morocco (Figure 1). These predictions are subsequently aggregated and visualized as dynamic wildfire risk maps, disseminated through an operational web-based geospatial platform.

The following sections detail the key methodological components: (i) spatio-temporal wildfire pressure modeling, (ii) temporal feature engineering for heatwave characterization, (iii) scalable spatial data abstraction and high-performance computing, and (iv) explainable artificial intelligence for operational interpretability.

3.3 Spatio-Temporal Feature Engineering for Wildfire Pressure Modeling

Encoding historical wildfire occurrences as raw geographic coordinates is insufficient for machine learning models, as such representations fail to capture spatial dependency structures and neighborhood interactions (Kuter et al., 2011). Similarly, traditional kernel density estimation (KDE) approaches, while effective for spatial smoothing, are limited in capturing temporal dynamics critical for short-term and seasonal forecasting.

To overcome these limitations, we introduce a novel Spatio-Temporal Wildfire Pressure (ST-WFP) index, explicitly designed to represent both the spatial connectivity and temporal memory of wildfire events (Figure 2). For each grid point p_t, wildfire pressure is computed as a function of surrounding historical events within multiple spatial buffers and temporal lags:

$$ST\_WildfirePressure(p_t) = \{st \mid st = Buffer(p_t, b, l), b_{min} < b < b_{max}, l_{min} < l < l_{max}\}$$

where $b$denotes the spatial buffer radius and $l$represents the temporal lag.

Spatial buffer values of 500 m, 1 km, 2 km, 3 km, and 5 km, combined with temporal lags of 3, 6, 9, 12, and 24 months, were selected based on an exhaustive grid search optimizing both predictive performance and computational efficiency. This formulation enables each grid cell to inherit a bi-dimensional memory kernel, accounting for both local fire clustering and longer-term regional fire recurrence patterns.

This spatio-temporal encoding allows the model to dynamically learn fire contagion mechanisms, spatial persistence, and temporal decay effects, significantly enhancing predictive capability.

Additional spatial predictors, particularly socio-economic variables, are derived through advanced spatial operations, including spatial joints, nearest-neighbor queries, and distance-based aggregation, implemented in a distributed computing environment (Zhou et al., 1998).

Figure 2. Spatio-temporal weather data cube construction

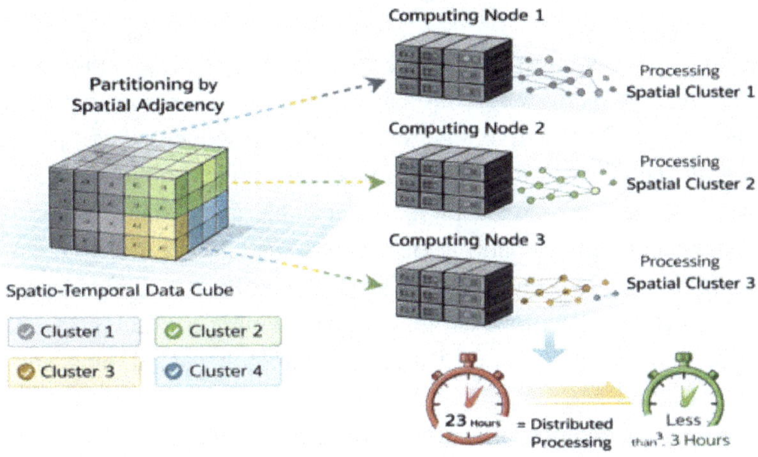

3.4 Temporal Feature Engineering for Heatwave Characterization

Heatwaves play a critical role in wildfire ignition by inducing fuel desiccation, atmospheric instability, and sustained high evaporative demand, thereby significantly increasing fire susceptibility (Hulley et al., 2020; Guion et al., 2022).

To explicitly capture these dynamics, we construct a set of aggregated temporal indicators describing short-term heat stress patterns. Specifically, within a rolling weekly window, we compute:

- The number of days exceeding temperature thresholds of 25°C, 30°C, 35°C, and 40°C, and
- The number of days with relative humidity below 20%, 25%, 30%, and 35%.

The temporal aggregation window and threshold values were optimized via grid search to maximize predictive accuracy. This formulation enables the model to capture cumulative thermal stress effects and persistent drought conditions, which are key drivers of extreme wildfire behavior.

3.5 Scalable Spatial Data Abstraction and High-Performance Computing

The operational constraints of AI4Fire require daily ingestion, processing, and inference over massive spatio-temporal datasets, encompassing more than one million grid cells and over 100 predictive features. Traditional sequential computing strategies proved infeasible, with processing times exceeding 23 hours per daily cycle.

To address this challenge, we implement a scalable spatial data abstraction layer based on Dask-GeoPandas, enabling distributed spatial operations across multiple computing nodes (Rocklin et al., 2015; Dask-GeoPandas, 2024; Dugré et al., 2023). The workflow is decomposed into embarrassingly parallel tasks, including spatio-temporal buffering, heatwave feature computation, spatial joins, and feature aggregation.

Furthermore, an efficient spatial data partitioning strategy is employed to cluster spatially adjacent grid points, thereby minimizing inter-node communication and optimizing spatial query performance (Zhou et al., 1998). This distributed architecture reduces total daily computation time from 23 hours to less than 3 hours, enabling near real-time wildfire risk monitoring (Figure 3).

Figure 3. Spatial partitioning and distrubution processing for predecting forest fire

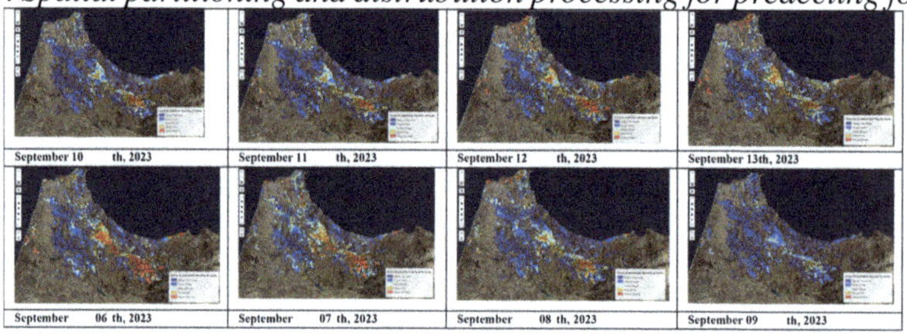

3.6 Explainable Artificial Intelligence for Operational Decision Support

Beyond predictive accuracy, operational wildfire management requires transparent and interpretable model outputs to foster trust and support informed decision-making. To this end, AI4Fire integrates a local explainability framework based on decision-path decomposition for tree-based models (ELI5, 2024).

For each grid point prediction $\hat{f}(x)$, feature contributions are computed as:

$$\hat{f}(x) = \bar{y} + \sum_{d=1}^{D} Split.contrib(d,x) + \sum_{j=1}^{P} feat.contrib(j,x)$$

where $\bar{y}$denotes the global model bias, $Split.contrib(d,x)$represents node-level contributions, and $feat.contrib(j,x)$quantifies the marginal influence of feature $j$for instance x.

This formulation enables the generation of daily explainability maps, highlighting the dominant drivers of wildfire risk at each spatial location. In addition, global model interpretability analyses are conducted to assess overall feature importance, with particular focus on the newly introduced spatio-temporal wildfire pressure and heatwave indicators.

3.7 Model Training Strategy and Validation Protocol

The operational deployment of machine learning models for wildfire risk prediction requires not only high predictive performance, but also *robust training strategies and rigorous validation protocols* to ensure generalization across time, space, and varying climatic conditions. In the AI4Fire framework, particular attention was paid to the design of the training workflow in order to address the *spatio-temporal nature of wildfire data*, the strong class imbalance between fire and non-fire observations, and the operational constraints associated with daily inference.

3.7.1 Training Dataset Construction and Class Imbalance Handling

The training dataset was constructed by associating each regular grid point with its corresponding feature vector and a binary target variable indicating wildfire occurrence within the grid cell on a given day. Given the rarity of wildfire ignition events relative to the total number of grid-day observations, the resulting dataset exhibits a *strong class imbalance*, a common characteristic of wildfire prediction problems.

To mitigate this issue, several complementary strategies were implemented. First, temporal stratification of samples was applied to preserve seasonal variability and ensure that the model adequately captures temporal patterns associated with wildfire occurrence. Second, a controlled undersampling of non-fire observations was performed to reduce the dominance of the negative class and improve class balance within the training dataset. Finally, a cost-sensitive learning approach was introduced through class weighting, assigning higher penalties to the misclassification of wildfire events during model optimization. This combination of techniques helps address class imbalance while maintaining the representativeness of the underlying spatio-temporal patterns.

These strategies aim to enhance model sensitivity to extreme-risk conditions while maintaining acceptable false-positive rates, which is critical for operational usability.

3.7.2 Temporal Cross-Validation and Generalization Assessment

Unlike conventional random cross-validation, which may lead to information leakage in time-dependent datasets, AI4Fire employs a temporal cross-validation strategy. Training and validation splits are performed along the temporal dimension, ensuring that model evaluation reflects true forecasting conditions.

Specifically, historical data were divided into sequential temporal blocks, where models were trained on earlier periods and validated on subsequent fire seasons. This approach allows for the assessment of model robustness under interannual climate variability, including years characterized by extreme heatwaves and drought conditions.

Such temporal validation is essential to verify that the learned spatio-temporal patterns are not overfitted to specific years, but instead capture generalizable wildfire risk drivers.

3.7.3 Hyperparameter Optimization and Model Stability

Model training relies on a gradient boosting framework based on XGBoost, selected for its strong predictive performance, scalability, and compatibility with explainable AI techniques. Hyperparameters, including tree depth, learning rate, number of estimators, and regularization coefficients, were optimized using a grid search strategy constrained by computational feasibility.

To ensure model stability, hyperparameter tuning was performed using aggregated performance metrics across multiple temporal folds rather than a single validation period. This procedure reduces sensitivity to anomalous fire seasons and improves robustness under changing environmental conditions.

3.7.4 Evaluation Metrics and Operational Thresholds

Model performance was evaluated using a combination of threshold-independent and threshold-dependent metrics to provide a comprehensive assessment of predictive capability. These included the Area Under the Receiver Operating Characteristic Curve (AUC) to measure the model's overall discriminative ability, as well as precision–recall analysis, which is particularly informative under class imbalance conditions. Additionally, detection rates at high-risk probability thresholds (e.g., >0.80 and >0.90) were analyzed to assess the model's effectiveness in identifying critical wildfire risk situations. Finally, false alarm ratios were examined to evaluate the operational reliability of the system and its suitability for decision-making in wildfire management contexts.

Rather than optimizing for a single global threshold, AI4Fire adopts adaptive risk thresholds, allowing wildfire managers to adjust alert sensitivity based on operational capacity, meteorological outlook, and available resources.

3.7.5 Continuous Learning and Model Updating

To maintain relevance under evolving climatic and land-use conditions, the AI4Fire framework supports periodic model retraining and performance monitoring. New wildfire observations and updated environmental data are incrementally integrated into the training dataset, enabling continuous learning and gradual adaptation to emerging fire regimes.

This strategy ensures that the predictive model remains responsive to long-term trends, such as climate-driven aridification and changes in ignition patterns, while preserving operational continuity.

3.8 Operational Deployment Architecture and End-to-End System Integration

Beyond methodological soundness, the effectiveness of a wildfire risk prediction system critically depends on its operational deployment architecture, including data ingestion, computational orchestration, model inference, and information dissemination. The AI4Fire framework was explicitly designed as an end-to-end operational system, capable of delivering daily wildfire risk and explainability maps in near real time, without manual intervention.

3.8.1 End-to-End Data Pipeline

The AI4Fire operational pipeline follows a fully automated workflow composed of four main stages: data acquisition, feature generation, model inference, and dissemination. Meteorological forecasts, satellite-derived indicators, and ancillary geospatial datasets are ingested daily from external providers through scheduled data synchronization processes. Historical wildfire databases and static geographic layers are stored locally within the system to minimize data access latency.

All incoming datasets are harmonized spatially and temporally using a common reference grid and coordinate system. This standardization ensures consistency across heterogeneous data sources and enables seamless feature generation. Automated validation checks are applied at each ingestion step to detect missing values, corrupted files, or anomalous observations, thereby enhancing system robustness and data reliability.

3.8.2 High-Performance Computing and Task Orchestration

Given the scale of daily processing,over one million grid cells and more than 100 predictive features, the AI4Fire system is deployed on a high-performance computing (HPC) infrastructure. Computational workloads are decomposed into modular tasks and executed in parallel using a distributed task scheduling framework.

Spatio-temporal feature computation, including wildfire pressure indices and heatwave indicators, is implemented as embarrassingly parallel operations, enabling efficient scaling across multiple compute nodes. Spatially aware data partitioning strategies are applied to minimize inter-process communication during spatial joins and buffering operations. This architecture allows AI4Fire to reduce daily processing times from more than 20 hours under sequential execution to less than 3 hours, thereby satisfying operational timeliness requirements.

3.8.3 Model Inference and Explainability Generation

Once feature vectors are generated, the trained machine learning model performs batch inference over the entire spatial grid. Prediction scores are computed independently for each grid point, ensuring scalability and fault tolerance. In parallel, the explainable AI module computes local feature contribution scores for each prediction, enabling the generation of daily interpretability maps.

The joint production of risk and explainability outputs is a central design principle of AI4Fire. Rather than treating interpretability as a post hoc analytical step, explainability is embedded directly into the operational pipeline, ensuring that transparency is preserved at scale and under real-time constraints.

3.8.4 Dissemination and Decision-Support Interfaces

Prediction and explainability outputs are automatically transformed into geospatial layers compatible with standard web-based geographic information systems. These layers are published through an interactive web mapping portal, allowing users to visualize daily wildfire risk levels, inspect dominant risk drivers, and explore spatio-temporal trends.

The platform supports multiple user profiles, including forest engineers, emergency coordinators, and decision-makers, each with tailored visualization and interaction capabilities. Risk thresholds, alert levels, and temporal aggregation options can be dynamically adjusted to reflect operational priorities and available response capacity.

3.8.5 Automation, Reliability, and System Resilience

A key operational requirement of AI4Fire is the elimination of human intervention in routine processing. All components of the system are scheduled and monitored through automated orchestration services, ensuring continuous operation during critical fire seasons. Logging and monitoring mechanisms track system performance, data availability, and computational health, enabling rapid detection and resolution of failures.

To enhance resilience, the architecture incorporates redundancy at multiple levels, including data storage, task execution, and output generation. In the event of partial data unavailability or delayed inputs, the system is capable of producing degraded but usable risk assessments, preserving operational continuity.

3.8.6 Toward Scalable and Transferable Operational Systems

The modular design of the AI4Fire deployment architecture ensures scalability and transferability beyond the initial study area. The system can be extended to other regions by adapting data sources, recalibrating model parameters, and adjusting spatial resolution, without fundamental architectural changes.

This operational architecture demonstrates how Big Data infrastructures, spatio-temporal machine learning, and explainable AI can be effectively integrated into a coherent, real-world wildfire intelligence system. The following section evaluates the predictive performance and operational impact of AI4Fire through multi-year validation and case-study analysis.

4. RESULTS AND PERFORMANCE ASSESSMENT

4.1 Spatio-Temporal Patterns of Wildfire Risk

Figure 4 illustrates representative daily wildfire risk maps generated during mid-September 2023, a period characterized by persistent heatwave conditions and elevated atmospheric instability. The predicted risk patterns exhibit pronounced spatial heterogeneity, with high-risk clusters consistently concentrated in mountainous and forest-dominated regions of Northern Morocco.

The spatial distribution of risk is strongly associated with areas combining dense vegetation cover, complex topography, low fuel moisture content, and recurrent historical wildfire activity. The temporal continuity of high-risk zones across consecutive days further indicates that the proposed framework effectively captures both short-term meteorological forcing and medium-term landscape susceptibility.

Moreover, the progressive evolution of daily risk patterns closely follows synoptic-scale atmospheric dynamics, confirming the physical consistency and temporal coherence of the model outputs. This spatio-temporal stability is particularly valuable for operational deployment, as it enables anticipation of critical fire danger periods and supports proactive resource allocation strategies.

Figure 4. Daily wildfire risk maps generated during mid-September 2023

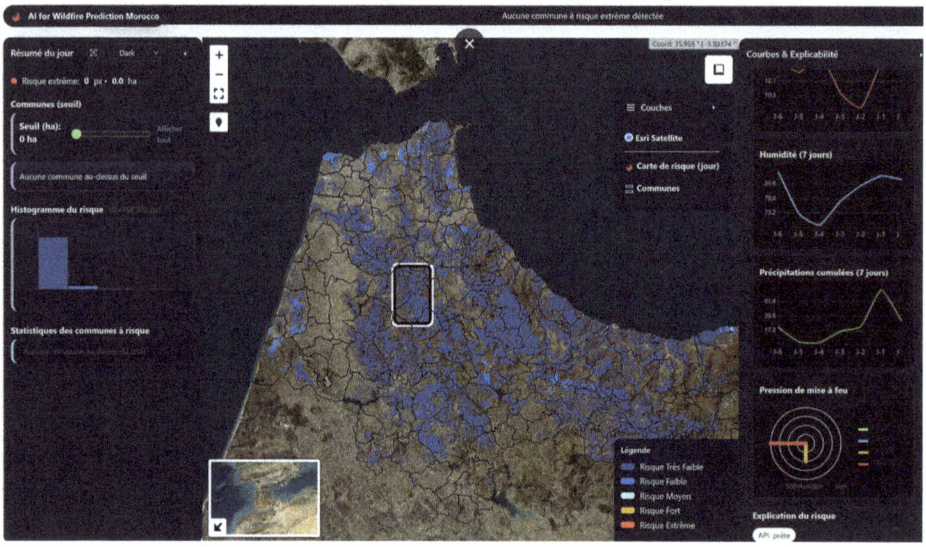

4.2 Quantitative Evaluation of Predictive Performance (2021–2023)

To quantitatively evaluate the predictive capability of AI4Fire, daily prediction outputs were systematically compared with observed wildfire ignition events recorded between January 2021 and September 2023. Each recorded fire event was spatially matched to the corresponding grid-cell prediction score on the day of occurrence.

The analysis reveals a high level of predictive discrimination. Approximately 80.3% of wildfire events occurred in grid cells assigned predicted risk scores exceeding 0.90, while 89.7% were located in areas with scores above 0.80. Conversely, fewer than 5% of wildfire events were associated with low predicted risk levels (< 0.60).

The Receiver Operating Characteristic (ROC) analysis yields an Area Under the Curve (AUC) exceeding 0.92, indicating excellent model performance. These results confirm the system's strong capacity to discriminate between high-risk and low-risk conditions, which is particularly critical in operational wildfire early-warning systems.

Compared to classical physically based fire danger indices, such as the Fire Weather Index (FWI), which typically achieve AUC values ranging between 0.65 and 0.80, the proposed spatio-temporal machine learning framework demonstrates substantial performance gains. These improvements can primarily be attributed to the integration of historical wildfire memory, spatial neighborhood interactions, and advanced temporal aggregation of heatwave dynamics.

4.3 Interannual Robustness and Temporal Generalization

To assess temporal generalization, model performance was analyzed separately for the 2021, 2022, and 2023 fire seasons. The results demonstrate remarkable temporal stability, despite substantial interannual variability in climatic conditions (Table 1).

Table 1. Comparative analysis of the predicted fire

Year	% Fires predicted > 0.90	% Fires predicted > 0.80
2021	78.4%	88.1%
2022	81.6%	90.5%
2023*	82.7%	91.3%

(*January – September 2023)

Notably, 2022 and 2023 were characterized by exceptional heatwave intensity and prolonged drought episodes, yet the predictive performance remained consistently high. This stability highlights the robustness and adaptability of the learned spatio-temporal representations, even under extreme climatic forcing.

4.4 Contribution of Spatio-Temporal and Heatwave Features

Feature contribution analysis reveals that the spatio-temporal wildfire pressure indices introduced in this work play a dominant role in driving model predictions. These features account for approximately 34–41% of the total explanatory power, while heatwave-derived temporal indicators contribute 21–27%. In contrast, classical meteorological variables, topographic descriptors, and land cover features collectively account for the remaining 32–45%.

An ablation study further confirms the critical role of these novel predictors. Removing the wildfire pressure features leads to:

- A reduction in AUC of 11–14%, and
- A decrease exceeding 18% in the detection rate of high-risk events (score > 0.90).

These results demonstrate that explicitly encoding spatial contagion processes and temporal persistence of past fire activity substantially enhances predictive performance, outperforming traditional hazard modeling paradigms.

4.5 Explainable Risk Mapping and Operational Interpretability

Figure 5 illustrates an example of the local explainability output for a grid point predicted with a wildfire risk score of 95%. The dominant positive contributors include short- and medium-range wildfire pressure indices (CB500 and CB3000), low fuel moisture levels, and persistent high-temperature anomalies during preceding days.

This combination reflects the synergistic effect of spatio-temporal fire memory and short-term meteorological stress, thereby reinforcing the ecological plausibility and physical consistency of the learned model behavior.

From an operational perspective, the availability of daily explainability maps substantially enhances model transparency and fosters greater user trust in the system. Feedback from forest management services indicates that these interpretability outputs contribute to improved situational awareness by highlighting the key drivers of predicted wildfire risk. They also support more effective prioritization of surveillance and patrol operations, facilitate the optimized deployment of firefighting resources, and strengthen the coordination of preventive logistics. Overall, these explainability tools bridge the gap between advanced predictive modeling and practical decision-making in wildfire management.

Thus, explainability transforms AI4Fire from a purely predictive model into a transparent, actionable, and decision-oriented wildfire intelligence system.

Figure 5. Most positives and negatives features contributing to a selected wildfire risk score

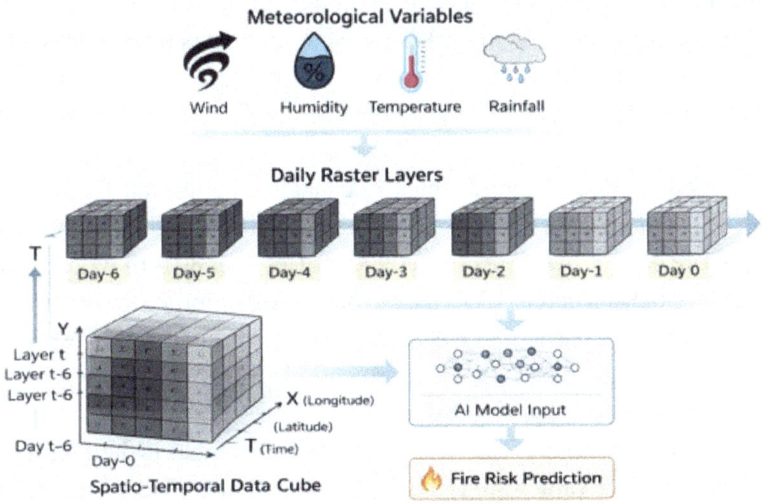

4.6 Comparative Evaluation Against Classical Fire Danger Indices

Classical fire danger indices, such as the Fire Weather Index (FWI) and its associated components, have long been used as reference tools for wildfire risk assessment and operational decision-making. These indices are grounded in physically based formulations that relate meteorological conditions—particularly temperature, relative humidity, wind speed, and precipitation—to fuel moisture dynamics and potential fire behavior. While widely adopted and operationally robust, such indices were not originally designed to capture the complex spatio-temporal interactions between climate, vegetation, historical fire activity, and human influence that characterize modern wildfire regimes.

To assess the added value of the AI4Fire framework, a comparative evaluation was conducted between daily wildfire risk predictions generated by AI4Fire and classical fire danger indicators derived from the Fire Weather Index system. For each wildfire event recorded during the 2021–2023 fire seasons, the corresponding daily FWI values were extracted at the spatial location of ignition. These values were then compared against AI4Fire risk scores computed for the same grid cells and dates.

Performance comparison focused on the ability of each approach to discriminate high-risk conditions, using metrics relevant to operational contexts, including detection rate of wildfire events at high-risk thresholds, false alarm ratios, and overall discriminatory power as measured by the Area Under the Receiver Operating Characteristic Curve (AUC).

Results indicate a clear performance advantage of the AI4Fire framework over classical fire danger indices. While FWI-based assessments demonstrate moderate discriminatory power, with AUC values typically ranging between 0.65 and 0.80, AI4Fire consistently achieves AUC values exceeding 0.92 across the evaluated fire seasons.

At high-risk thresholds, which are most relevant for operational decision-making, the difference becomes even more pronounced. Using thresholds calibrated to represent extreme fire danger, AI4Fire correctly identifies over 80% of wildfire events at risk levels above 0.90, whereas FWI-based thresholds capture a substantially lower proportion of events. In particular, FWI tends to underestimate wildfire risk in areas where non-meteorological factors, such as accumulated historical fire pressure, fuel continuity, or human activity, play a dominant role.

One of the key limitations of classical fire danger indices lies in their limited spatial contextualization. FWI values are typically computed at coarse spatial resolutions and do not explicitly account for spatial neighborhood effects or the memory of past fire events. As a result, FWI-based risk maps often exhibit broad, homogeneous risk patterns that lack the fine-scale differentiation required for targeted surveillance and prevention planning.

Furthermore, classical indices are inherently reactive to current or short-term meteorological conditions, with limited capacity to integrate medium-term drought effects or longer-term landscape susceptibility. In contrast, the spatio-temporal machine learning approach adopted by AI4Fire explicitly encodes both short-term weather dynamics and longer-term fire history, enabling a more nuanced representation of wildfire risk.

From an operational perspective, the comparative analysis highlights important differences in how risk information can be used by wildfire management agencies. While FWI remains a valuable indicator for general fire danger awareness, its limited spatial resolution and lack of interpretability at the local scale constrain its usefulness for fine-grained operational decision-making.

AI4Fire, by contrast, provides spatially explicit, high-resolution, and explainable risk assessments, enabling forest managers to prioritize specific zones, optimize patrol routes, and allocate resources more effectively. The integration of explainable AI further enhances this advantage by allowing users to understand why certain areas are classified as high risk, thereby supporting trust and informed action.

It is important to note that the objective of AI4Fire is not to replace classical fire danger indices, but rather to complement and enhance existing operational tools. Physically based indices such as FWI offer robustness, interpretability, and global applicability, making them valuable baseline indicators. When combined with data-driven, spatio-temporal machine learning approaches, these indices can contribute to hybrid wildfire intelligence systems that leverage the strengths of both paradigms.

Overall, the comparative evaluation demonstrates that AI4Fire provides substantial added value over classical fire danger indices, particularly in complex Mediterranean landscapes where wildfire risk emerges from the interaction of climatic, ecological, and human drivers. This added value translates directly into improved operational effectiveness and supports the transition toward more proactive and data-driven wildfire risk management strategies.

5. CONCLUSIONS

This study introduced AI4Fire, an operational spatio-temporal and explainable machine learning framework for daily wildfire risk prediction in Northern Morocco. By integrating large-scale geospatial data processing, advanced feature engineering, high-performance computing, and explainable artificial intelligence, the proposed system establishes a comprehensive wildfire early-warning and decision-support platform.

Quantitative evaluation over three consecutive fire seasons (2021–2023) demonstrates robust and consistent predictive performance, with approximately 80% of wildfire events correctly identified in areas assigned very high risk levels (> 0.90). These results confirm the reliability, scalability, and operational relevance of the proposed framework.

A major methodological contribution lies in the formulation of spatio-temporal wildfire pressure indices, which explicitly encode the memory and connectivity of past fire events, significantly enhancing predictive performance. Furthermore, the deployment of a distributed spatial big data architecture enables near real-time inference over more than one million grid points, meeting the stringent computational demands of operational wildfire management systems.

The integration of local explainability mechanisms ensures transparency and interpretability, thereby fostering trust among practitioners and facilitating informed tactical and strategic decision-making.

Future developments will focus on the integration of deep spatio-temporal learning architectures, particularly Spatio-Temporal Graph Neural Networks (ST-GNNs), which have demonstrated superior performance across multiple dynamic geospatial forecasting tasks. Such models may enable end-to-end learning of complex spatial interactions and temporal dependencies, further enhancing predictive accuracy and generalization.

Additional research directions include the assimilation of high-frequency satellite time series to better capture rapid environmental dynamics, as well as the integration of real-time observations from IoT sensors and UAV platforms to enhance data timeliness and spatial resolution. Future work should also focus on the development of probabilistic forecasting approaches and uncertainty quantification frameworks, enabling more robust risk assessment and decision support. Furthermore, advancing the coupling between ignition prediction models and fire spread and impact simulations would provide a more comprehensive framework for anticipating wildfire behavior and supporting proactive fire management strategies.

Through these advancements, AI4Fire aims to evolve into a fully integrated wildfire intelligence platform, supporting sustainable forest management, climate adaptation strategies, and disaster risk reduction policies in Mediterranean ecosystems.

REFERENCES

Alipour, M., La Puma, I., Picotte, J., Shamsaei, K., Rowell, E., Watts, A., Kosovic, B., Ebrahimian, H., & Taciroglu, E. (2023). A multimodal data fusion and deep learning framework for large-scale wildfire surface fuel mapping. *Fire (Basel, Switzerland), 6*(2), 36. DOI: 10.3390/fire6020036

Assali, F., Mharzi Alaoui, H., Rouchdi, M., & Badraoui, M. (2016). Modelling and mapping forest fire ignition risk in North-West Morocco (Chefchaouen–Ouazzane Region). [La Terre et la Vie]. *Revue d'Écologie, 71*, 111–128. DOI: 10.3406/revec.2016.1834

Bhamra, J. K., Anantha Ramaprasad, S., Baldota, S., Luna, S., Zen, E., Ramachandra, R., Kim, H., Schmidt, C., Arends, C., Block, J., Perez, I., Crawl, D., Altintas, I., Cottrell, G. W., & Nguyen, M. H. (2023). Multimodal wildland fire smoke detection. *Remote Sensing (Basel), 15*(11), 2790. DOI: 10.3390/rs15112790

Bishop, C. M. (2006). *Pattern recognition and machine learning*. Springer.

Breiman, L. (2001). Random forests. *Machine Learning, 45*(1), 5–32. DOI: 10.1023/A:1010933404324

Cardil, A., Monedero, S., Selegue, P., Navarrete, M. Á., de-Miguel, S., Purdy, S., Marshall, G., Chavez, T., Allison, K., Quilez, R., Ortega, M., Silva, C. A., & Ramirez, J. (2023). Performance of operational fire spread models in California. *International Journal of Wildland Fire, 32*(11), 1492–1502. DOI: 10.1071/WF22128

Chen, T., & Guestrin, C. (2016). XGBoost: A scalable tree boosting system. In *Proceedings of the 22nd ACM SIGKDD International Conference on Knowledge Discovery and Data Mining* (pp. 785–794). DOI: 10.1145/2939672.2939785

Chen, T., Kornblith, S., Norouzi, M., & Hinton, G. (2020). A simple framework for contrastive learning of visual representations. In *Proceedings of the 37th International Conference on Machine Learning* (pp. 1597–1607). doi:10.5555/3524938.3525087

Dask-GeoPandas. (2024). *Dask-GeoPandas documentation*. https://dask-geopandas.readthedocs.io

Dugré, M., Hayot-Sasson, V., & Glatard, T. (2023). Performance comparison of Dask and Apache Spark on HPC systems for neuroimaging. *Concurrency and Computation, e7635*(21), e7635. DOI: 10.1002/cpe.7635

ELI5. (2024). *ELI5 documentation*. https://eli5.readthedocs.io

Elia, M., D'Este, M., Ascoli, D., Giannico, V., Spano, G., Ganga, A., Colangelo, G., Lafortezza, R., & Sanesi, G. (2020). Estimating the probability of wildfire occurrence in Mediterranean landscapes using artificial neural networks. *Environmental Impact Assessment Review*, *85*, 106474. DOI: 10.1016/j.eiar.2020.106474

Fernandes, S., Niknejad, M., Barata, C., & Bernardino, A. (2023). Forest fires identification using self-supervised learning. In Rousseau, J.-J., & Kapralos, B. (Eds.), *Pattern Recognition, Computer Vision, and Image Processing* (LNCS Vol. 13644). Springer. https://doi.org/DOI: 10.1007/978-3-031-37742-6_17

Friedman, J. H. (2001). Greedy function approximation: A gradient boosting machine. *Annals of Statistics*, *29*(5), 1189–1232. DOI: 10.1214/aos/1013203451

Gandomi, A., & Haider, M. (2015). Big data concepts, methods, and analytics. *International Journal of Information Management*, *35*(2), 137–144. DOI: 10.1016/j.ijinfomgt.2014.10.007

Guion, A., Turquety, S., Polcher, J., Pennel, R., Bastin, S., & Arsouze, T. (2022). Droughts and heatwaves in the Western Mediterranean: Impact on vegetation and wildfires using the coupled WRF-ORCHIDEE regional model. *Climate Dynamics*, *58*(9–10), 2881–2903. DOI: 10.1007/s00382-021-05938-y

Hastie, T., Tibshirani, R., & Friedman, J. (2009). *The elements of statistical learning: Data mining, inference, and prediction* (2nd ed.). Springer. DOI: 10.1007/978-0-387-84858-7

Hulley, G. C., Dousset, B., & Kahn, B. H. (2020). Rising trends in heatwave metrics across southern California. *Earth's Future, 8*(7), e2020EF001480. DOI: 10.1029/2020EF001480

Intergovernmental Panel on Climate Change (IPCC). (2021). *Climate Change 2021: The Physical Science Basis – Summary for Policymakers*. Cambridge University Press. https://www.ipcc.ch/report/ar6/wg1

Jain, P., Coogan, S. C. P., Subramanian, S. G., Crowley, M., Taylor, S., & Flannigan, M. D. (2020). A review of machine learning applications in wildfire science and management. *Environmental Reviews*, *28*(4), 478–505. DOI: 10.1139/er-2020-0019

Kuter, N., Yenilmez, F., & Kuter, S. (2011). Forest fire risk mapping by kernel density estimation. *Croatian Journal of Forest Engineering*, *32*(2), 599–610.

McCormick, J. (2020, October 1). California firefighters tap AI for an edge in battling wildfires. *The Wall Street Journal.*

Mharzi Alaoui, H., Assali, F., Hajji, H., Lahssini, S., Aadel, T., & Moukrim, S. (2019). Machine learning approach for predicting seasonal risk of forest fires in Morocco. *Biodiversidade Brasileira, 9*(1). doi:10.37002/biodiversidadebrasileira.v9i1.1116

Mharzi Alaoui, H., Assali, F., Rouchdi, M., Tahiri, D., Lahssini, S., Aafi, A., & Moukrim, S. (2017). Cartographie de l'intensité du feu et de la sensibilité aux feux de cimes en forêts méditerranéennes. Cas de la province de Chefchaouen au nord du Maroc. [La Terre et La Vie]. *Revue d'Écologie*, *72*(4), 387–409. DOI: 10.3406/revec.2017.1900

Niknejad, M., & Bernardino, A. (2021). Attention on classification for fire segmentation. In *Proceedings of the 20th IEEE International Conference on Machine Learning and Applications* (pp. 616–621). doi:10.48550/arXiv.2111.03129DOI: 10.1109/ICMLA52953.2021.00104

Qiao, Y., Jiang, W., Wang, F., Su, G., Li, X., & Jiang, J. (2023). FireFormer: An efficient transformer to identify forest fire from surveillance cameras. *International Journal of Wildland Fire*, *32*(9), 1364–1380. DOI: 10.1071/WF22220

Randerson, J. T., Chen, Y., van der Werf, G. R., Rogers, B. M., & Morton, D. C. (2012). Global burned area and biomass burning emissions from small fires. *Journal of Geophysical Research*, *117*(G4), G04012. Advance online publication. DOI: 10.1029/2012JG002128

Rocklin, M. (2015). Dask: Parallel computation with blocked algorithms and task scheduling. In *Proceedings of the 14th Python in Science Conference* (pp. 130–136). DOI: 10.25080/Majora-7b98e3ed-013

Ruffault, J., Curt, T., Moron, V., Trigo, R. M., Mouillot, F., Koutsias, N., Pimont, F., Martin-StPaul, N., Barbero, R., Dupuy, J.-L., Russo, A., & Belhadj-Khedher, C. (2020). Increased likelihood of heat-induced large wildfires in the Mediterranean Basin. *Scientific Reports*, *10*(1), 13790. DOI: 10.1038/s41598-020-70069-z PMID: 32796945

Van Der Werf, G. R., Randerson, J. T., Giglio, L., van Leeuwen, T. T., Chen, Y., Rogers, B. M., Mu, M., van Marle, M. J. E., Morton, D. C., Collatz, G. J., Yokelson, R. J., & Kasibhatla, P. S. (2017). Global fire emissions estimates during 1997–2016. *Earth System Science Data*, *9*(2), 697–720. DOI: 10.5194/essd-9-697-2017

Yu, M., Masrur, A., & Blaszczak-Boxe, C. (2023). Predicting hourly PM2.5 concentrations in wildfire-prone areas using a spatio-temporal transformer model. *The Science of the Total Environment*, *860*, 160446. DOI: 10.1016/j.scitotenv.2022.160446 PMID: 36436649

Zhou, X., Abel, D. J., & Truffet, D. (1998). Data partitioning for parallel spatial join processing. *GeoInformatica*, *2*(2), 175–204. DOI: 10.1023/A:1009755931056

Chapter 6
Wildfire Research and Governance in Morocco Within a Shifting Mediterranean Context:
Toward Integrated and Resilient Fire Management

Sanaa Noury
Faculty of Arts and Humanities of Beni Mellal, Morocco

Abderrahim Ettaqy
http://orcid.org/0000-0003-4662-5770
Polydisciplinary Faculty of Beni Mellal, Morocco

Hicham Bouzekraoui
http://orcid.org/0000-0002-3980-4378
Faculty of Arts and Humanities of Beni Mellal, Morocco

Moussa Couachi
http://orcid.org/0009-0006-6134-0000
Faculty of Arts and Humanities of Beni Mellal, Morocco

ABSTRACT

Wildfires are becoming more intense across North Africa, yet research on this topic in Morocco remains limited and uneven despite rising climatic pressure and increasing fire recurrence. This chapter delivers the first bibliometric analysis on Moroccan wildfire studies, examining publication trends, collaboration networks and themat-

DOI: 10.4018/979-8-2600-0783-9.ch006

ic areas. The most notable progress concerns satellite monitoring, burn-severity evaluation and spatial modelling, while governance, socio-economic vulnerability, institutional coordination, community participation and post-fire restoration are still rarely addressed. Compared with other Mediterranean countries that have developed more integrated and prevention-oriented wildfire systems, Morocco continues to face shortcomings in scientific cooperation, regional integration and links between research and policy. Improving data interoperability and coordinated research strategies will be crucial to support climate-resilient wildfire management in North Africa.

1. BACKGROUND

Forests are central to Earth's life-support systems. They regulate hydrological cycles, stabilize soils, moderate climate, sustain biodiversity, and store large amounts of carbon that contribute to climate mitigation and human well-being (Bonan, 2008; Feng et al., 2024; Pan et al., 2011). Across the Mediterranean basin and North Africa, however, these functions are increasingly threatened by accelerating disturbances. Among them, wildfires have emerged as one of the most recurrent and damaging pressures on ecosystem integrity, human safety, and sustainable land management (Ford et al., 2021; Gallo Granizo et al., 2025; Graus et al., 2024; Mauri et al., 2023).

Recent wildfire activity signals more than a rise in fire occurrence. It reflects a structural shift in the ecological and socio-environmental conditions that shape fire regimes in Mediterranean and North African landscapes. Climate variability, fuel accumulation linked to land-use change, and multiple forms of anthropogenic pressure are interacting in ways that depart from historical dynamics (Gallo Granizo et al., 2025; Graus et al., 2024; Jones et al., 2022; Mauri et al., 2023). Over roughly four decades, fire seasons have tended to lengthen, burned areas have expanded, and extreme events have become more consequential (Mauri et al., 2023). Warming trends and prolonged drought increase fuel aridity, while landscape homogenization and altered management can promote fuel continuity, together enabling larger and more severe fires (Allen et al., 2015; Hantson et al., 2015; Mauri et al., 2023). This shift is particularly problematic for forest systems historically shaped by periodic, relatively low-intensity fires, which are now more frequently exposed to high-severity events that can overwhelm recovery processes.

Multiple regional indicators support this diagnosis. Long-term syntheses report that, since the early 1970s, the average annual number of fires across the Mediterranean has roughly doubled to around 50,000 events per year, and mean annual burned area has risen to nearly 600,000 hectares, about twice the levels reported for the 1970s and 1980s (Alexandrian et al., 1999). More recent assessments similarly point to substantial fire activity and strong spatial concentration: between 2010 and 2023, 22,370 fires were recorded, with Western Mediterranean countries accounting for 53 percent of all fires, and more than 5.5 million hectares burned overall, with close to half of this area in the Western Mediterranean (Gallo Granizo et al., 2025). At the same time, harmonized satellite-based analyses suggest that contemporary fire regimes are increasingly shaped by sharp interannual variability and short-lived extreme fire-weather episodes rather than stable long-term averages (San-Miguel-Ayanz et al., 2025).

In the Maghreb, long time series confirm sustained exposure to wildfire disturbance. In Morocco, forest fire records for 1960 to 2019 report 17,711 ignitions and nearly 178,773 hectares burned, corresponding to annual averages of about 295 fires and 2,980 hectares affected, with particularly severe years reaching around 11,000 hectares in 1983 and 8,660 hectares in 2004 (Département des Eaux et Forêts, 2020). In Algeria, more than 42,500 fires affected approximately 910,000 hectares between 1985 and 2010 (Meddour-Sahar et al., 2012). In Tunisia, more than one hundred fires occurred annually between 1985 and 2000, burning an average of around 1,500 hectares per year (Chriha & Sghari, 2013). Together, these records situate current concerns within a multi-decadal disturbance history, while also suggesting that recent climatic and land-use conditions may be intensifying impacts.

Ecological vulnerability is not uniform across North African forest types. Historically, dominant ecosystems were associated with different fire regimes shaped by species traits and land-use practices. Cork oak woodlands often experienced low- to moderate-intensity surface fires and can recover relatively rapidly through resprouting. Cedar forests in the Middle and High Atlas, by contrast, tended to burn less frequently but are highly sensitive to severe events, given slower regeneration and long recovery times. Pine-dominated formations and mixed Mediterranean forests have frequently been exposed to recurrent fires, often linked to human activities, with outcomes that depend strongly on intensity and recurrence. Under intensifying climatic stress and shifting land use, these ecosystem-specific regimes are increasingly disrupted, exposing less fire-tolerant systems to fire frequencies and severities beyond their adaptive capacity. In this context, wildfire resilience should be framed not only as resistance to burning, but also as the capacity to absorb disturbance, avoid irreversible degradation, and support post-fire recovery through natural regeneration or adapted restoration strategies.

The broader climatic backdrop reinforces these risks. The Mediterranean basin is widely described as a climate-change hotspot (IPCC, 2021). Observational evidence and dendroclimatic reconstructions indicate that aridification has accelerated since the 1980s, contributing to reduced forest productivity and radial growth (Touchan et al., 2008; Sarris et al., 2014). Projections suggest additional warming of 2 to 4 °C by the end of the century, combined with declining precipitation and more frequent droughts, particularly in the western Mediterranean and North Africa (Bazza et al., 2018). These climatic trends interact with demographic growth and agricultural expansion that can intensify deforestation and fragmentation, thereby increasing both ignition likelihood and fire spread potential (Ameztegui et al., 2024; Bazza et al., 2018; Bowring et al., 2024; Peñuelas & Sardans, 2021). Recent assessments accordingly report a rising regional fire risk under combined biophysical and socio-economic pressures (Gallo Granizo et al., 2025).

North African forest distribution and risk patterns further reflect regional climate gradients. Forest cover decreases from west to east, broadly tracking precipitation. Morocco retains the largest forest area, estimated at nearly 9 million hectares, dominated by oaks, cedar, thuya, and argan systems, yet these ecosystems are increasingly stressed by recurrent drought, overgrazing, and land conversion (Serbouti et al., 2023). Wildfire recurrence has become a major concern, particularly in the Rif and Middle Atlas (Jones et al., 2022; Seddouki et al., 2022; Serbouti et al., 2023). Algeria and Tunisia face comparable vulnerabilities, with annual burned areas fluctuating at the scale of several thousand hectares despite investments in prevention and suppression capacity (Belhadj-Khedher et al., 2018; San-Miguel-Ayanz et al., 2025).

Policy responses vary across the Maghreb. In Morocco, the institutional framework for wildfire management has been substantially modernized since the early 2000s, culminating in the National Intersectoral Strategy for Integrated Forest-Fire Management 2020 to 2030, which structures prevention, suppression, and post-fire restoration through strengthened inter-agency coordination (Département des Eaux et Forêts, 2020). Governance is ensured through an interministerial committee involving the Ministry of Interior, the National Agency for Water and Forests, the Ministry of Defense, the Ministry of Equipment, and operational forces (Département des Eaux et Forêts, 2020). This modernization includes expanded automated meteorological monitoring and real-time fire-risk modelling through the Forest and Climate Risk Research Center (CRCF), supporting the spatial planning of fuel breaks, water points, and access infrastructure (Département des Eaux et Forêts, 2020). Operational response mobilizes aerial and ground resources and integrates satellite-based monitoring (MODIS, Sentinel). Post-fire rehabilitation is guided by severity diagnostics and stand survival assessments, combining short-term erosion control and access restrictions with longer-term strategies based on natural regeneration or planting when needed (ANEF, n.d.).

In Algeria and Tunisia, governance structures remain more centralized, with recurring limits in coordination and community-level integration. In Algeria, the General Directorate of Forests (DGF) and Civil Protection services lead prevention, surveillance, and suppression, and although multi-tier committees exist across administrative levels, assessments report uneven implementation, limited technical capacity, and persistent coordination gaps (Banque mondiale, Direction générale des forêts & Délégation nationale aux risques majeurs, 2023; Meddour-Sahar, 2015; Meddour-Sahar et al., 2012). In Tunisia, wildfire management is similarly centralized and local involvement remains limited; while some development programmes encourage participation in natural-resource management, structured community-based initiatives dedicated to wildfire prevention and preparedness remain scarce (Belhadj-Khedher et al., 2018; Chriha & Sghari, 2013). Regional syntheses argue that early warning, long-term planning, and integrated strategies remain insufficiently developed, sustaining recurrent vulnerabilities (Gallo Granizo et al., 2025).

Against this operational backdrop, the scientific literature on wildfires in Morocco remains relatively sparse and fragmented. Existing research is often concentrated on remote sensing applications, including risk mapping, burn severity assessment, and vegetation monitoring, while fewer studies address governance, socio-economic impacts, or long-term ecological trajectories. The dispersion of contributions across institutions and disciplines further limits cumulative visibility and weakens research-to-policy uptake.

Bibliometric analysis offers a systematic way to characterize this knowledge landscape. By combining performance indicators with science mapping, bibliometrics can identify dominant themes, collaboration networks, and persistent gaps in a field where ecological processes, climatic drivers, and policy responses are increasingly interdependent (Aria & Cuccurullo, 2017; Ellegaard & Wallin, 2015; Van Eck et al., 2010). In this sense, bibliometric patterns can be read not only as measures of scientific output, but also as indirect signals of institutional priorities and the strength of research-policy interfaces. Applied to wildfire science in Morocco, this approach can help clarify what is well studied, what remains underexplored, and how future research could better align with evolving management needs.

This chapter provides a comprehensive assessment of wildfire-related research in Morocco, examining its structure, evolution, and thematic orientation within the broader Mediterranean context. It aims to: (1) analyze temporal trends, scientific productivity, and source dispersion; (2) identify major collaboration networks, leading authors, and institutional contributions; and (3) characterize the dominant, emerging, and evolving research themes that shape wildfire studies in the country. By doing so, the chapter provides an evidence-based understanding of how national scientific production can inform policy development, regional coordination, and adaptive management strategies to strengthen resilience to climate-driven fire risks.

2. MATERIALS AND METHODS

This study adopted a bibliometric approach to evaluate the evolution, structure, and thematic orientation of wildfire research in Morocco. Bibliometric methods are increasingly used in environmental sciences to synthesize large volumes of scientific output and to reveal structural patterns in the development of knowledge (Aria & Cuccurullo, 2017; Donthu et al., 2021; Ellegaard & Wallin, 2015). The present work combined performance indicators and scientific mapping to provide both quantitative and conceptual insights into Moroccan wildfire research, with a focus on identifying key contributors, thematic clusters, and emerging trends.

The bibliographic data were retrieved from the Scopus database, which is widely recognized for its extensive coverage of peer-reviewed journals and reliable metadata structure (Bar-Ilan, 2010; Martín-Martín et al., 2018; Jumilah et al., 2025). Scopus was preferred to other databases such as Web of Science due to its broader inclusion of regional and interdisciplinary journals, especially those dealing with Mediterranean and North African environmental issues (Bosman et al., 2006; Cabeza et al., 2020). The search was conducted in October 2025 and included all wildfire-related publications indexed in Scopus up to that date.

The search strategy employed a set of controlled keywords: *("forest fire*" OR "wildfire*" OR "wildland fire*" OR "vegetation fire*" OR "incendie de forêt" OR "feux de forêt" OR "incendies de forêt") AND ("Morocco" OR "Maroc" OR "Atlas Mountains" OR "Rif" OR "Moyen Atlas" OR "Haut Atlas" OR "Rif Mountains")*. No restrictions were applied regarding language or document type. Scopus was selected for its wide coverage of peer-reviewed literature and its relevance for bibliometric studies, as it encompasses a broad range of scientific fields (Ghani et al., 2022). Nonetheless, it may provide limited representation of publications written in languages other than English. Only peer-reviewed journal articles, book chapters, and conference papers explicitly addressing forest fires within the Moroccan context were retained. Publications written in French or Spanish were included when English abstracts and keywords were available. Duplicate records were removed, and metadata such as author names, institutional affiliations, and keywords were standardized to ensure consistency and to avoid double counting of publications or authors.

The final dataset was exported in BibTeX format (.bib) and analysed using the *Bibliometrix* R-package (Aria & Cuccurullo, 2017) via its graphical interface *Biblioshiny* in RStudio. *VOSviewer* software (Van Eck & Waltman, 2010) was used to generate network visualizations of co-authorship relationships, keyword co-occurrence, and thematic clusters. These complementary tools are commonly applied in bibliometric studies for their ability to combine quantitative performance metrics with structural mapping (Cobo et al., 2011).

The analysis followed three main steps. First, a performance analysis was carried out to quantify publication and citation trends, authorship patterns, and institutional productivity. Second, collaboration networks were examined to assess national and international partnerships and to visualize the degree of scientific cooperation.

Finally, a conceptual structure analysis was performed through co-word mapping, allowing the identification of dominant themes, emerging topics, and their relative development measured by Callon's centrality–density indices (Cobo et al., 2011). Lotka's Law (Kawamura et al., 2000) was applied to model author productivity, reflecting the concentration of contributions within the research field.

Temporal keyword analysis was also conducted to trace shifts in thematic emphasis over time, particularly the growing presence of topics such as climate change, drought, remote sensing, and fire ecology. This diachronic view provides an understanding of how research priorities have evolved alongside climatic and institutional developments. All bibliometric data used in this study are publicly accessible from Scopus and comply with ethical standards of reproducibility and transparency recommended in bibliometric research (Donthu et al., 2021).

The final dataset comprised 41 publications drawn from 32 different publication sources and authored by 179 researchers affiliated with national and international institutions. Table 1 presents the main bibliometric characteristics of the dataset, which forms the analytical basis for the subsequent sections.

Table 1. Descriptive statistics of the bibliographic dataset

Description	**Results**
MAIN INFORMATION ABOUT DATA	
Timespan	2013:2025
Sources (Journals, Books, etc)	32
Documents	41
Annual Growth Rate %	2.43
Document Average Age	4.27
Average citations per doc	14.22
References	375
DOCUMENT CONTENTS	
Keywords Plus (ID)	364
Author's Keywords (DE)	158
AUTHORS	
Authors	179
Authors of single-authored docs	0
AUTHORS COLLABORATION	
Single-authored docs	0
Co-Authors per Doc	5.66
International co-authorships %	46.34

continued on following page

Table 1. Continued

Description	Results
DOCUMENT TYPES	
Article	34
Conference paper	6
Conference review	1

3. RESULTS

3.1 Scientific production trends

The timeline of wildfire-related publications in Morocco shows that scientific work on this subject emerged only recently. The earliest traceable papers appear in 2013, and output stays low until around 2016. A clearer upward trend takes shape between 2016 and 2019, when yearly production varies between one and five articles. The strongest growth starts in 2021 and continues through 2022 to 2024, when annual output reaches six to eight publications, marking the most productive phase in the dataset (Figure 1). This sudden increase aligns with the severe wildfire seasons that affected Morocco and with the growing institutional attention given to fire-risk evaluation, forest monitoring and climate-adaptation efforts.

The overall trend from 2013 to 2025 is positive, although moderate. When considering only years with non-zero output, the correlation between publication year and scientific production ($r \approx 0.58$) suggests a clear upward tendency, consistent with the progressive structuring of wildfire research as a distinct scientific field in Morocco. Periods with zero output (2014 and 2020) indicate that the field is still in a consolidating phase rather than fully mature.

Citation patterns provide additional insights into the evolution of scientific visibility. Early publications, which have accumulated longer citable windows, exhibit moderate mean citation counts per article (e.g., 5.33 in 2013; 13.00 in 2016). Some years display higher per-article citation performance, such as 2018 (22 citations per article) and especially 2021 (127 citations per article), driven by studies that have gained substantial international attention. When normalized by exposure time (TC/year), most years remain in the range of 1 to 3 citations per article per year, reflecting typical citation accumulation rates for an emerging field. The exceptionally high TC/year value observed in 2021 (25.40) corresponds to specific high-impact contributions rather than a systemic shift across the field. Recent years (2022–2024) display a combination of increased publication volume and moderate citation intensity, which is expected given their short citation windows.

Figure 1. Evolution of forest fire research publications related to Morocco

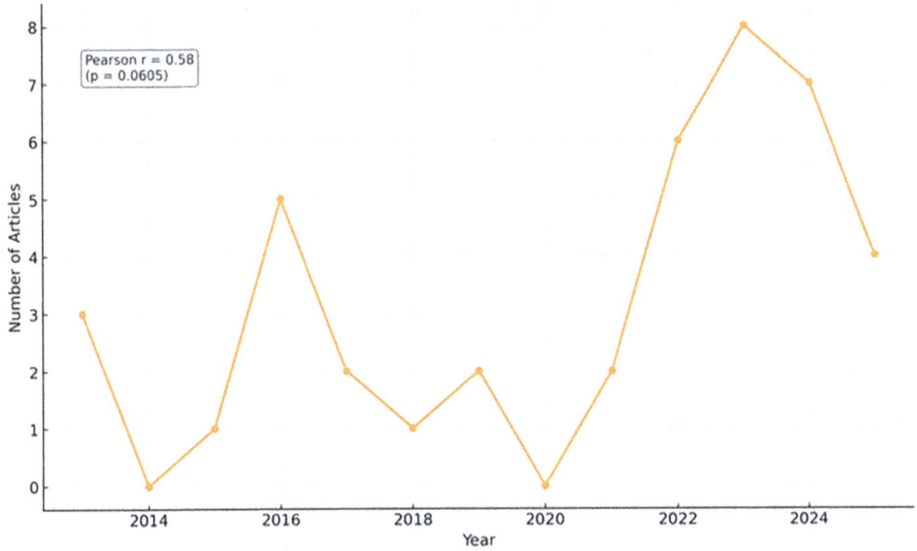

3.2 Core journals and source dispersion

Bradford's Law was applied to assess the distribution of publications across journals and to identify the core sources contributing to Moroccan wildfire research. The resulting curve (Figure 2) shows a highly concentrated structure. Three journals form the core zone: *Revue d'Écologie (La Terre & La Vie)*, *Euro-Mediterranean Journal for Environmental Integration*, and *Fire*. These sources each publish between two and four articles and represent the highest-frequency journals within the dataset.

Beyond this core, the number of articles per journal decreases sharply. Several additional outlets appear in the subsequent zones with only one publication each, including the *Journal of Environmental Management*, *Lecture Notes in Computer Science*, and *Lecture Notes in Networks* and *Systems*. The distribution thus follows a classical Bradford pattern, characterized by a small nucleus of journals with relatively higher productivity and a larger number of peripheral sources with minimal contributions.

The visualization also illustrates the ranking of journals along the logarithmic scale, highlighting the rapid transition from the core to the peripheral zones. No clear intermediate zone appears in this dataset. The overall pattern indicates that wildfire-related research linked to Morocco is published across a wide range of journals, but only a very limited number of these sources exhibit repeated contributions.

Figure 2. Bradford's Law distribution of journals publishing wildfire research related to Morocco

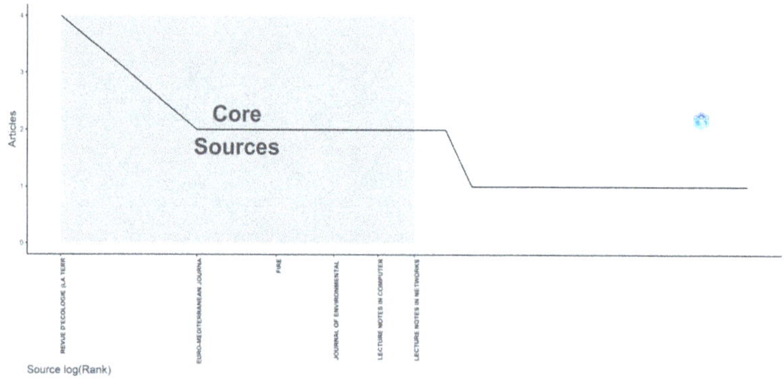

3.3 Authorship patterns and leading contributors

The distribution of first-author publications reveals that wildfire research in Morocco is driven by a relatively small group of recurrent contributors. Table 2 identifies the ten most active authors, whose work spans a diverse but complementary set of themes, reflecting the emerging structure of the field.

Table 2. Leading first authors in Moroccan wildfire research

Rank	Author	Number of First-Author Publications
1	Boubekraoui, H.	4
2	Essaghi, S.	3
3	Jellouli, O.	3
4	Mharzi-Alaoui, H.	2
5	Jadouli, A.	2
6	Badda, H.	2
7	Zidane, I.E.	2
8	El Mazi, M.	2
9	El Amarty, F.	1
10	Cherki, K.	1

The thematic orientation of these contributions highlights four major research profiles:

- A first group concentrates on fire-risk assessment and spatial susceptibility mapping, focusing on the role of fuel continuity, topography and land-use pressure in shaping fire-prone landscapes. Their work relies heavily on GIS-based multicriteria analysis and spatial modelling.
- A second group specializes in fuel characteristics and fire behavior, investigating the flammability of dominant Mediterranean species, the structure of fine fuels, and traits influencing ignition potential and combustion dynamics.
- A third set of contributors advances computational and predictive modelling, developing fire-spread simulations, cellular automata frameworks and deep-learning architectures for ignition prediction and early-warning systems.
- A final group works on burned-area monitoring, fire severity and post-fire responses, using multisensor remote sensing to characterize ecological impacts on vegetation, soils and hydrological functioning.

The timeline of author publications (Figure 3) shows that wildfire research in Morocco has developed through two distinct phases. A first group of contributors appears between 2016 and 2019, marking the initial establishment of research activity in the field. Their publication trajectories extend over several years, indicating sustained engagement and early structuring of wildfire scholarship. A second group of authors enters the field more recently, with publications concentrated after 2021. This recent influx corresponds to the acceleration of wildfire research observed at the national scale and reflects growing scientific interest triggered by the severe fire seasons of the early 2020s.

The authors tend to publish about one paper a year, and only a few show a more active rhythm. This suggests that the field is moving forward through small, individual contributions rather than through a high yearly output. The color scale linked to citations per year also points to a low to moderate citation rate for most papers, which makes sense given that many of them were published recently and that research on wildfires in Morocco is still gaining visibility in the international scientific community.

Figure 3. Top contributing authors in Moroccan forest fire research and their publication timelines (2016–2025)

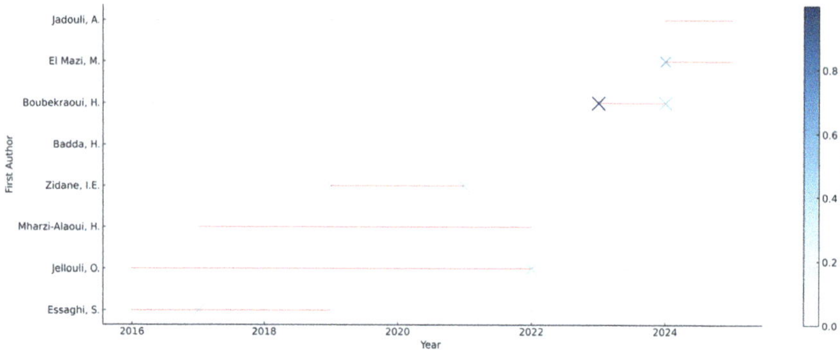

3.4 Collaboration networks and institutional landscape

The co-authorship network (Figure 4) reveals a research landscape characterized by clear fragmentation and the presence of multiple small, internally connected clusters. Collaboration patterns are organized around a series of discrete groups, each corresponding to research teams that operate largely independently. These clusters display dense internal linkages but only limited connections with one another, indicating that wildfire research in Morocco is structured more around localized or institution-specific initiatives than around a cohesive national network.

The temporal gradient shown by node colors highlights a progressive diversification of contributors over time. Early publications form smaller and less interconnected groups, while more recent years display an expansion of nodes and the emergence of new clusters, reflecting the growing number of researchers entering the field after 2020. However, despite this increase in participation, cross-cluster collaborations remain weak, suggesting that knowledge exchange and methodological integration across institutions are still limited.

The prevalence of isolated nodes and dyadic links further illustrates the developmental stage of the field. These solitary or near-solitary positions correspond to one-off contributions or short-lived collaborations that have not yet evolved into stable research partnerships.

Figure 4. Co-authorship network of researchers involved in Moroccan forest fire studies

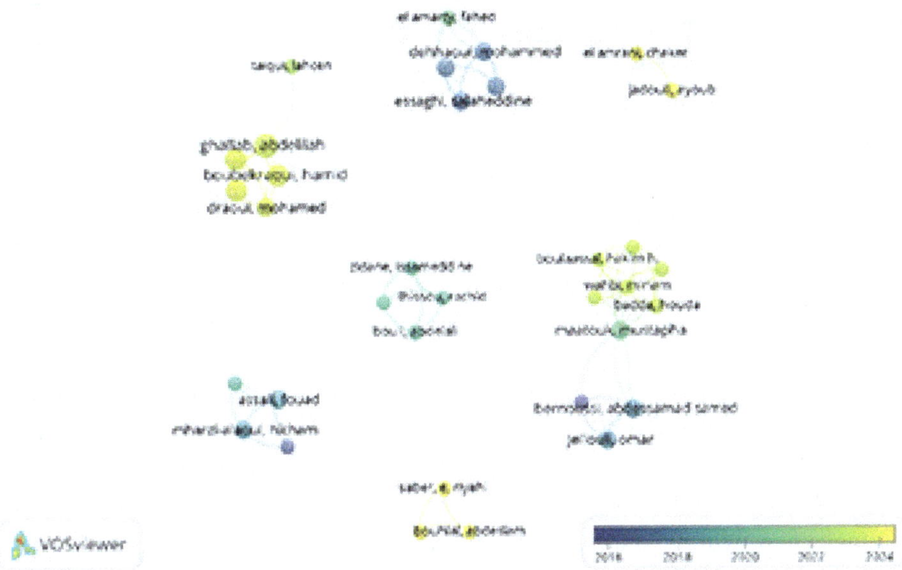

The institutional landscape of Moroccan forest fire research is dominated by Abdelmalek Essaadi University (UAE), which alone accounts for fifteen publications. Its position as the leading contributor is closely linked to its geographical location in the country's northern region, an area that faces the highest recurrence and intensity of forest fires. The School of Architecture in Tetouan reinforces this regional focus through research connecting spatial planning and fire-prone landscapes.

The Hassan II Institute of Agronomy and Veterinary Medicine in Rabat and Moulay Ismaïl University (UMI) in Meknes, each contributing six publications, represent additional centers of expertise supporting environmental and ecological research on fire prevention and forest resilience. Beyond the national framework, the collaboration with the Instituto de Sistemas e Robótica in Lisbon underscores a growing integration with European research networks, particularly those of Portugal and Spain, where similar Mediterranean fire dynamics prevail.

This configuration illustrates that Moroccan forest fire research is largely structured around northern academic institutions exposed to recurrent fire events, while gradually expanding through partnerships that link local scientific capacity with regional and international research initiatives (Table 3).

Table 3. Main institutional contributors to forest fire research in Morocco

Affiliation	Articles
Abdelmalek Essaadi University (UAE)	15
Hassan II Institute of Agronomy and Veterinary Medicine, Rabat	6
Moulay Ismaïl University (UMI), Meknes	6
National School of Forest Engineers (ENFI), Salé	4
Sidi Mohamed Ben Abdellah University (USMBA), Fez	4
Sultan Moulay Slimane University (USMS), Beni Mellal	3
Ibn Tofail University (UIT), Kenitra	2
School of Architecture, Tetouan	2
Instituto de Sistemas e Robótica, Lisbon, Portugal	2

At the international level, the country collaboration map (Figure 5) highlights the international orientation of Moroccan forest fire research, with strong partnerships established mainly with European and North American institutions. The most active collaborations link Morocco with Spain, Portugal, and France, reflecting both geographical proximity and shared Mediterranean fire regimes. These partnerships are often grounded in joint studies on fire risk modeling, remote sensing, and forest management, where Mediterranean countries face similar ecological and climatic pressures. Additional but less frequent links with the United States, Canada, and China indicate an expanding yet selective global engagement.

What stands out, however, is the absence of significant scientific collaboration within North Africa itself. No visible research links appear with neighboring Maghreb countries such as Algeria, Tunisia, or Libya, despite their comparable environmental conditions and growing exposure to wildfires. This lack of intra-regional cooperation suggests that forest fire research across North Africa remains nationally fragmented, with limited coordination in data sharing, regional monitoring, or policy-oriented research.

In the context of southern Mediterranean forest governance, this pattern underlines a missed opportunity for a coordinated North African scientific framework. Strengthening collaboration across the Maghreb would not only enhance comparative understanding of fire dynamics but also contribute to shared strategies for prevention, resilience, and sustainable forest management under increasing climatic stress.

Figure 5. International collaboration network in Moroccan forest fire research

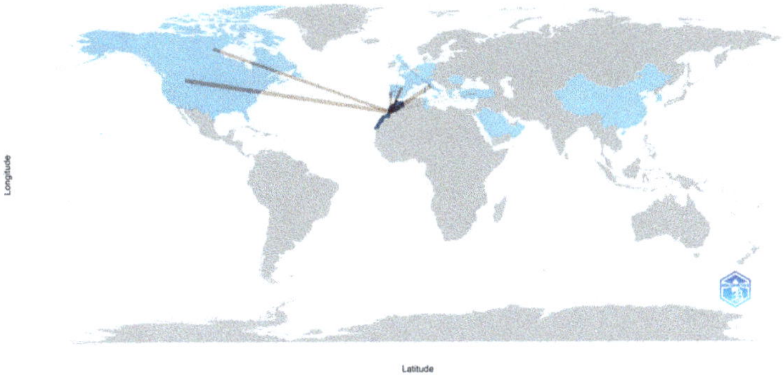

3.5 Authorship, productivity, and citation dynamics

The analysis of citation performance and author productivity provides complementary insights into the structure and visibility of Moroccan forest fire research. Morocco leads in publication volume with 206 citations and an average of 7.4 citations per article, reflecting a growing but still nationally focused research base. In contrast, Portugal and the United States, despite fewer publications, display higher citation averages (18.5 and 11.0, respectively), indicating greater international impact and stronger integration into high-visibility scientific networks.

Author productivity through Lotka's Law (Figure 6) confirms that the field follows a classic pattern of concentration, where most authors have produced only a single paper, while a small core group accounts for multiple contributions. This distribution suggests that Moroccan forest fire research remains in a developmental stage, driven by a limited number of highly active researchers whose collaborations—particularly with European partners—enhance both scientific output and citation impact.

Figure 6. Author productivity through Lotkas Law

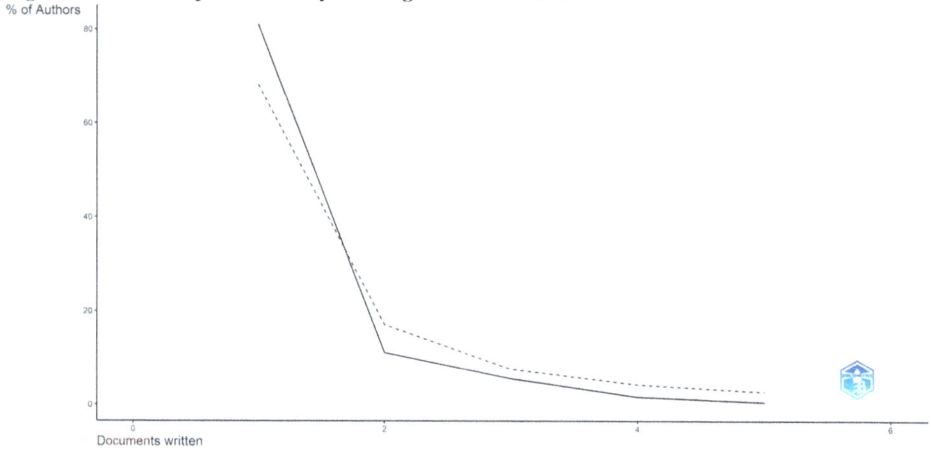

3.6 Conceptual and thematic structure

The keyword co-occurrence network (Figure 7) provides a detailed view of the conceptual structure of Moroccan forest fire research. Three main thematic clusters can be identified. The red cluster centers on *forest fire*, *fire hazards*, *deforestation*, and *climate change*, reflecting the dominant focus on the causes, spread, and environmental impacts of wildfires. It captures the intersection between climatic drivers, land-use change, and fire dynamics—core issues in understanding fire behavior in Mediterranean ecosystems.

The green cluster, organized around *Morocco*, *wildfire*, and *forest management*, represents the applied and territorial dimension of research. This cluster links local case studies and management practices with regional characteristics, particularly in the *northern Morocco* and *Mediterranean region*, where fire frequency is highest.

The blue cluster gathers methodological approaches such as *remote sensing*, *GIS*, *NDVI*, and *machine learning*, showing the integration of technological and analytical tools to monitor burned areas, assess risk, and model fire behavior. The growing prominence of these terms indicates a clear trend toward data-driven and spatially explicit research frameworks.

These interconnected themes illustrate an increasingly multidisciplinary and system-oriented research field that combines environmental, technological, and management perspectives to address wildfire challenges within Morocco's Mediterranean forest ecosystems.

Figure 7. Keyword co-occurrence network of forest fire research related to Morocco

The temporal trend of keyword occurrences (Figure 8) shows the progressive diversification and consolidation of research themes related to forest fires in Morocco. The term Morocco dominates throughout the period, reflecting the central geographic focus of the studies. Since 2016, there has been a noticeable rise in keywords such as *deforestation, remote sensing, forest ecosystem, and climate change*, indicating a growing integration of environmental and technological dimensions into wildfire research. The increasing frequency of GIS and risk assessment after 2020 suggests enhanced methodological sophistication and stronger links between fire modeling, landscape monitoring, and risk management. Overall, the pattern highlights an expanding and more structured research landscape that aligns with the intensification of wildfire events and policy responses in Morocco over the past decade.

Figure 8. Temporal evolution of the most frequent keywords in wildfire research in Morocco (2005–2025)

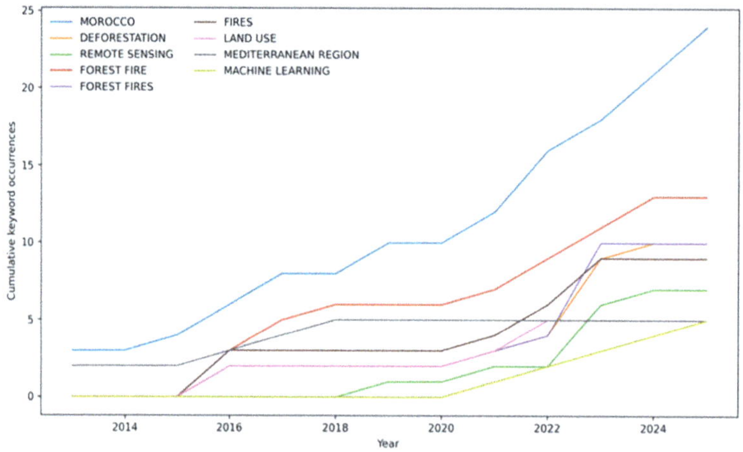

The thematic coupling analysis, based on Callon's centrality and density indices (Figure 9), provides insight into the conceptual and thematic structure of Moroccan forest fire research. The thematic map shows a differentiated structure in Moroccan wildfire research. Advanced computational methods such as *machine learning* and *deep learning* appear as the most influential and internally coherent themes, signalling a shift toward data-intensive approaches for fire detection, severity assessment, and predictive modeling. Classical remote-sensing tools, including *forest fire monitoring* and *cellular automata*, remain central because they provide the methodological base for mapping and simulation, although their internal development is less dynamic than that of AI-based methods.

Themes such as *northern Morocco* or *autocorrelation* form a distinct group of regionally focused studies with strong internal consistency but limited integration into broader research directions. In contrast, topics like *wildfire* or *modis* occupy peripheral positions with low density and centrality, reflecting either declining use or the replacement of older data sources by more recent spatial products.

Intermediate themes including *spectral indices, burned areas, random forest,* and *climate-related modeling* bridge the gap between remote sensing and advanced analytics. Their placement indicates ongoing consolidation and the potential to become more structuring elements in future research.

This configuration highlights a field in methodological transition, where traditional earth-observation tools continue to structure scientific practice while newer computational paradigms increasingly shape the direction and depth of wildfire research in the Moroccan and Southern Mediterranean context.

Figure 9. Thematic coupling analysis of Moroccan forest fire research based on Callon's centrality and density indices

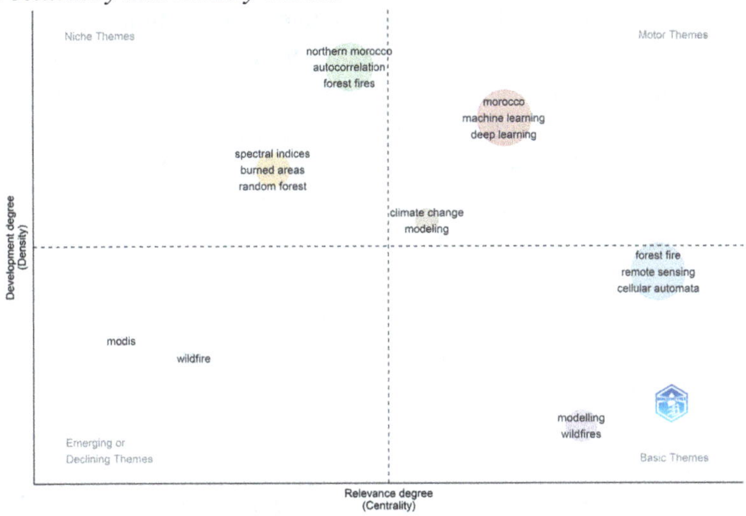

4. DISCUSSION

4.1 Scientific advances and methodological evolution in Moroccan wildfire research

The results of this bibliometric and thematic analysis show that wildfire research in Morocco has expanded significantly since 2016 and especially after 2021, reflecting its growing relevance as a socio-environmental challenge in North Africa. This trajectory parallels broader Mediterranean tendencies, where intensifying drought and fuel desiccation have driven more severe fire regimes (Bowman et al., 2020; Turco et al., 2017). Moroccan contributions increasingly move beyond descriptive assessments toward approaches that integrate risk governance and ecosystem resilience under climate stress.

A major factor behind this shift is the rapid adoption of geospatial tools and satellite-derived indicators for vegetation monitoring, burn detection and spectral change analysis (Bouta et al., 2024; El Amarty et al., 2025; Zidane et al., 2021). Several studies also explore post-fire vegetation dynamics, where recovery paths depend strongly on severity levels and landscape configuration (El Mazi et al., 2024). The widespread mobilization of spectral indices reinforces the alignment of Moroccan research with global advances in quantitative fire ecology (Chuvieco et al., 2019; Giglio et al., 2018). Machine-learning and hybrid models are increasingly used to predict ignition and potential fire spread (Jadouli & El Amrani, 2025; 2024), although these frameworks still rarely incorporate socio-institutional variables.

Land-use change emerges as a consistent determinant of wildfire susceptibility. Urban expansion, agriculture and forest clearing reshape fuel patterns and exacerbate fire vulnerability, particularly in human-impacted landscapes of the Rif and Mediterranean zones (Boubekraoui et al., 2024a; 2024b). Other contributions analyze ecological and infrastructural impacts, including soil alterations (El Mazi et al., 2025) and increasing exposure of transport and development infrastructure in chronically affected areas (El Maachi et al., 2024). Despite these advances, research on governance, social vulnerability and long-term socio-ecological transitions remains scarce, confirming international observations that these dimensions remain underrepresented in wildfire science (Tedim et al., 2018).

The analysis of source dispersion through Bradford's Law provides a useful perspective on how Moroccan wildfire research is structured. Only a small group of journals can be considered central to the field, while many others have published just a single article. This suggests that the field still lacks stable publication platforms that could help build cumulative knowledge over time. In more mature research areas, core journals serve as places where methods are standardized, debates unfold and research programs gain continuity. The situation here is quite different. The strong dispersion of publications points to a fragmented landscape in which studies are spread across a wide variety of outlets that often have little thematic connection. This aligns with the early stage of development highlighted by the co-authorship networks and thematic maps, where research groups remain relatively small and only loosely connected. A more deliberate publication strategy, especially one directed toward journals specializing in Mediterranean studies, fire ecology or remote sensing, could increase the visibility of this work, allow findings to be compared more reliably and help bring together a more cohesive scientific community around wildfire research in Morocco.

A similar fragmentation pattern appears when examining author productivity. The concentration of scientific output in a small group of recurrent contributors, as revealed by fractionalized publication counts, indicates that Moroccan wildfire research relies heavily on a limited number of active researchers. While this nucleus provides continuity and drives methodological advancement, the narrow distribution of contributions also limits the diversification of expertise and reduces the system's capacity to generate interdisciplinary perspectives. In combination with the dispersed journal landscape revealed by Bradford's Law, this concentration of authorship suggests that Moroccan wildfire research remains structurally dependent on a few individual research groups rather than on a broad, interconnected scientific community. Expanding the pool of contributors, strengthening institutional training, and encouraging inter-university collaboration would therefore be essential for consolidating the field and sustaining long-term scientific development.

4.2 Structural fragmentation of research collaboration and regional disparities

Collaboration structures show that Moroccan wildfire research remains concentrated in universities located near fire-prone regions such as the Rif and the Middle Atlas, reflecting the geographical proximity between researchers and affected landscapes. However, the resulting co-authorship networks remain fragmented and short-lived, which limits national coordination and prevents the emergence of long-term multi-institutional programmes capable of generating shared datasets and methodological consistency. By contrast, southern Europe has developed a more consolidated scientific landscape, where wildfire research is supported by permanent institutions and integrated monitoring infrastructures. The European Forest Fire Information System (EFFIS), for instance, provides harmonised fire information across participating countries and has become a central tool used by European research communities to support comparative analyses, long-term data series and joint scientific programmes (Chuvieco et al., 2019; San-Miguel-Ayanz et al., 2025). Although EFFIS has progressively expanded its monitoring activities to include parts of North Africa, its most sustained scientific integration has been driven by European organisations with established institutional frameworks.

International collaborations are still mostly directed toward Europe, especially Spain, France and Portugal, which supports the sharing of methods and comparative research. However, the absence of scientific cooperation with Algeria and Tunisia remains a serious drawback. These two countries experience similar climate challenges and have comparable forest ecosystems according to previous studies (Belhadj-Khedher et al., 2018; Meddour-Sahar, 2015; Meddour-Sahar et al., 2012). Efforts such as building a shared fire-severity database, adopting common monitoring procedures and conducting comparative post-fire studies could make wildfire research in the Maghreb far more robust. Regional entities like the Union for the Mediterranean and FAO's Silva Mediterranea promote collaboration, yet concrete progress is still limited in practice.

This structural fragmentation of research collaboration not only limits cumulative scientific progress but also constrains the translation of scientific knowledge into coherent governance frameworks, as fragmented evidence bases tend to produce fragmented policy responses.

4.3 Governance models, science–policy interface, and regional contrasts

Keyword trajectories suggest a conceptual shift from post-event documentation to resilience-oriented, risk-assessment perspectives (Pyne et al., 2019). However, the weak presence of governance, institutional coordination, community perception and economic impacts indicates that the science–policy interface is still insufficiently developed. Across Mediterranean-type regions, a continued reliance on suppression keeps reinforcing structural fuel accumulation, which in turn reduces the controllability of wildfires under extreme weather (Moreira et al., 2020). Evaluating policies primarily through burned area obscures their actual capacity to prevent socio-ecological loss.

In contrast, southern Europe has advanced toward integrated fire-management systems that combine landscape-scale planning, fuel reduction, rural development and structured community engagement (MITECO, 2022; AGIF, 2020). Within North Africa, governance models follow differentiated trajectories. Morocco displays the most structured system, based on interministerial coordination, territorial action plans and operational protocols linking national strategies to local implementation (Département des Eaux et Forêts, 2020; Assali et al., 2019; Mharzi Alaoui et al., 2017). Algeria's framework is legally established but continues to face substantial limitations in interagency coordination and monitoring capacity (Meddour-Sahar, 2015; Meddour-Sahar et al., 2012). Tunisia maintains a broad disaster-management architecture, yet wildfire strategies remain weakly integrated and largely centralized, with response capacities outweighing long-term planning and limited involvement of local communities in prevention and preparedness (European Commission, 2019; Mauri et al., 2023). Mediterranean assessments consistently highlight the same regional priorities: interoperable data infrastructures, early-warning capacities, fuel-management strategies, coordinated territorial planning and stronger community engagement (FAO, 2006; Mauri et al., 2023). Recent evaluations further emphasize that integrated and multi-level governance is essential to confront increasingly severe fire regimes across the Mediterranean forests (Gallo Granizo et al., 2025).

4.4 Outstanding gaps and future directions for wildfire research in Morocco

The thematic-coupling map shows that drought research and spatiotemporal approaches are beginning to gain attention, but they are still not fully developed. Reducing these gaps will require combining long climatic records such as ERA5-Land and CHIRPS with vegetation time-series data to examine how temperature anomalies, soil-moisture deficits and fuel availability interact. These factors play a

central role in driving extreme fire events in Mediterranean regions (Jia et al., 20 Turco et al., 2018). Adding more ground observations through forest invento would also make predictive models more reliable.

Globally, sustainability research increasingly emphasizes post-fire recovery cesses, including carbon losses, regeneration failures, and biodiversity shifts (Ke & Pausas, 2022; Zuo et al., 2024). In Morocco, however, these dimensions re weakly represented in the scientific literature. Post-fire management practice still largely discussed in operational terms, with limited scientific assessment respective roles of natural regeneration and artificial planting. Inappropriate ration strategies, including poorly adapted monocultures or insufficient con invasive species, may compromise long-term ecosystem resilience and alter pc trajectories. Strengthening research on restoration ecology and post-fire mon would therefore be essential to support climate-resilient forest management

Moroccan wildfire research is advancing in conceptual and methodc sophistication but still lacks the interdisciplinary depth, sustained institutic ordination and regional integration that underpin advanced Mediterranean s Developing cross-border research platforms, expanding open-access wildfi bases and aligning scientific output with national climate adaptation fran represent essential steps toward strengthening both scientific visibility long-term resilience of Moroccan forest ecosystems. Beyond these thema several broader structural and socio-ecological considerations further cont the development of wildfire science in Morocco and help interpret the research patterns.

A broader reflection on the evolution of wildfire science in Mor highlights the need to consider wildfires not as isolated events but as co of larger socio-ecological transitions unfolding across Mediterranean African landscapes. In regions where climate change accelerates aridific degradation, and vegetation stress, fire increasingly interacts with long-te ical processes such as species turnover, shifts toward drought-tolerant s and progressive loss of forest resilience. These emerging transitions documented in parts of southern Europe, where repeated high-severit triggered conversions from forest to low-stature vegetation types, with consequences for biodiversity, carbon storage, and rural economies. Morc similar vulnerabilities, particularly in the Rif and Middle Atlas, yet literature still rarely examines wildfires through the lens of long-ter trajectories or irreversible ecological thresholds. Integrating these pers future research would enrich the analysis of Moroccan fire regimes and work with global advances in socio-ecological resilience theory.

Another dimension that deserves closer attention is the relation wildfire risk and rural socio-economic dynamics. Across the Medi

sin, processes such as depopulation of mountain areas, abandonment of traditional agro-sylvo-pastoral systems, and the expansion of unmanaged shrublands have contributed to increased fuel continuity and fire propagation. In Morocco, the progressive weakening of traditional management systems, such as the Agdal, has reduced collective regulation of grazing and resource use, indirectly favoring fuel accumulation in forest and rangeland landscapes. Combined with agricultural burning, honey harvesting, charcoal production, and other resource-extraction practices, these changes contribute to heterogeneous ignition patterns and fire vulnerability. Beyond their ecological implications, these dynamics entail tangible socio-economic consequences affecting forest-based livelihoods and ecosystem services. Yet the role of traditional management practices in fuel regulation and wildfire prevention remains largely overlooked in Moroccan wildfire research.

Finally, the discussion of regional contrasts suggests that North African countries could benefit from a more explicit integration of wildfire risk into climate adaptation, land-use planning, and regional cooperation frameworks. Mediterranean assessments consistently emphasize the importance of harmonized fire-danger indices, interoperable datasets, and joint monitoring infrastructures. Although Morocco has made significant progress in institutional coordination, the absence of shared protocols and cross-border scientific collaboration limits the development of a coherent regional perspective on evolving fire regimes. Strengthening scientific bridges across the Maghreb would not only improve comparative analyses but also support the emergence of shared knowledge platforms capable of anticipating extreme events and informing coordinated prevention strategies.

Taken together, these broader socio-ecological and regional considerations highlight the need for continued refinement of wildfire research frameworks in Morocco, while also clarifying the methodological constraints that shape the present analysis.

4.5 Limitations of the Study

This bibliometric assessment provides a structured overview of wildfire research in Morocco, but several limitations should be acknowledged. First, reliance on the Scopus database excludes national strategies, technical reports, and French-language institutional publications, which contain valuable operational insights but are not captured in the quantitative dataset.

Second, bibliometric indicators depend on the consistency of metadata. Variations in how authors report affiliations, keywords, or research themes may influence co-occurrence patterns and thematic clustering, particularly in interdisciplinary fields such as wildfire science. The resulting networks should therefore be interpreted as indicative rather than exhaustive.

Third, the 2013–2025 publication window captures major trends but does not fully reflect the most recent advances following the extreme 2021–2022 fire seasons, notably in AI-based detection, fuel dynamics, and risk-governance frameworks. Given the rapid evolution of the field, updated analyses will be necessary.

Finally, bibliometric approaches do not capture experiential and community-based knowledge held by firefighters, forest rangers, local authorities, and rural populations. Integrating these perspectives through qualitative or mixed methods would offer a more comprehensive understanding of wildfire dynamics and management challenges in Morocco.

These limitations do not undermine the validity of the analysis but instead help identify priority gaps in data coverage, thematic focus, and institutional integration. The following recommendations are therefore framed as direct responses to these gaps, with the aim of strengthening both scientific production and wildfire governance in Morocco.

5. RECOMMENDATIONS AND ROADMAP

The bibliometric analysis shows that Moroccan wildfire research is entering a phase of consolidation, yet remains structurally and thematically unbalanced. The following recommendations propose a roadmap (Figure 10) that links scientific priorities, governance reforms and regional cooperation, in order to move from a fragmented and largely reactive system toward an integrated, resilience oriented wildfire regime.

5.1 Consolidating data infrastructures and operational monitoring

A first strategic priority concerns the consolidation of wildfire information systems. The current landscape is characterized by scattered datasets, heterogeneous formats and limited interoperability between institutions. Building on existing geospatial and modelling efforts, Moroccan research and management agencies should work toward a national wildfire data infrastructure that integrates at least four components: systematically georeferenced fire occurrence records, climate indicators, land use and fuel information, and post fire impact assessments.

From a research perspective, this implies that universities and research centers move beyond case specific analyses and develop standardized protocols for fire detection, severity mapping and vegetation recovery monitoring, based for instance on NDVI, NBR and related indices already used in recent work. Operationally, National Agency for Water and Forests (ANEF) and Civil Protection could gradually align their monitoring systems with these scientific standards, ensuring that the same datasets serve both academic analysis and real time decision support. In practical terms, this would also require solid data-sharing agreements and the establishment of a national wildfire observatory, where time series from diverse climate, precipitation and satellite datasets are routinely processed to supply early warning systems and seasonal risk bulletins.

The development of such an infrastructure is not merely a technical issue. It conditions the possibility of robust trend analysis, comparative studies across regions and the evaluation of policy measures. Without a shared and cumulative data foundation, the gains brought by advanced modelling and machine learning remain confined to isolated experiments that are difficult to translate into stable management instruments.

5.2 Rebalancing the research portfolio toward socio ecological resilience and post fire trajectories

The bibliometric profile of Moroccan wildfire research shows a strong emphasis on biophysical determinants, with comparatively little attention to governance, social vulnerability and long term recovery. To move toward a resilience oriented agenda, the research portfolio needs to be rebalanced in at least three directions.

First, socio economic and cultural dimensions of wildfire should be addressed systematically. This includes the analysis of household dependence on forest resources, perceptions of fire risk, prevention practices, conflicts around land use and the differentiated impacts of fire on rural livelihoods and local economies, including tourism. Such work would allow the identification of social groups and territories where targeted prevention and compensation measures are most needed, and would provide an empirical basis for community centred fire management.

Second, post fire ecology and restoration require a much more central position in the research agenda. International experience shows that high severity and recurrent fires can drive shifts in soil carbon stocks, vegetation composition and regeneration regimes that compromise long term forest sustainability. In Morocco, similar processes are likely to occur, yet they remain poorly documented. Future studies should therefore link burn severity, soil and vegetation dynamics, and management interventions, in order to identify thresholds beyond which active restoration becomes necessary and to design cost effective restoration strategies adapted to local ecological conditions.

Third, it is necessary to link these ecological and social analyses through integrated socio ecological frameworks. Rather than treating fire solely as a disturbance of forest cover, research should examine how repeated fires interact with demographic change, agricultural dynamics and territorial development policies, and how these interactions shape resilience or degradation pathways at landscape scale. This implies closer collaboration between ecologists, geographers, economists and social scientists, and the design of longitudinal studies that follow affected territories over time.

5.3 Strengthening the science policy interface and North African cooperation

The comparison with southern European experiences and the governance gradient within North Africa suggest that the limits of Moroccan wildfire research are not only thematic but also institutional. The consolidation of a science policy interface is therefore a key condition for the effective use of knowledge in decision making.

At the national level, this interface could take the form of a permanent platform bringing together ANEF, Civil Protection, universities and regional authorities. Its mandate would be to co define priority research questions, ensure that scientific results are translated into operational tools and guidelines, and monitor the integration of wildfire considerations into territorial and sectoral policies, including spatial planning, rural development and climate adaptation strategies. Such a platform would also help reduce the current fragmentation of research networks, by encouraging multi institutional projects and shared doctoral and post-doctoral programmes focused on fire.

At the regional scale, the absence of structured collaboration with Algeria and Tunisia is a major missed opportunity, given the ecological and climatic similarities across the Maghreb. The initiatives launched under the Union for the Mediterranean and Silva Mediterranea provide an institutional entry point, but they will only produce concrete effects if they are backed by joint research projects, shared databases and common methodological frameworks. A pragmatic first step would be to build a North African fire atlas based on harmonized indicators of fire occurrence, severity and impacts, complemented by a series of comparative case studies on governance and community vulnerability. Over time, this could evolve into a regional knowledge hub that supports collective negotiation of international climate and biodiversity agendas.

5.4 A phased roadmap for an integrated and climate adapted wildfire regime

Taken together, the previous elements suggest a phased roadmap. In the short term, efforts should focus on improving the quality, accessibility and interoperability of wildfire data, and on embedding existing geospatial and modelling tools into operational monitoring and early warning systems. In the medium term, the research agenda should expand toward socio economic impacts, governance mechanisms and post fire trajectories, with a clear orientation toward socio ecological resilience. In the longer term, the objective should be to establish an integrated wildfire regime where scientific knowledge, territorial planning and community based prevention mutually reinforce each other at national and regional scales.

Such a transition requires sustained political commitment, investment in human and technical capacities, and a deliberate opening of national research systems toward regional and international cooperation. For Morocco, which already possesses a relatively structured governance framework compared with its neighbours, the challenge is not only to produce more scientific output, but to orient this output toward the levers that can effectively reduce risk and support resilient forest landscapes under increasingly severe climate pressures.

Figure 10. Conceptual roadmap for advancing wildfire science and governance in Morocco

6. CONCLUSION

This chapter provides a comprehensive overview of wildfire science in Morocco by combining bibliometric analysis, thematic mapping and a comparative assessment of Mediterranean governance models. The sharp rise in publications over the last decade reflects a growing recognition of wildfires as a central issue for environmental research and landscape management in the country. Progress in satellite-based monitoring, modelling frameworks and climate–fire interactions demonstrates clear methodological advances and an increasing alignment with international approaches in geospatial fire ecology.

Despite these advances, the scientific landscape remains uneven. Governance, socio-economic vulnerability and long-term ecological recovery are still marginal in Moroccan wildfire research, although these dimensions form the foundation of resilient fire-management systems in Mediterranean countries. The limited national collaboration among research teams and the insufficient integration of scientific evidence into operational decision-making reduce the potential impact of existing knowledge. Comparisons with Southern Europe show that structured science–policy interfaces and multi-institutional programs can substantially accelerate progress.

The evidence gathered highlights three priorities for strengthening wildfire science in Morocco: developing interoperable national datasets, promoting interdisciplinary research on socio-ecological resilience, and expanding cooperation across North Africa to align monitoring systems and adaptation strategies. Integrating ecological, social and governance dimensions in future research is essential to support a transition from reactive suppression toward proactive, risk-based landscape management.

By situating Moroccan wildfire research within the wider Mediterranean context, this chapter underscores the importance of shared scientific infrastructures and regional cooperation to confront converging climate pressures and build resilient forest landscapes across the southern rim of the basin.

REFERENCES

AGIF. (2020). *Landscape fire governance framework*. Agência para a Gestão Integrada de Fogos Rurais.

Alexandrian, D., Esnault, F., & Calabri, G. (1999). Forest fires in the Mediterranean area. *Unasylva*, *50*(197), 35–41.

Allen, C. D., Breshears, D. D., & McDowell, N. G. (2015). On underestimation of global vulnerability to tree mortality and forest die-off from hotter drought in the Anthropocene. *Ecosphere*, *6*(8), 1–55. DOI: 10.1890/ES15-00203.1

Ameztegui, A., Coll, L., De Cáceres, M., & Morán-Ordóñez, A. (2024). Disturbance impacts on Mediterranean forests across climate and management scenarios. *Journal of Environmental Management*, *371*, 123193. DOI: 10.1016/j.jenvman.2024.123193 PMID: 39536583

ANEF. (n.d.). *Incendies: stratégie de gestion*. http://www.eauxetforets.gov.ma/Pages/Incendies-strategie-de-gestion.aspx

Aria, M., & Cuccurullo, C. (2017). bibliometrix: An R-tool for comprehensive science mapping analysis. *Journal of Informetrics*, *11*(4), 959–975. DOI: 10.1016/j.joi.2017.08.007

Assali, F., D'Avezac de Castera, H., Mharzi Alaoui, H., Taberkant, S., & Nabaha, A. (2019). *Integrated forest fires management in Morocco: New national strategy and action plan 2019–2028*. https://doi.org/DOI: 10.37002/biodiversidadebrasileira.v9i1.1119

Banque mondiale, Direction générale des forêts & Délégation nationale aux risques majeurs. (2023). *Note sur les forêts algériennes: Gestion durable des forêts pour lutter contre les feux de forêts*. World Bank.

Bar-Ilan, J. (2010). Citations to the "Introduction to informetrics" indexed by WOS, Scopus and Google Scholar. *Scientometrics*, *82*(3), 495–506. DOI: 10.1007/s11192-010-0185-9

Bazza, M., Kay, M., & Knutson, C. (2018). Drought characteristics and management in North Africa and the Near East. FAO Water Reports No. 45. Food and Agriculture Organization of the United Nations.

Belhadj-Khedher, C., Koutsias, N., Karamitsou, A., El-Melki, T., Ouelhazi, B., Hamdi, A., Nouri, H., & Mouillot, F. (2018). A revised historical fire regime analysis in Tunisia (1985–2010) from a critical analysis of the national fire database and remote sensing. *Forests*, *9*(2), 59. DOI: 10.3390/f9020059

Bonan, G. B. (2008). Forests and climate change: Forcings, feedbacks, and the climate benefits of forests. *Science*, *320*(5882), 1444–1449. DOI: 10.1126/science.1155121 PMID: 18556546

Bosman, J., Mourik, I., Rasch, M., & Sieverts, E. (2006). Scopus reviewed and compared: The coverage and functionality of the citation database Scopus. *Online Information Review*, *30*(1), 4–21.

Boubekraoui, H., Attar, Z., Maouni, Y., Ghallab, A., & Draoui, B. (2024b). Forest Loss Drivers and Landscape Pressures in a Northern Moroccan Protected Areas' Network: Introducing a Novel Approach for Conservation Effectiveness Assessment. *Conservation*, *4*(3), 452–485. DOI: 10.3390/conservation4030029

Boubekraoui, H., Maouni, Y., Ghallab, A., Draoui, B., & Attar, Z. (2024a). Deforestation drivers in northern Morocco: An exploratory spatial data analysis. *Environmental Research Communications*, *6*(7), 071005. DOI: 10.1088/2515-7620/ad5ad6

Bouta, A., Laaribya, S., Brache, O., & Dindaroglu, T. (2024). Geospatial modelling of forest fire risk and mitigation strategies in Moroccan Mediterranean regions. *International Journal of Global Warming*. Advance online publication. DOI: 10.1504/IJGW.2024.142423

Bowman, D. M. J. S., Kolden, C. A., Abatzoglou, J. T., Johnston, F. H., van der Werf, G. R., & Flannigan, M. (2020). Vegetation fires in the anthropocene. *Nature Reviews. Earth & Environment*, *1*(10), 500–515. DOI: 10.1038/s43017-020-0085-3

Bowring, S. P. K., Li, W., Mouillot, F., Rosan, T. M., & Ciais, P. (2024). Road fragment edges enhance wildfire incidence and intensity, while suppressing global burned area. *Nature Communications*, *15*(1), 9176. DOI: 10.1038/s41467-024-53460-6 PMID: 39448625

Cabeza, L. F., Chàfer, M., & Mata, É. (2020). Comparative Analysis of Web of Science and Scopus on the Energy Efficiency and Climate Impact of Buildings. *Energies*, *13*(2), 409. DOI: 10.3390/en13020409

Chriha, S., & Sghari, A. (2013). Les incendies de forêt en Tunisie. Séquelles irréversibles de la révolution de 2011. *Méditerranée. Revue géographique des pays méditerranéens/Journal of Mediterranean geography*, (121), 87-93.

Chuvieco, E., Mouillot, F., van der Werf, G. R., San Miguel, J., Tanase, M., Koutsias, N., García, M., Yebra, M., Padilla, M., Gitas, I., Heil, A., Hawbaker, T. J., & Giglio, L. (2019). Historical background and current developments for mapping burned area from satellite Earth observation. *Remote Sensing of Environment, 225*, 45–64. DOI: 10.1016/j.rse.2019.02.013

Cobo, M. J., López-Herrera, A. G., Herrera-Viedma, E., & Herrera, F. (2011). Science mapping software tools: Review, analysis, and cooperative study among tools. *Journal of the American Society for Information Science and Technology, 62*(7), 1382–1402. DOI: 10.1002/asi.21525

Département des eaux et Eaux et Forêts. (2020). *Stratégie nationale de gestion intégrée des incendies de forêt 2020–2030.*

Donthu, N., Kumar, S., Mukherjee, D., Pandey, N., & Lim, W. M. (2021). How to conduct a bibliometric analysis: An overview and guidelines. *Journal of Business Research, 133*, 285–296. DOI: 10.1016/j.jbusres.2021.04.070

El Amarty, F., Chakir, A., Purohit, S., Elhag, A., & Taleb, M. A. (2025). Comparative evaluation of machine learning models for forest fire prediction in Morocco. *Earth Systems and Environment.* Advance online publication. DOI: 10.1007/s41748-025-00823-8

El Maachi, S. E., Saadane, R., Abdellah, A., et Jamal, D. (2024). Navigating change: enhancing transportation resilience in fire-prone Moroccan areas. DOI: 10.1109/WFPST58552.2024.00030

El Mazi, M., Bouhlal, A., Saber, E.-R., Hmamouchi, M., & Ait Haddou, Y. (2025). Effects of wildfire severity on the soil properties of Mediterranean forest ecosystems in Morocco. *Euro-Mediterranean Journal for Environmental Integration.* Advance online publication. DOI: 10.1007/s41207-025-00896-x

El Mazi, M., Boutallaka, M., Saber, E.-R., Chafik, N., & Hmamouchi, M. (2024). Forest fire risk modeling in Mediterranean forests of Morocco using remote sensing and geospatial tools. *Euro-Mediterranean Journal for Environmental Integration.* Advance online publication. DOI: 10.1007/s41207-024-00591-3

Ellegaard, O., & Wallin, J. A. (2015). The bibliometric analysis of scholarly production: How great is the impact? *Scientometrics, 105*(3), 1809–1831. DOI: 10.1007/s11192-015-1645-z PMID: 26594073

European Commission. (2019). *Peer review report Tunisia 2018.* Directorate-General for European Civil Protection and Humanitarian Aid Operations (DG ECHO).

FAO. (2006). *Fire management: voluntary guidelines. Principles and strategic actions* (Fire Management Working Paper 17). Food and Agriculture Organization of the United Nations.

Feng, Q., Li, Y., Zhang, H., & Chen, X. (2024). Interdisciplinary perspectives on forest ecosystems and their functions in hydrology, climate and biodiversity. *Environmental Reviews*. Advance online publication. DOI: 10.1139/er-2024-0010

Ford, A. E. S., Coughlan, M. R., & Meentemeyer, R. K. (2021). modelling human–fire interactions: Combining alternative perspectives and approaches. *Frontiers in Environmental Science*, *9*, 649835. DOI: 10.3389/fenvs.2021.649835

Gallo Granizo, C., Bacciu, V., Eden, J., Hernández Paredes, E., García Feced, C., Bedhief, S., Oom, D., Goldammer, J., Dieppois, B., Belen, İ., Mitri, G., Turhan, Ü., Ascoli, D., Alfonso, L., & San-Miguel-Ayanz, J. (2025). fire management in the Mediterranean region. In *The status of Mediterranean forests 2025* (pp. 83-103). Food and Agriculture Organization of the United Nations.

Ghani, N. A., Teo, P. C., Ho, T. C., Choo, L. S., Kelana, B. W. Y., Adam, S., & Ramliy, M. K. (2022). Bibliometric analysis of global research trends on higher education internationalization using Scopus database: Towards sustainability of higher education institutions. *Sustainability (Basel)*, *14*(14), 8810. DOI: 10.3390/su14148810

Giglio, L., Boschetti, L., Roy, D. P., Humber, M. L., & Justice, C. O. (2018). The Collection 6 MODIS burned area mapping algorithm and product. *Remote Sensing of Environment*, *217*, 72–85. DOI: 10.1016/j.rse.2018.08.005 PMID: 30220740

Graus, S., Ferreira, T. M., Vasconcelos, G., & Ortega, J. (2024). Changing Conditions: Global Warming-Related Hazards and Vulnerable Rural Populations in Mediterranean Europe. *Urban Science (Basel, Switzerland)*, *8*(2), 42. DOI: 10.3390/urbansci8020042

Hantson, S., Pueyo, S., & Chuvieco, E. (2015). Global fire size distribution is driven by human impact and climate. *Global Ecology and Biogeography*, *24*(1), 77–86. DOI: 10.1111/geb.12246

IPCC. (2021). *Climate Change 2021: The Physical Science Basis*. Cambridge University Press.

Jadouli, A., & El Amrani, C. (2024). *Advanced wildfire prediction in Morocco: development of a geospatial machine learning framework*. IEEE., DOI: 10.1109/ACCESS.2024.3516784

Jadouli, A., & El Amrani, C. (2025). *Hybrid parallel architecture integrating FFN, CNN, and MLP models for wildfire prediction*. Lecture Notes in Networks and Systems., DOI: 10.1007/978-3-031-88653-9_16

Jia, S., Zhang, X., & Yin, G. (2024). Hydrological responses to global warming intensification. *Nature, 629*, 137–143. DOI: 10.1038/s41586-025-09047-2

Jones, M. W., Abatzoglou, J. T., Veraverbeke, S., Andela, N., Lasslop, G., Forkel, M., Smith, A. J. P., Burton, C., Betts, R. A., van der Werf, G. R., Sitch, S., Canadell, J. G., Santín, C., Kolden, C., Doerr, S. H., & Le Quéré, C. (2022). Global and regional trends and drivers of fire under climate change. *Reviews of Geophysics, 60*(3), Article e2020RG000726. https://doi.org/DOI: 10.1029/2020RG000726

Jumilah, J., Nadi, S., Eko, H., & Friny, N. (2025). Bibliometrics analysis of environmental sustainability issues: Trends, challenges, and future directions. *E3S Web of Conferences, 640*, 01012. https://doi.org/DOI: 10.1051/e3sconf/202564001012

Kawamura, M., Thomas, C. D., Tsurumoto, A., Sasahara, H., & Kawaguchi, Y. (2000). Lotka's law and productivity index of authors in a scientific journal. *Journal of Oral Science*, *42*(2), 75–78. DOI: 10.2334/josnusd.42.75 PMID: 10989588

Keeley, J. E., & Pausas, J. G. (2022). Evolutionary ecology of fire. *Annual Review of Ecology, Evolution, and Systematics*, *53*(1), 203–225. DOI: 10.1146/annurev-ecolsys-102320-095612

Martín-Martín, A., Orduna-Malea, E., Thelwall, M., & Delgado López-Cózar, E. (2018). Google Scholar, Web of Science, and Scopus: A systematic comparison. *Journal of Informetrics*, *12*(4), 1160–1177. DOI: 10.1016/j.joi.2018.09.002

Mauri, E., Hernández Paredes, E., Núñez Blanco, I., & García Feced, C. (2023). *Key recommendations on wildfire prevention in the Mediterranean*. European Forest Institute., DOI: 10.36333/rs6

Meddour-Sahar, O. (2015). Wildfires in Algeria: Problems and challenges. *iForest - Biogeosciences and Forestry, 8*(6), 818–826. https://doi.org/DOI: 10.3832/ifor1279-007

Meddour-Sahar, O., González-Cabán, A., Meddour, R., & Derridj, A. (2012). Wildfire management policies in Algeria: Present and future needs. In *Proceedings of the Fourth International Symposium on Fire Economics, Planning, and Policy: Climate Change and Wildfires* (pp. 382–395).

Mharzi Alaoui, H., Assali, F., Rouchdi, M., Tahiri, D., Lahssini, S., Aafi, A., & Moukrim, S. (2017). Cartographie de l'intensité du feu et de la sensibilité aux feux de cimes en forêts méditerranéennes. Cas de la province de Chefchaouen au nord du Maroc. [La Terre et La Vie]. *Revue d'Écologie*, *72*(4), 387–409. DOI: 10.3406/revec.2017.1900

MITECO. (2022). *Strategic guidelines for wildland fire management*. Ministerio para la Transición Ecológica y el Reto Demográfico.

Moreira, F., Ascoli, D., Safford, H., Adams, M. A., Moreno, J. M., Pereira, J. M. C., Catry, F. X., Armesto, J., Bond, W., González, M. E., Curt, T., Koutsias, N., McCaw, L., Price, O., Pausas, J. G., Rigolot, E., Stephens, S., Tavşanoğlu, C., Vallejo, V. R., & Fernandes, P. M. (2020). Wildfire management in Mediterranean-type regions: Paradigm change needed. *Environmental Research Letters*, *15*(1), 011001. DOI: 10.1088/1748-9326/ab541e

Pan, Y., Birdsey, R. A., Fang, J., Houghton, R., Kauppi, P. E., Kurz, W. A., Phillips, O. L., Shvidenko, A., Lewis, S. L., Canadell, J. G., Ciais, P., Jackson, R. B., Pacala, S. W., McGuire, A. D., Piao, S., Rautiainen, A., Sitch, S., & Hayes, D. (2011). A large and persistent carbon sink in the world's forests. *Science*, *333*(6045), 988–993. DOI: 10.1126/science.1201609 PMID: 21764754

Peñuelas, J., & Sardans, J. (2021). Global change and forest disturbances in the Mediterranean Basin: Breakthroughs, knowledge gaps, and recommendations. *Forests*, *12*(5), 603. DOI: 10.3390/f12050603

Pyne, S. J. (2019). *Fire: A brief history* (2nd ed.). University of Washington Press. DOI: 10.1515/9780295746197

San-Miguel-Ayanz, J., Durrant, T., Boca, R., Maianti, P., Liberta', G., Oom, D., Branco, A., De Rigo, D., Suarez-Moreno, M., Ferrari, D., Roglia, E., Scionti, N., Broglia, M., & Sedano, F. (2025). *Advance report on forest fires in Europe, Middle East and North Africa 2024* (JRC141505). Publications Office of the European Union. https://data.europea.eu/doi/10.2760/1264626, JRC141505

Sarris, D., Christopoulou, A., Angelonidi, E., Koutsias, N., Fulé, P. Z., & Arianoutsou, M. (2014). Increasing extremes of heat and drought associated with recent severe wildfires in southern Greece. *Regional Environmental Change*, *14*(3), 1257–1268. DOI: 10.1007/s10113-013-0568-6

Seddouki, M., Benayad, M., Aamir, Z., Tahiri, M., Maanan, M., & Rhinane, H. (2022). Using machine learning coupled with remote sensing for forest fire susceptibility mapping: Case study Tetouan Province, northern Morocco. *The International Archives of the Photogrammetry, Remote Sensing and Spatial Information Sciences, XLVIII-4*(W6), 333–2023.

Serbouti, S., Ettaqy, A., Boukcim, H., El Mderssa, M., El Ghachtouli, N., & Abbas, Y. (2023). Forests and woodlands in Morocco: Review of historical evolution, services, priorities for conservation measures and future research. *International Forestry Review, 25*(1), 121–145. DOI: 10.1505/146554823836838745

Tedim, F., Leone, V., Amraoui, M., Bouillon, C., Coughlan, M. R., Delogu, G. M., Fernandes, P. M., Ferreira, C., McCaffrey, S., McGee, T. K., Parente, J., Paton, D., Pereira, M. G., Ribeiro, L. M., Viegas, D. X., & Xanthopoulos, G. (2018). Defining Extreme Wildfire Events: Difficulties, Challenges, and Impacts. *Fire (Basel, Switzerland), 1*(1), 9. DOI: 10.3390/fire1010009

Touchan, R., Anchukaitis, K. J., Meko, D. M., Attalah, S., Baisan, C., & Aloui, A. (2008). Long-term context for recent drought in northwestern Africa. *Geophysical Research Letters, 35*(13), L13705. Advance online publication. DOI: 10.1029/2008GL034264

Turco, M., Rosa-Cánovas, J. J., Bedia, J., Jerez, S., Montávez, J. P., Llasat, M. C., & Provenzale, A. (2018). Exacerbated fires in Mediterranean Europe due to anthropogenic warming projected with non-stationary climate-fire models. *Nature Communications, 9*(1), 3821. DOI: 10.1038/s41467-018-06358-z PMID: 30279564

Turco, M., von Hardenberg, J., AghaKouchak, A., Llasat, M. C., Provenzale, A., & Trigo, R. M. (2017). On the key role of droughts in the dynamics of summer fires in Mediterranean Europe. *Scientific Reports, 7*(1), 81. Advance online publication. DOI: 10.1038/s41598-017-00116-9 PMID: 28250442

Van Eck, N. J., & Waltman, L. (2010). Software survey: VOSviewer, a computer program for bibliometric mapping. *Scientometrics, 84*(2), 523–538. DOI: 10.1007/s11192-009-0146-3 PMID: 20585380

Wunder, S., Calkin, D. E., Charlton, V., Feder, S., Martínez de Arano, I., Moore, P., Rodríguez y Silva, F., Tacconi, L., & Vega-García, C. (2021). Resilient landscapes to prevent catastrophic forest fires: Socioeconomic insights towards a new paradigm. *Forest Policy and Economics, 128*, 102458. DOI: 10.1016/j.forpol.2021.102458

Zidane, I. E., Lhissou, R., Ismaili, M., Bouli, A., & Mabrouki, M. (2021). Characterization of fire severity in the Moroccan Rif using Landsat-8 and Sentinel-2 satellite images. *International Journal on Advanced Science, Engineering and Information Technology, 11*(1), 72–83. DOI: 10.18517/ijaseit.11.1.10342

Zuo, F., Yao, Q., Shi, L., Wang, Z., Bai, M., Fang, K., Guo, F., Yuan, L., & Zhang, W. (2024). Research on wildfire and soil water: A bibliometric analysis from 1990 to 2023. *Fire (Basel, Switzerland), 7*(12), 434. DOI: 10.3390/fire7120434

Section 3
Climate Impacts and Environmental Processes

Chapter 7

A Dynamic Approach to Soil Erosion Modeling for Watershed Management:

Case Study of the N'Fiss Basin, Morocco

Hicham Mharzi Alaoui
http://orcid.org/0000-0003-2226-5133
Department of Climate and Environnemental Risks, ANEF, Morocco

Majdouline Belhouari
http://orcid.org/0009-0006-4241-5361
Department of Climate and Environnemental Risks, ANEF, Morocco

Zineb Loukili
National Agency for Water and Forest, Morocco

Youness Bouziani
National Agency for Water and Forest, Morocco

Fouad Assali
National Agency for Water and Forest, Morocco

Hicham Hajji
IAV Hassan II, Morocco

Nora Hammouda
Rural Development Directorate, Morocco

ABSTRACT

Soil erosion is a major environmental challenge in Mediterranean mountainous watersheds, especially under increasing climatic variability and anthropogenic pressure. This chapter presents an integrated and dynamic framework for soil erosion assessment and prioritization, combining RUSLE-based modeling, field-based validation, and SMART multi-criteria decision analysis, applied to the N'Fis watershed (Morocco), upstream of the Lalla Takerkoust Dam. Results reveal strong spatial

DOI: 10.4018/979-8-2600-0783-9.ch007

heterogeneity, with soil loss rates ranging from <3 to >22.7 $t{\cdot}ha^{-1}{\cdot}yr^{-1}$*. Field surveys highlight marked local contrasts related to erosion dynamics, vegetation degradation, land use, and infrastructure exposure. SMART-based multi-criteria analysis integrates biophysical and socio-economic factors into a unified prioritization framework, enabling refined identification of priority intervention zones. The proposed approach offers a robust, scalable, and transferable decision-support tool for sustainable watershed management, dam protection, and climate adaptation in semi-arid and mountainous regions.*

I. INTRODUCTION

Soil erosion by water is among the most critical forms of land degradation worldwide, particularly in semi-arid and mountainous environments, where climatic variability and anthropogenic pressures strongly interact. Over recent decades, watershed degradation has intensified under the combined effects of climate change, increasing land-use pressure, deforestation, and inadequate soil and water conservation practices (FAO & ITPS, 2015; Borrelli et al., 2017). These interacting drivers have accelerated erosion processes, leading to severe on-site and off-site impacts, including soil fertility loss, sediment transport, downstream flooding, and widespread reservoir siltation (Panagos et al., 2015a; Cerdan et al., 2010).

In Mediterranean and North African contexts, water-induced soil erosion represents a major environmental, economic, and social challenge. The increasing frequency and intensity of extreme rainfall events, attributed to climate change, exacerbate runoff generation and sediment mobilization, particularly in fragile mountainous watersheds (Labbaci et al., 2021, Nearing et al., 2017). One of the most critical consequences of this process is the progressive siltation of reservoirs, which drastically reduces their storage capacity, operational efficiency, and lifespan (Panagos et al., 2015b). In Morocco, where dams play a strategic role in drinking water supply, irrigation, flood regulation, and energy production, sedimentation caused by upstream erosion has become a major concern for sustainable water resources management (Aht-Rising, 2016, Markhi et al., 2019).

Despite the magnitude of this challenge, most watershed management strategies and erosion assessment studies still rely on static modeling approaches. Empirical models such as the Revised Universal Soil Loss Equation (RUSLE) are commonly applied using long-term average climatic and land-use conditions, producing erosion maps that remain unchanged over extended periods (Renard et al., 1997; Kouli et al., 2009; Djoukbala et al., 2019). However, soil erosion is inherently a dynamic process, governed by temporal variations in rainfall intensity, vegetation cover, soil properties, and human activities (Morgan et al., 1998; Nearing et al., 2017). Static representations therefore fail to capture short-term variability and long-term trends, limiting the capacity of decision-makers to anticipate, monitor, and effectively respond to evolving erosion patterns, particularly under rapidly changing climatic and land-use regimes (Panagos & Katsoyiannis, 2019).

Furthermore, a significant proportion of watershed management plans currently implemented in Morocco are based on studies conducted several decades ago, in some cases dating back to the mid-1990s. While these early investigations provided valuable baseline knowledge, their continued application without periodic updating neglects profound changes in climatic conditions, land-use dynamics, population pressure, and socio-economic development (Boardman et al., 2003; Itzkin et al., 2021). As a result, many existing erosion control interventions are no longer optimally located, adequately designed, or sufficiently effective.

In this context, dynamic modeling of water-induced soil erosion emerges as both a scientific necessity and an operational imperative. By integrating spatio-temporal variability and enabling continuous updating of erosion risk maps, dynamic approaches provide a more realistic representation of erosion processes (Nearing et al., 2017; Panagos & Katsoyiannis, 2019). They support adaptive watershed management, facilitate the identification of priority intervention zones, and enable the optimization of soil and water conservation strategies under changing environmental conditions (Borrelli et al., 2017).

This chapter focuses on the N'Fiss watershed, located upstream of the Lalla Takerkoust Dam in central Morocco. This mountainous basin is particularly vulnerable to erosion due to its steep slopes, fragile lithology, heterogeneous soils, and increasing pressures from agricultural expansion, grazing, and infrastructure development (Markhi et al., 2019). Sedimentation of the Lalla Takerkoust reservoir constitutes a major threat to regional water security, irrigation systems, and socio-economic activities. Moreover, observed forest degradation and agricultural land deterioration within the watershed are largely attributed to insufficient soil conservation practices, further disrupting the hydrological balance and sediment dynamics (Boardman et al., 2003; Itzkin et al., 2021). The N'Fiss River is also known for its devastating floods, particularly the event of October 28, 1999 (with a peak discharge of 1,575 m^3/s), which caused considerable damage, including the destruction of agricultural land due to severe erosion, siltation of dam reservoirs, and flooding of road infrastructure.

The general objective of this chapter is to contribute to the protection of the Lalla Takerkoust reservoir and its surrounding environment through the development and application of a dynamic soil erosion modeling and decision-support framework. Specifically, this work aims to:

- Dynamically model soil erosion processes within the N'Fiss watershed, integrating spatio-temporal variability and recent Earth observation data;
- Develop a generic and transferable tool applicable to other watersheds, enabling objective prioritization and improved decision-making for soil and water conservation interventions (Panagos & Katsoyiannis, 2019; Padarian et al., 2020).

The main scientific contributions of this study are twofold: (i) the implementation of a dynamic, multi-source erosion modeling framework in a critical Moroccan watershed threatened by sedimentation and ecological degradation, and (ii) the development of an operational, scalable, and transferable methodology that supports evidence-based watershed management and climate-resilient land-use planning at regional and national scales.

II. STUDY AREA

The study area corresponds to the N'Fis watershed, a major sub-basin of the Tensift hydrological basin in central Morocco. The Tensift basin covers approximately 31,827 km^2 and is characterized by highly contrasted topography, with altitudes ranging from sea level to 4,167 m in the High Atlas Mountains. It is bounded to the south by the crest line of the High Atlas, to the north by the Precambrian massif of Jbilet, to the east by the watershed divide separating it from the Tessaout basin, and to the west by the Atlantic Ocean (Markhi et al., 2019).

The N'Fis watershed drains an area of about 1,380 km^2 (138,000 ha) at the level of the Lalla Takerkoust Dam, which represents the downstream control point of sediment and water fluxes. The main river extends over approximately 152 km before joining the Tensift River, and the hydrographic network reaches a total length of nearly 1,100 km, reflecting a dense and highly structured drainage system. This network plays a crucial role in sediment routing from mountainous upstream areas toward the downstream reservoir.

II.1- Topography and Geomorphology:

The watershed exhibits strong altitudinal gradients, with elevations ranging from approximately 600 m to over 4,000 m, generating steep slopes, long hillslopes, and highly dissected terrain. These geomorphological conditions strongly favor surface runoff concentration and sediment transport, making topography one of the dominant controls of soil erosion processes (Markhi et al., 2019; Desmet & Govers, 1996; Panagos et al., 2015a).

Upstream sectors, located in the High Atlas Mountains, are characterized by steep slopes (>30%), narrow valleys, and fragile lithological formations, mainly composed of schists, marls, and limestones. These materials exhibit variable resistance to weathering, with schistose formations being particularly prone to mechanical disintegration and erosion (Markhi et al., 2019; Cerdan et al., 2010). Downstream areas display gentler slopes and wider valley floors, where sediment deposition and agricultural activities are more dominant.

II.2- Climate and Hydrological Regime:

The climate of the N'Fis watershed ranges from semi-arid to sub-humid Mediterranean, with strong spatial and temporal variability driven by altitude and orographic effects. Mean annual precipitation varies from less than 250 mm in downstream lowlands to over 1,000 mm in mountainous zones, with rainfall mainly concentrated between October and April (Aht-Rising, 2016).

Precipitation events are often short, intense, and highly erosive, particularly during autumn storms, generating flash floods and intense sediment transport (Aht-Rising, 2016). Temperature regimes also exhibit strong altitudinal gradients, with hot summers in lowlands and cold winters in high-altitude zones, where snowfall may significantly contribute to seasonal runoff, further enhancing hydrological variability and erosion risk (Markhi et al., 2019).

II.3- Soils and Land Cover

Soils within the N'Fis watershed are highly heterogeneous, reflecting variations in parent material, topography, and land use. In upstream areas, soils are generally shallow, skeletal, and poorly developed, often characterized by low organic matter content and limited water-holding capacity, making them highly vulnerable to detachment and transport (Markhi et al., 2019; FAO & ITPS, 2015). In contrast, deeper alluvial soils dominate valley bottoms, where agricultural activities are concentrated.

Land cover is dominated by forest ecosystems, rangelands, and traditional agriculture. Forested areas, mainly composed of holm oak (*Quercus ilex*), juniper, and cedar species, provide partial protection against erosion, although degradation due to overgrazing, wood harvesting, and climate stress is increasingly observed (Vrieling, 2006; Boardman et al., 2003). Agricultural practices mainly involve terraced cultivation along watercourses and slopes, aimed at reducing runoff velocity and conserving soil moisture. However, the progressive abandonment of terraces and inadequate maintenance have reduced their effectiveness, contributing to enhanced erosion (Markhi et al., 2019; Itzkin et al., 2021).

II.4 Anthropogenic Pressures and Infrastructure

The watershed is subjected to increasing anthropogenic pressure, driven by population growth, agricultural expansion, grazing, road construction, and infrastructure development (Boardman et al., 2003; Itzkin et al., 2021). Several national and regional roads, as well as dense networks of rural tracks, cross the basin, locally modifying natural drainage patterns and accelerating soil degradation. Sediment inputs from road works and poorly managed construction activities are directly discharged into watercourses, intensifying downstream sediment transport (Baghdadi et al., 2007; Vrieling, 2006).

Rural settlements (douars) are scattered throughout the basin, particularly along valley floors and near water resources, increasing local vulnerability to erosion and flooding hazards. The Lalla Takerkoust Dam, located at the outlet of the N'Fis watershed, represents a strategic hydraulic infrastructure for drinking water supply, irrigation, and energy production. Its progressive siltation due to upstream erosion constitutes a major threat to regional water security and socio-economic development (Markhi et al., 2019).

II.5 Environmental and Socio-Economic Stakes

The N'Fis watershed represents a critical socio-ecological system, where land degradation directly affects agricultural productivity, forest sustainability, water resources, and rural livelihoods (FAO & ITPS, 2015). Accelerated erosion not only leads to soil fertility loss but also generates high sediment loads that threaten downstream hydraulic infrastructures and irrigation schemes (Markhi et al., 2019; Cerdan et al., 2010).

Protecting the watershed against erosion is therefore essential for sustaining agricultural production, preserving forest ecosystems, extending the operational lifespan of the Lalla Takerkoust Dam reservoir, and enhancing resilience to climate variability.

III. SOIL EROSION MODELS OVERVIEW

Soil erosion is an inherently dynamic process, driven by temporal variations in rainfall, runoff, soil properties, land cover, and human activities (Morgan et al., 1998; Nearing et al., 2017). Dynamic water erosion reflects not only the intensity and frequency of erosive events but also the changing conditions of the watershed over time. This temporal dimension is crucial for understanding the evolution of sediment transport, the formation of rills and gullies, and the resulting impacts on soil fertility and downstream reservoirs (Cerdan et al., 2010; Boix-Fayos et al., 2006). In mountainous and semi-arid watersheds, such as N'Fiss in Morocco, these variations are particularly pronounced due to steep slopes, heterogeneous soils, and sporadic but intense rainfall events (Meusburger et al., 2012).

The main physical processes involved in water-induced soil erosion include:

- Surface runoff generated by rainfall events and snowmelt, which mobilizes soil particles (Nearing et al., 2017);
- Rill and gully formation, which concentrate flow and accelerate local erosion (Cerdan et al., 2010; Boix-Fayos et al., 2006);
- Sediment transport both on the surface and within the stream network, contributing to downstream deposition (Morgan et al., 1998; Schmitter et al., 2010);
- Soil–rain–vegetation interactions, which determine infiltration, protection against raindrop impact, and sediment detachment (Vrieling, 2006; Zhou et al., 2008);
- Anthropogenic influences, such as deforestation, agriculture, and construction, which modify land cover and increase susceptibility to erosion (Boardman et al., 2003; Itzkin et al., 2021).

Understanding these processes is essential for the selection of an appropriate modeling approach. Over the past decades, three main families of methods have been developed to quantify and predict soil erosion (Panagos & Katsoyiannis, 2019):

III.1. Empirical or Statistical Models

These approaches rely on observed correlations between key factors such as rainfall, slope, soil characteristics, and measured soil loss (Wischmeier & Smith, 1978; Renard et al., 1997). Their main advantages are simplicity and low data requirements, making them suitable for preliminary analyses and regional-scale applications (Kouli et al., 2009; Djoukbala et al., 2019). However, they often fail to represent complex erosion processes such as gully erosion, and their predictive accuracy may decrease when applied outside their original calibration environments (Cerdan et al., 2010; Naipal et al., 2015).

III.2. Process-Based Models (Physical Models)

These models explicitly simulate the physical mechanisms of runoff generation, sediment detachment, transport, and deposition (Morgan et al., 1998; Arnold et al., 1998). They provide a detailed and mechanistic understanding of erosion processes and allow testing of conservation measures and climate change scenarios (Arnold et al., 2012). However, their application is often limited by high data requirements, computational intensity, and complex calibration procedures, particularly in data-scarce environments.

III.3. Semi-empirical Models

Remote sensing platforms, including multispectral, SAR, and LiDAR systems, enable repeated, non-destructive, and high-resolution monitoring of topography, soil moisture, and vegetation cover (Jensen, 2000; Baghdadi et al., 2007). GIS methods and digital elevation models (DEM) are indispensable for visualizing spatial patterns of erosion, extracting topographic parameters, and integrating soil and land cover databases such as SoilGrids and the Harmonized World Soil Database (Hengl et al., 2017; FAO et al., 2012; Vrieling, 2006). These technologies have significantly enhanced the spatial explicitness and temporal monitoring capabilities of erosion modeling (Gitas, 2009; Padarian et al., 2020).

Although several modeling approaches exist, including physically based models such as WEPP and SWAT, the Revised Universal Soil Loss Equation (RUSLE) was selected for this study due to its simplicity, robustness, limited data requirements, and proven applicability in Mediterranean and semi-arid environments (Renard et al., 1997; Kouli et al., 2009; Djoukbala et al., 2019; Rajabi et al., 2022). Numerous studies have demonstrated the suitability of RUSLE for erosion assessment in North African and Moroccan contexts, particularly when integrated within GIS environments (Markhi et al., 2019; Gitas, 2009) (Table 1).

Table 1. Comparative analysis of soil erosion models

Criterion	Empirical Models	Process-Based Models	Semi-empirical Models
Precision	Approximate / General	High (after calibration)	Approximate
Data requirements	Limited / Simple	Complex	Sensor-dependent
Ideal scale	Regional / Preliminary	Watershed /Plot/field	Multi-scale
Cost / complexity	Low	High	Variable

Moreover, RUSLE can be adapted for dynamic applications by incorporating time-series rainfall products, updated land cover datasets, and high-resolution topographic inputs, allowing for periodic recalculation of erosion risk maps (Naipal et al., 2015; Panagos & Katsoyiannis, 2019; Nearing et al., 2017). This dynamic implementation enhances the capacity of RUSLE to support adaptive watershed management and decision-making processes.

IV. DATA AND MATERIALS

IV.1. Overview of Data Sources

The dynamic assessment of water-induced soil erosion in the N'Fiss watershed was based on the integration of multi-source spatial, climatic, and environmental datasets. These datasets were selected to capture both the spatial heterogeneity of erosion-controlling factors and their temporal variability, which is essential for dynamic erosion modeling (Vrieling, 2006; Panagos & Katsoyiannis, 2019).

The data used in this study were grouped into four main categories: (i) climatic data, (ii) topographic data, (iii) land use and vegetation data, and (iv) soil-related data. All datasets were harmonized in terms of spatial reference system, resolution, and temporal coverage prior to analysis, ensuring methodological consistency and reliable spatial integration (Jensen, 2000; Padarian et al., 2020).

IV.2. Climatic and Rainfall Data

Rainfall constitutes the primary driving force of water-induced soil erosion, as it controls both runoff generation and sediment detachment. To capture the temporal variability of rainfall erosivity, precipitation data were obtained from meteorological stations and gridded datasets provided by the Moroccan Meteorological Agency and the National Agency for Forest and Water, covering the period 2010-2025. These datasets provide monthly rainfall records suitable for analyzing interannual variability and extreme rainfall events (Table 2).

Rainfall data were processed to derive rainfall erosivity indicators required for dynamic erosion modeling. The use of time-series rainfall data allowed the identification of temporal fluctuations in erosive rainfall, which are often masked when long-term averages are applied. This approach improves the representation of short-term climatic variability and enhances the predictive capacity of erosion models under changing climatic conditions.

Table 2. Characteristics of rainfall monitoring stations in the N'Fiss watershed

Station	Station No.	Mean Rainfall (mm)	X	Y	Elevation Z (m)
Lalla Takerkoust Area	8968	275	239,400	80,200	636
Amezmiz	1009	485	228,80	71,600	1120
Tizigui	8664	473	234,10	469,50	1410
Tizi Ghouran	8468	507	226,000	70,000	1150
Ouirgane	5913	453	244,65	67,730	950
Talat N'os	7608	325	239,55	458,550	1240
Ijoukak	4328	352	236,90	446,70	1150
Idni	4284	626	223,25	437,90	1670
Aghbar	1232	624	216,20	434,75	1750
Tizin'Test	8512	532	215,60	433,70	2100
N'Kouris	4299	230	238,90	453,800	1100
Imin El Hammam	4432	393	241,40	72,40	770
Takerkoust Dam	8969	264	239,50	88,200	630

IV.3. Topographic Data

Topographic characteristics of the watershed were derived from a Digital Elevation Model (DEM) with a spatial resolution of 12.5 m. The DEM was used to compute slope gradient, slope length, flow accumulation, and other morphometric parameters influencing runoff generation and erosion processes.

These topographic derivatives constitute the basis for estimating the topographic factor (LS) and for analyzing spatial patterns of erosion susceptibility across the watershed. High-resolution DEM data significantly improve the representation of complex mountainous terrains, allowing more accurate simulation of slope-driven erosion dynamics.

IV.4. Land Use and Vegetation Cover Data

Land use and land cover (LULC) information was obtained from Sentinel-2 multispectral imagery, processed for multiple dates to capture temporal changes in vegetation cover. Vegetation indices, particularly the Normalized Difference Vegetation Index (NDVI), were calculated to characterize vegetation dynamics and seasonal variability (Jensen, 2000; Vrieling, 2006).

These data were used to dynamically estimate the vegetation cover factor (C), reflecting changes in land cover conditions over time rather than assuming static land use patterns. Numerous studies have demonstrated that the integration of

NDVI-based indicators significantly enhances the accuracy of erosion modeling by better representing vegetation protection effects (Zhou et al., 2008; Gitas, 2009).

IV.5. Soil Data

Due to the limited number of pedological soil profiles available in the study area, soil data used in this research were derived from a combination of local field observations and global soil datasets provided by the SoilGrids platform (Hengl et al., 2017). SoilGrids is based on more than 230,000 georeferenced soil profile observations worldwide, compiled within the WoSIS database, and represents one of the most comprehensive global sources of harmonized soil information (FAO et al., 2012; Hengl et al., 2017).

The SoilGrids framework integrates soil profile observations with a wide range of environmental covariates, including vegetation characteristics, terrain morphology, climatic variables, geological context, and hydrological conditions. These covariates are combined using advanced digital soil mapping and machine learning techniques to estimate and spatially predict soil properties at high spatial resolution (Hengl et al., 2017; Padarian et al., 2020).

This approach enables the estimation and mapping of soil physical and chemical properties at the regional scale, with a sufficiently fine spatial resolution to support watershed-scale erosion modeling. The soil parameters considered in this study include soil organic carbon, total nitrogen, coarse fragments, soil pH (H_2O), cation exchange capacity, bulk density, and soil texture fractions, extracted for multiple depth intervals down to 200 cm.

Although soil properties generally exhibit limited temporal variability compared to climatic or vegetation factors, their spatial heterogeneity plays a critical role in controlling erosion processes (FAO & ITPS, 2015; Cerdan et al., 2010). The integration of SoilGrids data therefore provides a robust and spatially explicit basis for estimating soil erodibility in data-scarce regions, while ensuring methodological consistency and transferability to other watersheds.

IV.6 Soil Erosion Model Framework

The Revised Universal Soil Loss Equation (RUSLE) estimates the average annual soil loss (A, $t \cdot ha^{-1} \cdot yr^{-1}$) as the product of five factors representing the influence of climate (R), soil properties (K), topography (LS), land cover (C), and human practices (P) (Wischmeier & Smith, 1978; Renard et al., 1997):

$$A = R \times K \times LS \times C \times P$$

Each factor was calculated as a raster layer using Google Earth Engine (GEE) and GIS software, allowing spatially explicit modeling across the N'Fiss watershed. This approach enables dynamic updates of soil erosion maps by integrating temporal variations in rainfall and land cover (Table 3).

IV.7 Data Acquisition and Factor Computation

IV.7.1 Rainfall Erosivity Factor (R)

The rainfall erosivity factor (R, MJ·mm·ha^{-1}·h^{-1}·yr^{-1}) quantifies the kinetic energy of precipitation, which drives soil detachment and runoff. Due to the absence of dense hourly rainfall data, R was estimated from monthly and annual precipitation data obtained from the CHIRPS database for the period 2000–2023.

The empirical formula proposed by Arnoldus (1977), adapted for Moroccan conditions, was used:

$$R = \sum_{i=1}^{12} \left[\left(\frac{P_i}{P} \right)^{1.5} \times 100 \right]$$

Where:

- P= annual precipitation (mm)
- P_i= precipitation of month i (mm)

This method allows capturing temporal variability of rainfall for dynamic modeling.

IV.7.2 Soil Erodibility Factor (K)

The soil erodibility factor (K, t·ha·h·ha^{-1}·MJ^{-1}·mm^{-1}) represents the intrinsic susceptibility of soils to erosion. It was calculated using Wischmeier and Smith's equation (1978), which accounts for soil texture, organic matter content, and permeability. Soil data were extracted from the SoilGrids global database at 250 m resolution.

$$K = 0.1317 \times \frac{[2.1 \times 10^{-4} \times M^{1.14} \times (12 - OM) + 3.25 \times (S - 2) + 2.5 \times (P - 3)]}{100}$$

Where:

- M= texture factor (% silt + % very fine sand) × (100 – % clay)

- OM= organic matter content (%)
- S= soil structure class (1 = very fine granular, 2 = fine granular, 3 = medium/ coarse granular, 4 = block/lamellar/massive)
- P= soil permeability

IV.7.3 Topography Factor (LS)

The topographic factor (LS, dimensionless) accounts for slope length and steepness, which strongly influence runoff and soil transport. LS was calculated from a 12.5 m DEM derived from Sentinel-1 data, using the method of Desmet and Govers (1996):

$$LS = \left(\frac{AS}{22.13}\right)^{0.4} \times \left(\frac{\sin\beta}{0.0896}\right)^{1.3}$$

Where:

- AS= specific contributing area (unit area flowing into a given cell)
- β= slope gradient (radians)

IV.7.4 Facteur de couverture et de gestion du sol (C)

The cover-management factor (C) reflects the protective effect of vegetation and land use. It was estimated from NDVI derived from Sentinel-2 images (2023) using the Zhou et al. (2008) relation:

$$C = \exp\left(\left|\frac{a \cdot NDVI}{\beta - NDVI}\right)$$

Where $a$and $\beta$are parameters controlling the shape of the NDVI-C relationship.

IV.7.5 Conservation Practices Factor (P)

In the absence of detailed spatial data on soil and water conservation practices, and following regional studies (Zouagui et al., 2018), a uniform value of P = 1 was assigned across the watershed. This assumption reflects the low implementation of anti-erosive practices in the N'Fiss basin.

Table 3. Description of erosion factors

Type of factor	Factor	Role	Units	Typical Range	Dimensionality
Dynamic	R	Erosive	MJ·mm·ha^{-1}·h^{-1}	0 – ∞	[P][L^{-1}]
	C	Protective	–	0 – 1	–
Static	K	Erosive	t·ha·h·ha^{-1}·MJ^{-1}·mm^{-1}	0 – 0.1	[M][L^{-2}][P^{-1}]
	LS	Erosive	–	0 – 20	–
	P	Protective	–	1 – 2	–

The methodological framework adopted in this study is fundamentally based on the Google Earth Engine (GEE) platform, which constitutes the central pillar of the analytical approach. GEE provides a cloud-based computing environment specifically designed for large-scale geospatial data processing and Earth observation analysis, offering substantial advantages for soil erosion assessment and dynamic monitoring (Jensen, 2000; Vrieling, 2006).

One of the main strengths of GEE lies in its ability to provide direct access to a wide range of real-time and historical geospatial datasets without requiring local downloading or storage. This significantly reduces constraints related to data handling, computational resources, and storage capacity. The platform integrates multi-source datasets, including satellite-derived vegetation indices, precipitation products, digital elevation models, land cover classifications, and soil databases, all of which are essential for dynamic erosion modeling (Hengl et al., 2017; Padarian et al., 2020; Panagos & Katsoyiannis, 2019).

Moreover, GEE offers powerful parallel computing capabilities, enabling the efficient processing of large geospatial datasets and the implementation of robust, spatially explicit RUSLE models. This computational architecture allows the continuous updating of erosion risk maps and supports near real-time monitoring of spatio-temporal changes in erosion drivers, particularly rainfall erosivity and vegetation cover dynamics (Naipal et al., 2015; Nearing et al., 2017; Gitas, 2009).

Beyond its analytical advantages, the use of GEE enables the development of a generic, scalable, and fully automated erosion assessment tool, capable of being applied to any watershed with minimal parameter adjustment. By standardizing data acquisition, preprocessing, modeling workflows, and output generation, this tool can be readily transferred and deployed across multiple watersheds at regional and national scales.

Such a generic tool offers an integrated and coherent framework for the automated retrieval and processing of geospatial datasets, the dynamic computation of RUSLE factors, the real-time updating of erosion risk maps, the integration of field observations and decision-support indicators, and the generation of standardized outputs to support planners and policymakers in informed decision-making.

This approach supports the establishment of a national dynamic soil erosion monitoring and prioritization system, enabling consistent evaluation, spatial comparison, and hierarchical ranking of erosion risk across Moroccan watersheds. Such a system represents a strategic asset for optimizing investments in soil and water conservation, enhancing reservoir protection strategies, and strengthening climate change adaptation policies (Borrelli et al., 2017; FAO & ITPS, 2015; Panagos et al., 2015b).

Furthermore, the modular structure of the proposed framework allows the integration of advanced modeling techniques, including machine learning, deep learning, and scenario-based simulations, which significantly enhance predictive capabilities and long-term planning efficiency (Ghorbanzadeh et al., 2019; Padarian et al., 2020).

Overall, the GEE-based methodological framework developed in this study constitutes a robust, operational, and transferable decision-support system, capable of guiding sustainable watershed management, erosion mitigation strategies, and adaptive land-use planning under changing climatic and socio-economic conditions.

V. MULTI-CRITERIA ANALYSIS FOR SUB-WATERSHED PRIORITIZATION

The confrontation of modeling results with field observations was conducted using a multi-criteria decision analysis (MCDA) framework, aimed at identifying and ranking priority sub-watersheds according to their overall vulnerability to water-induced soil erosion. This integrated approach combines biophysical and anthropogenic criteria, allowing the assessment to move beyond a sole reliance on estimated soil loss and to incorporate key environmental and socio-economic dimensions of erosion risk (Boardman et al., 2003; Panagos & Katsoyiannis, 2019; Itzkin et al., 2021).

The RUSLE-derived soil loss factor (A) was incorporated as one criterion among others, serving as a synthetic indicator of erosion intensity, but not as a stand-alone decision variable. Although soil loss estimates provide valuable quantitative insight into erosion processes, several studies emphasize that erosion vulnerability and management priorities must also consider land cover conditions, erosion dynamics, human pressures, and infrastructure exposure (Boix-Fayos et al., 2006; Cerdan et al., 2010; Borrelli et al., 2017).

Consequently, three main groups of criteria were considered in the prioritization framework:

- Erosion-related factors, including soil loss rate (A factor), differential erosion patterns, and observed erosion dynamics (rill and gully activity, sediment deposition), which reflect both modeled and field-based indicators of erosion processes (Renard et al., 1997; Cerdan et al., 2010; Nearing et al., 2017);

- Vegetation cover and agricultural practices, incorporating vegetation density, forest cover, and dominant agricultural systems, which strongly influence soil protection, infiltration capacity, and runoff generation (Vrieling, 2006; Zhou et al., 2008; Schmitter et al., 2010);
- Infrastructure and territorial assets, including road networks, hydraulic crossing structures, rural settlements, dams, watercourses, and hill lakes, representing critical exposure elements and socio-economic vulnerability to erosion-induced impacts, These infrastructure data were obtained from the OpenStreetMap database (OpenStreetMap Contributors, 2023).

Table 4. Multicritera analysis factors

Analysis Category	**Sub-factor**	**Weight**
Infrastructure and Equipment	Road Network and Ponds	0.10
	Douars	0.10
	Dams	0.15
	Presence of Watercourses	0.10
	Hill Reservoirs	0.02
Erosion Factors	Soil Loss Rate	0.12
	Differential Erosion	0.10
	Erosion Dynamics	0.15
Vegetation Cover and Agricultural Practices	Vegetation Density	0.10
	Legal Nature of Land (DF)	0.03
	Agricultural Practices	0.02
	Protected Areas	0.01

Decision criteria were weighted according to their relative importance using the SMART (Simple Multi-Attribute Rating Technique) methodology, which enables the transformation of heterogeneous qualitative and quantitative indicators into normalized utility scores and their aggregation into a single vulnerability index (Goodwin & Wright, 2014; Panagos & Katsoyiannis, 2019). This weighting procedure ensures logical consistency, transparency, and reduced subjectivity in the decision-making process (Table 4).

Each sub-watershed was evaluated based on spatial datasets derived from remote sensing and GIS analysis, combined with detailed field observations, allowing a comprehensive representation of erosion processes and vulnerability factors (Vrieling, 2006; Jensen, 2000). A global vulnerability score was then computed through weighted aggregation of the criteria, enabling the ranking of sub-watersheds and the identification of priority intervention zones for soil conservation and sustainable watershed management.

This integrated multi-criteria framework enhances the operational relevance of erosion assessment, providing decision-makers with a robust, transparent, and reproducible tool for targeting investments, optimizing soil and water conservation strategies, and improving long-term watershed resilience under changing climatic and socio-economic conditions (FAO & ITPS, 2015).

VI. RESULTS AND DISCUSSION

VI.1 Results of the RUSLE modeling

The results of the RUSLE modeling are presented through a series of spatially explicit maps illustrating the distribution and relative contribution of the main erosion-controlling factors across the N'Fis watershed. These maps provide a comprehensive overview of the spatial variability of rainfall erosivity (R), soil erodibility (K), topographic influence (LS), vegetation cover, and land management (C).

The combined analysis of these factors allows for a better understanding of the interactions between climatic, geomorphological, pedological, and anthropogenic drivers of soil erosion, while highlighting the zones most vulnerable to erosion processes (Cerdan et al., 2010; Boix-Fayos et al., 2006). The following sections interpret the observed spatial patterns in relation to local environmental conditions and land-use dynamics and discuss their implications for soil conservation and sustainable watershed management.

VI.1.1 Rainfall erosivity factor (R)

The R-factor map of the N'Fis watershed (fig.1) exhibits *pronounced spatial variability*, structured into five classes of rainfall erosivity that reflect the heterogeneity of local climatic conditions. R values range from 41.4 $MJ \cdot mm \cdot ha^{-1} \cdot h^{-1} \cdot yr^{-1}$ in the least erosive areas to values exceeding 86 $MJ \cdot mm \cdot ha^{-1} \cdot h^{-1} \cdot yr^{-1}$ in the most exposed zones, with a mean value of approximately 65 $MJ \cdot mm \cdot ha^{-1} \cdot h^{-1} \cdot yr^{-1}$.

This variability highlights the strong influence of rainfall aggressiveness on soil erosion risk and underlines the importance of integrating the spatial heterogeneity of erosivity into erosion assessment and watershed management strategies (Brown & Foster, 1987; Meusburger et al., 2012; Naipal et al., 2015). Similar ranges of rainfall erosivity have been reported in Mediterranean and semi-arid environments, where intense and short-duration precipitation events generate high erosive energy (Kouli et al., 2009; Djoukbala et al., 2019).

The diversity of erosivity classes emphasizes the complexity of climatic controls on erosion processes, particularly the role of local rainfall variability and orographic effects. The cartographic representation provides a clear visualization of areas exposed to high erosion risk driven primarily by rainfall intensity and distribution, notably in upstream mountainous zones (Nearing et al., 2017; Panagos & Katsoyiannis, 2019).

Figure 1. Rainfall erosivity factor (R) in MJ·mm·ha^{-1}·h^{-1}·yr^{-1}

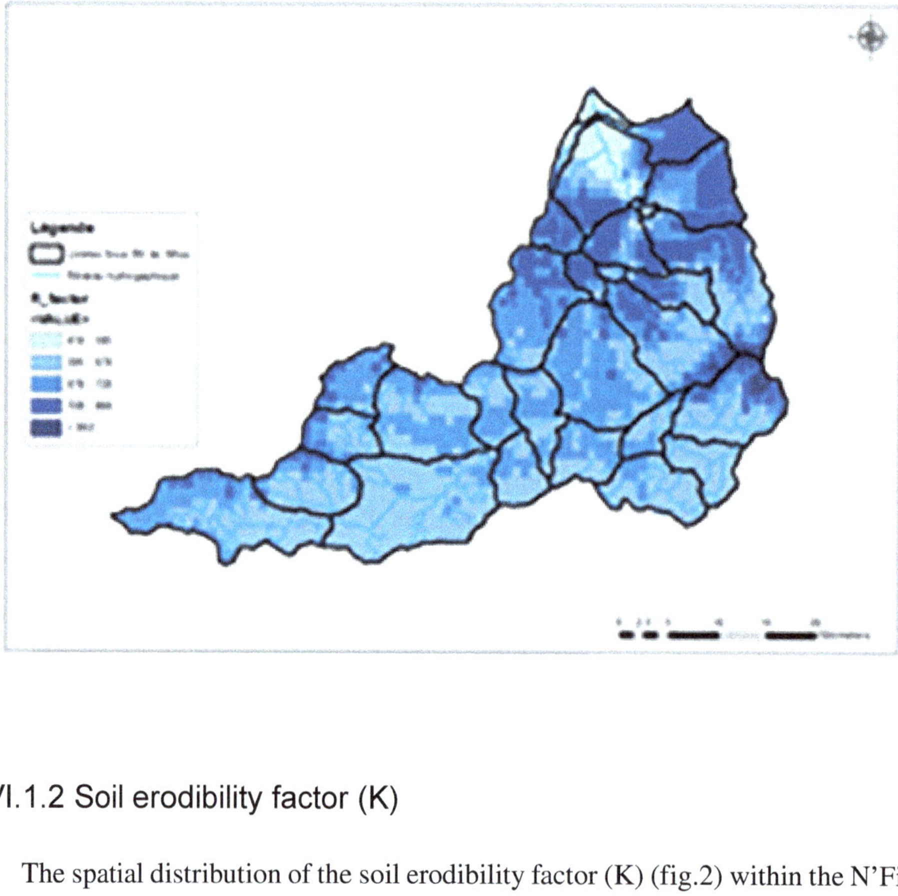

VI.1.2 Soil erodibility factor (K)

The spatial distribution of the soil erodibility factor (K) (fig.2) within the N'Fis watershed reveals marked heterogeneity, with values ranging from 0.015 t·ha·h·ha^{-1}·MJ^{-1}·mm^{-1} for the most erosion-resistant soils to 0.034 t·ha·h·ha^{-1}·MJ^{-1}·mm^{-1} for the most erodible soils. These values were classified into five distinct classes.

Low K values correspond to soils with greater resistance to detachment and transport, while higher values indicate increasing susceptibility to erosion. The mean K value of approximately 0.025 t·ha·h·ha^{-1}·MJ^{-1}·mm^{-1} reflects an overall moderate level of soil erodibility across the watershed. Comparable ranges have been reported in Mediterranean and North African watersheds characterized by shallow, skeletal, and structurally weak soils.

This spatial variability reflects differences in soil texture, organic matter content, structural stability, and permeability, emphasizing the necessity of accounting for pedological heterogeneity when assessing erosion risk (Hengl et al., 2017; Cerdan et al., 2010). The K-factor map therefore provides valuable insight into the spatial distribution of erosion-prone soils, supporting targeted soil conservation strategies.

Figure 2. Soil erodibility factor (K) in t·ha·h / (ha·MJ·mm)

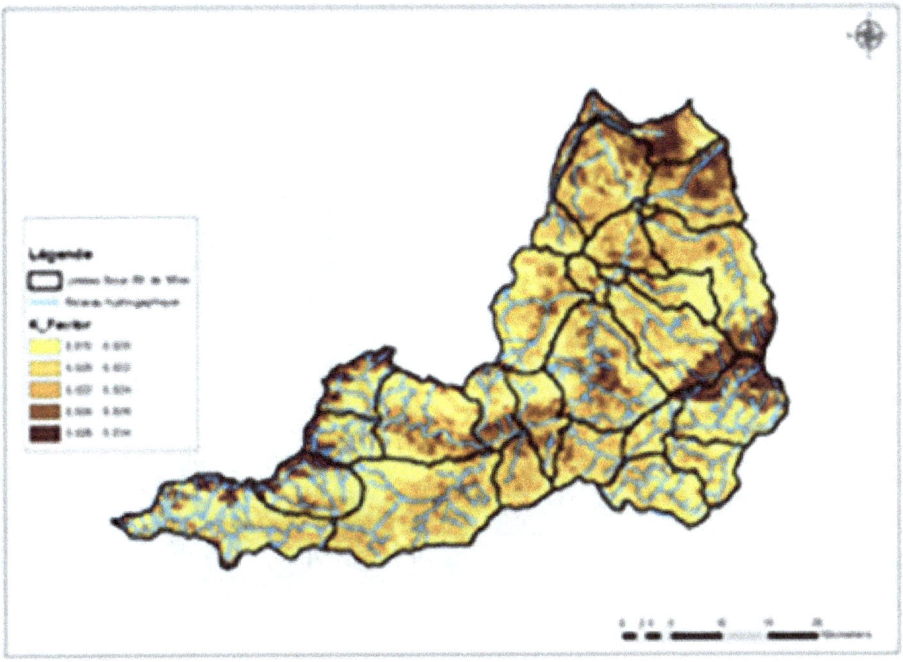

VI.1.3 Topographic factor (LS)

The analysis of the topographic factor (LS) (fig.3) highlights strong contrasts in slope length and steepness across the N'Fis watershed, with values ranging from 0.029 to 62.41. These values were grouped into five classes representing increasing levels of erosion sensitivity.

Low LS values (0.029–6.63) characterize areas with gentle slopes and limited erosion potential, while intermediate classes (6.63–12.99 and 12.99–20.09) indicate moderate to increased sensitivity. Higher values (20.09–28.40) correspond to zones with high erosion susceptibility, and the highest class (28.40–62.41) identifies very highly sensitive areas, mainly located on steep and elongated slopes in the upstream mountainous sectors.

Among all RUSLE factors, LS emerges as the dominant driver of erosion, reflecting the strong control exerted by topography on runoff concentration and sediment transport (Desmet & Govers, 1996; Nearing, 1997; Panagos et al., 2015a). Similar findings have been widely reported in mountainous Mediterranean environments, where steep slopes and long hillslopes significantly amplify erosion processes (Kouli et al., 2009; Djoukbala et al., 2019; Cerdan et al., 2010).

Figure 3. LS factor

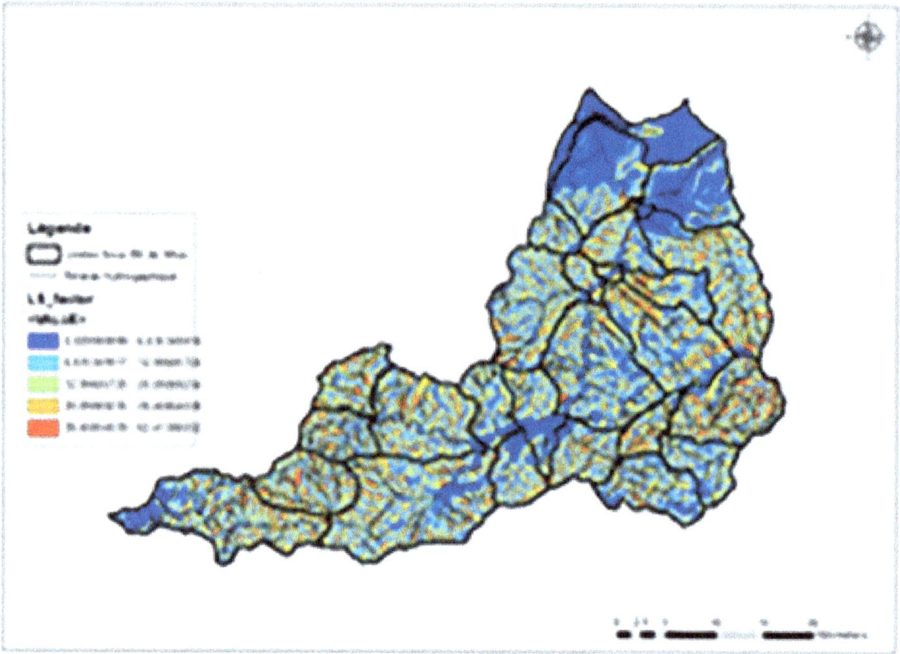

VI.1.4 Cover-management factor (C)

The spatial distribution of the cover-management factor (C) (fig.4) reveals substantial variability in land cover and vegetation density across the N'Fis watershed. Five classes were identified, reflecting different degrees of soil protection.

Very low C values (0–0.17) correspond to dense vegetation cover, primarily forested areas, which provide effective protection against soil erosion. Intermediate values (0.17–0.43 and 0.43–0.62) represent moderate to reduced vegetation cover, including croplands and orchards. Higher values (0.62–0.76) indicate sparse vegetation, while the highest class (0.76–1.59) characterizes bare or poorly vegetated soils that are highly vulnerable to erosion.

These results confirm the critical role of vegetation cover in mitigating soil erosion and demonstrate the negative impact of deforestation, overgrazing, and land degradation on erosion susceptibility (Vrieling, 2006; Zhou et al., 2008; Boardman et al., 2003). Similar spatial patterns have been documented in Mediterranean and mountainous watersheds, where vegetation loss rapidly translates into increased runoff and sediment production (Schmitter et al., 2010; Itzkin et al., 2021).

Figure 4. C factor

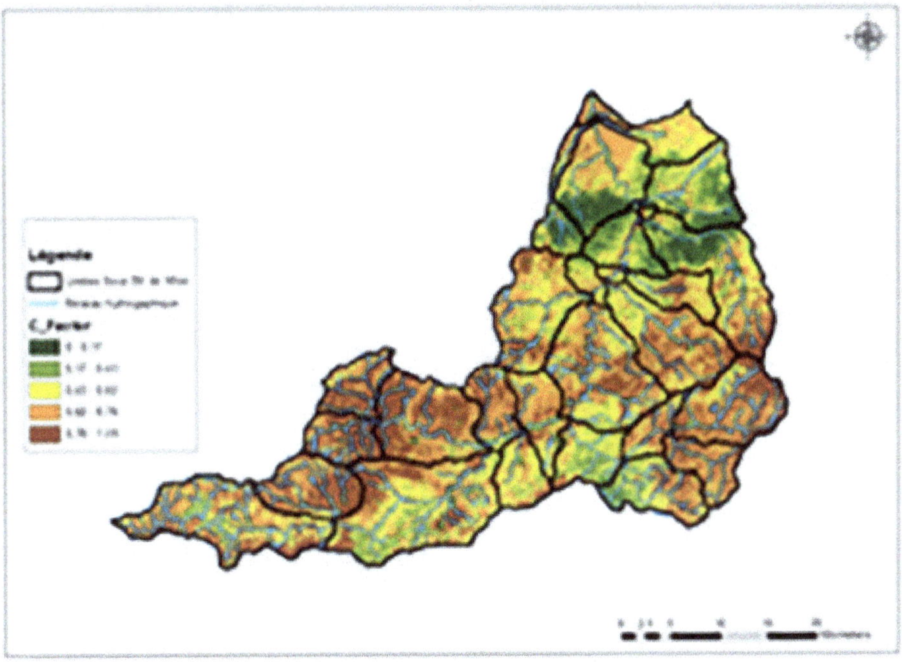

VI.1.5 Estimated soil loss (A-factor)

The spatial distribution of estimated annual soil loss (A-factor) (fig.5), classified into five erosion risk classes, reveals pronounced spatial variability across the N'Fis watershed. Areas with low erosion rates (< 3 t·ha^{-1}·yr^{-1}) account for approximately 16% of the basin, mainly corresponding to dense forest and stable land-use zones.

Moderate erosion levels (3–12.6 t·ha^{-1}·yr^{-1}) dominate the landscape, covering about 43% of the watershed, particularly in agricultural and transitional zones. Higher erosion rates (12.6–22.7 t·ha^{-1}·yr^{-1}) affect roughly 26% of the basin and correspond to areas with steep slopes, degraded vegetation cover, and shallow soils. Areas experiencing high to very high erosion (> 22.7 t·ha^{-1}·yr^{-1}) represent approximately 15% of the watershed and are mainly located in upstream mountainous regions characterized by steep slopes, active gully erosion, and intense rainfall.

These values are consistent with erosion rates reported in similar Mediterranean and semi-arid watersheds, confirming the severe erosion risk faced by mountainous basins in North Africa (Kouli et al., 2009; Djoukbala et al., 2019; Rajabi et al., 2022; Cerdan et al., 2010).

Overall, these results clearly identify erosion hotspots and provide a robust spatial basis for prioritizing soil conservation and watershed management interventions, particularly in zones contributing disproportionately to sediment yield and reservoir siltation (Panagos et al., 2015b; Borrelli et al., 2017; FAO & ITPS, 2015) (Figure 6).

Figure 5. Soil erosion in 10 -1 T/ha/an (A factor)

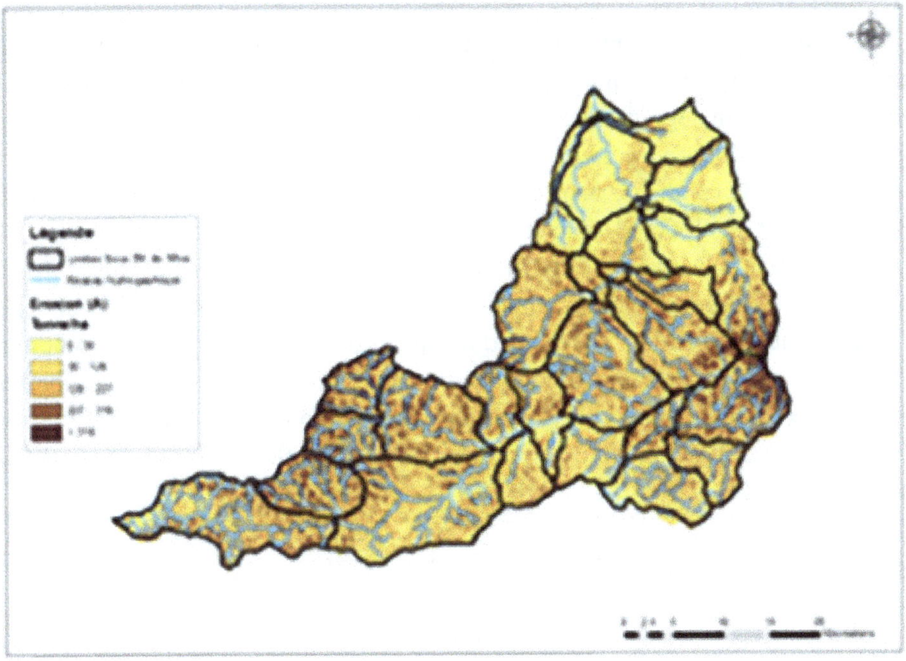

Figure 6. A factor by sub-watershed

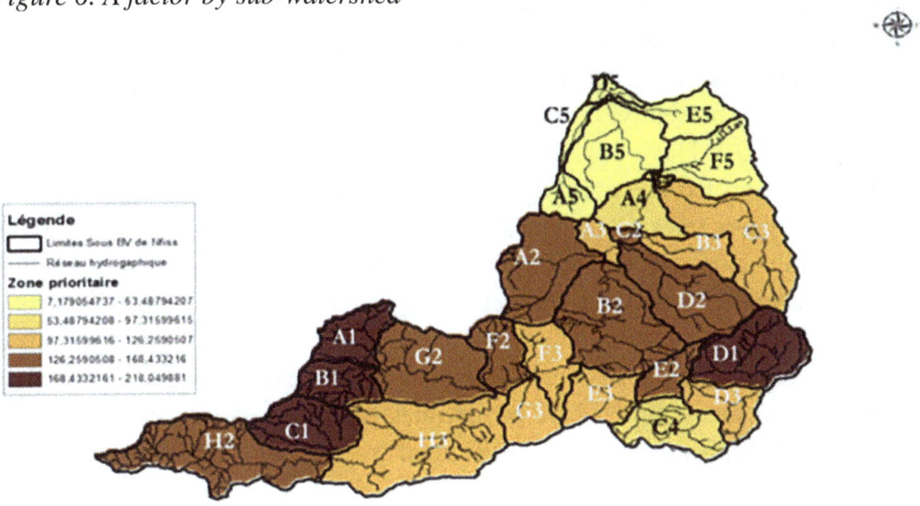

VI.2. Field Validation and Observed Erosion Dynamics

Field surveys were conducted to validate the RUSLE outputs and integrate additional vulnerability factors not explicitly captured by the model, including dynamic erosion, vegetation degradation, infrastructure exposure, and anthropogenic disturbances. Such field-based validation is essential to ensure the operational relevance of erosion modeling results, particularly in heterogeneous mountainous environments (Boix-Fayos et al., 2006; Cerdan et al., 2010; Boardman et al., 2003).

Overall, field observations confirmed the general consistency of the RUSLE-based erosion patterns, while simultaneously highlighting important local-scale contrasts that are not fully represented by spatial modeling alone. These findings are consistent with previous studies demonstrating the necessity of combining model-based assessments with in situ observations to capture fine-scale erosion processes and land degradation dynamics (Vrieling, 2006; Schmitter et al., 2010; Itzkin et al., 2021).

VI.2.1 Case of Sub-watershed C3

Sub-watershed C3 (fig.7) was classified as priority level 3 according to the RUSLE model. Although the entire area belongs to the forest domain, several factors observed during the field survey significantly increase its vulnerability beyond what is suggested by the modeled soil loss values.

A future dam, named Tissa Ouirgane, is currently under construction within this sub-watershed, considerably increasing the strategic importance of sediment control in this area. The presence of hydraulic infrastructures strongly amplifies the socio-economic and environmental consequences of upstream erosion, requiring a precautionary and preventive management approach (Panagos et al., 2015b; Borrelli et al., 2017; FAO & ITPS, 2015).

The sub-watershed is crossed by a national road, a regional road, and a dense network of unpaved tracks, which locally disturb surface runoff patterns, accelerate slope destabilization, and enhance sediment connectivity. Road construction activities have generated significant sediment deposits that are directly discharged into existing watercourses, thereby intensifying downstream sediment transfer. Similar impacts of road infrastructures on erosion dynamics have been widely documented in mountainous watersheds, where poorly managed road networks constitute major sources of sediment production (Boardman et al., 2003; Baghdadi et al., 2007; Cerdan et al., 2010).

Figure 7. Soil erosion in Sub bassin C3

Agricultural practices are limited and mainly consist of terraced cultivation along the main wadis, which traditionally contribute to runoff attenuation and soil conservation. However, the progressive degradation and partial abandonment of terraces, combined with insufficient maintenance, have reduced their protective efficiency, thereby increasing erosion susceptibility (Schmitter et al., 2010; Itzkin et al., 2021).

Slopes are generally moderate, and vegetation cover is moderately dense in several areas; nevertheless, field observations revealed the presence of both active and dormant gullies, indicating a *dynamic erosion system*. Such geomorphic features reflect unstable equilibrium conditions, where currently dormant erosion forms may be rapidly reactivated under intense rainfall events, leading to sudden increases in sediment yield (Boix-Fayos et al., 2006; Cerdan et al., 2010; Nearing et al., 2017).

The soils are predominantly schistose and relatively friable, making them particularly susceptible to detachment once vegetation cover is disturbed. Lithological fragility is a well-known driver of accelerated erosion in mountainous Mediterranean environments, where weathered schists and marls exhibit low structural stability and high erodibility (Cerdan et al., 2010; FAO & ITPS, 2015).

Although the RUSLE model assigns a moderate priority to this sub-watershed, the combination of active erosion processes, infrastructure development, sediment inputs from road works, and the presence of a strategic hydraulic structure suggests a significantly higher operational priority for intervention. This discrepancy highlights the limitations of relying solely on modeling outputs and demonstrates the critical role of field validation and multi-criteria analysis in refining erosion control strategies (Panagos & Katsoyiannis, 2019; Borrelli et al., 2017) (Figure 8).

Figure 8. Sub bassin F3

VI.2.2 Case of Sub-watershed F3

Sub-watershed F3 (Figure 8), classified as priority level 3 by the RUSLE model, is also entirely located within the forest domain but exhibits a markedly different vulnerability profile. This highlights the necessity of field-based validation and multi-criteria interpretation to refine erosion risk assessment beyond modeled outputs (Boix-Fayos et al., 2006; Cerdan et al., 2010; Panagos & Katsoyiannis, 2019).

The area is crossed by a national road and a sparse network of unpaved tracks, and several rural settlements (douars) are present, significantly increasing anthropogenic pressure, land-use intensity, and socio-economic exposure. Such infrastructure and settlement patterns locally disturb natural drainage networks, enhance surface runoff concentration, and accelerate soil destabilization, thereby amplifying erosion processes (Baghdadi et al., 2007) (Figure 9).

Agricultural activities are mainly represented by terraced cultivation along the wadis, which traditionally contributes to reducing runoff velocity and conserving soil moisture. However, the limited maintenance and partial degradation of terraces observed during field surveys reduce their protective role, leading to enhanced erosion susceptibility, especially during high-intensity rainfall events.

Slopes are generally moderate, while vegetation cover is locally sparse, particularly near infrastructure corridors and settlement areas. This spatial fragmentation of vegetation creates preferential runoff pathways and promotes rill and gully initiation. Field observations revealed the presence of both active and dormant gullies, as well as significant sediment deposits, indicating a spatially heterogeneous yet dynamically active erosion system.

The soils are predominantly friable, which further facilitates particle detachment once surface protection is reduced. Similar lithological and pedological conditions have been widely associated with accelerated erosion rates in mountainous Mediterranean environments (FAO & ITPS, 2015).

Although F3 is ranked as a moderate-priority area by the RUSLE model, the coexistence of active erosion processes, infrastructure development, agricultural activities, and rural settlements substantially increases its overall vulnerability and potential downstream impacts. This combination therefore justifies a higher operational priority for soil conservation and erosion control measures compared to other sub-watersheds presenting similar modeled soil loss but lower exposure levels.

These observations clearly demonstrate that similar RUSLE priority classes may correspond to very different real-world situations in terms of erosion dynamics and socio-economic vulnerability. Consequently, integrating field validation and multi-criteria analysis is essential to ensure accurate prioritization and effective targeting of soil conservation interventions.

VI.3. Contribution of the SMART multi-criteria analysis

To reduce subjectivity in sub-watershed prioritization and integrate both modeled and observed factors, a multi-criteria decision analysis based on the SMART methodology was implemented. This approach allowed the weighting and aggregation of erosion-related indicators, land cover and agricultural practices, and infrastructure exposure.

The SMART analysis highlighted those sub-watersheds with moderate modeled soil loss but significant exposure of infrastructures and settlements may require higher priority for intervention than forested areas with higher erosion rates but limited socio-economic stakes. In this context, sub-watershed F3 (IJOUKAK) was ranked higher than suggested by RUSLE alone due to the presence of roads, douars, and agricultural terraces vulnerable to sedimentation and gully expansion.

Conversely, sub-watershed B2, despite exhibiting active erosion processes, was assigned a lower operational priority due to its forested status and absence of human or hydraulic infrastructures.

The combined use of RUSLE modeling, field surveys, and SMART-based multi-criteria analysis provides a comprehensive framework for erosion risk assessment and action prioritization. While RUSLE remains an effective tool for identifying erosion hotspots, it is not sufficient on its own to support decision-making for soil and water conservation interventions.

Field observations are essential to capture erosion dynamics, land-use practices, and local vulnerability, while the SMART methodology ensures a transparent and reproducible integration of heterogeneous factors. This combined approach reduces reliance on expert judgment alone and enhances the robustness of prioritization outcomes.

Overall, the results demonstrate that effective erosion control planning must move beyond soil loss estimates and explicitly incorporate exposure and vulnerability criteria. The proposed methodology offers a scalable and transferable decision-support tool for watershed management, particularly in Mediterranean and semi-arid contexts where erosion processes and land-use pressures are highly heterogeneous.

Figure 9. Priority of intervention against soil erosion by Sub-watershed priority

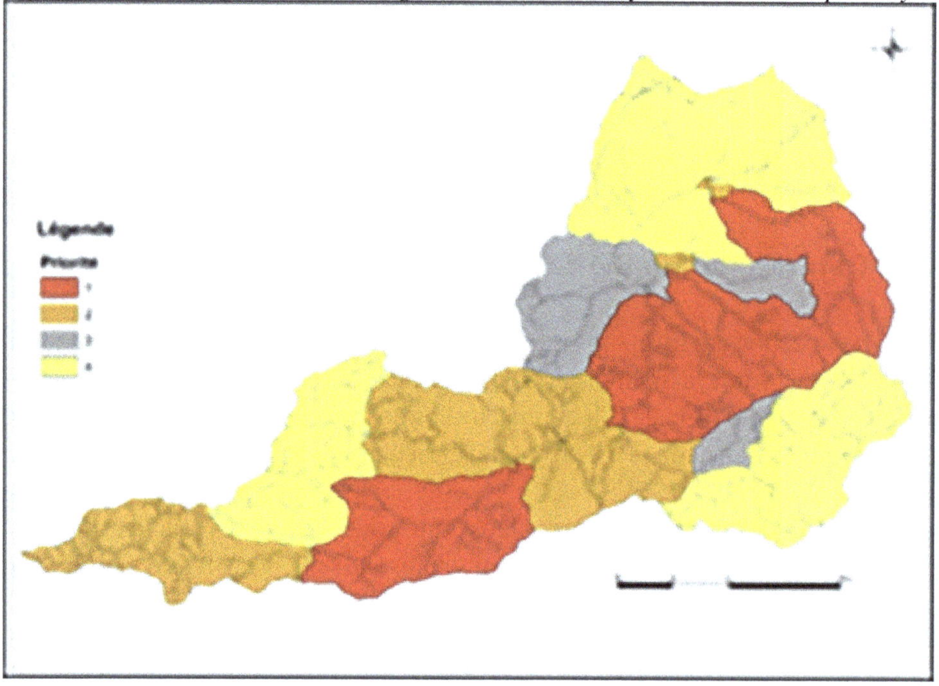

The sub-watershed-based prioritization framework highlights the need for differentiated and targeted soil erosion control strategies, adapted to the level of environmental vulnerability and socio-economic exposure. Three main priority levels were identified.

High-priority sub-watersheds are characterized by high to very high soil loss rates, active erosion processes, degraded vegetation cover, and significant exposure of infrastructures and rural settlements. These areas require urgent and intensive interventions, combining mechanical erosion control structures, biological restoration measures, and improved agricultural practices. Actions in these zones are expected to substantially reduce sediment production, limit reservoir siltation, and protect socio-economic assets.

Moderate-priority sub-watersheds exhibit moderate erosion levels and partial vegetation degradation, often associated with expanding agricultural or grazing pressures. These areas represent strategic zones for preventive management, where early interventions can effectively halt degradation trajectories. Recommended measures include vegetation reinforcement, sustainable agricultural practices, and controlled grazing, aiming to stabilize slopes and reduce long-term restoration costs.

Low-priority sub-watersheds, largely dominated by stable forest cover and minimal human disturbance, require primarily conservation and monitoring actions. Management efforts in these areas focus on preserving ecosystem stability, preventing degradation, and detecting early signs of erosion activation under climate variability.

Overall, this hierarchical prioritization framework enables optimized allocation of resources, enhanced effectiveness of erosion control programs, and improved protection of hydraulic infrastructures, contributing to sustainable watershed management and long-term environmental resilience (Figure 9 and Table 5).

Table 5. Summary of the prioritization framework and management implications

Priority Level	*Characteristics of Sub-Watersheds*	*Main Objectives*	*Recommended Management Actions*	*Expected Impacts*
Priority 1 – High	• High to very high soil loss rates (> 12.6 $t \cdot ha^{-1} \cdot yr^{-1}$) • Active erosion forms (rills, gullies) • Degraded vegetation cover • Strong anthropogenic pressure • Presence of infrastructures, roads, settlements, dams	• Rapid reduction of sediment production • Protection of infrastructures and populations • Reduction of sediment inflow to reservoirs	• Mechanical measures: check dams, gabions, gully plugs, channel stabilization • Biological measures: reforestation, slope revegetation, assisted natural regeneration • Agricultural practices: terrace rehabilitation, contour farming, mulching • Infrastructure protection: drainage improvement, slope stabilization	• Significant decrease in erosion rates • Reduced reservoir siltation • Increased lifespan of dams • Enhanced socio-economic resilience
Priority 2 – Moderate	• Moderate soil loss (3–12.6 $t \cdot ha^{-1} \cdot yr^{-1}$) • Moderate vegetation degradation • Limited infrastructures • Expanding agricultural or grazing activities	• Prevent transition toward high erosion risk • Maintain vegetation cover and soil stability • Promote sustainable land management	• Preventive biological measures: enrichment planting, agroforestry • Sustainable agriculture: improved terracing, crop rotation, soil cover • Controlled grazing and forest management	• Stabilization of slopes • Reduction of long-term restoration costs • Improved land productivity
Priority 3 – Low	• Low soil loss (< 3 $t \cdot ha^{-1} \cdot yr^{-1}$) • Dense and stable vegetation cover • Minimal anthropogenic pressure • Predominantly forested areas	• Preserve ecological stability • Prevent land degradation • Maintain ecosystem services	• Forest conservation • Grazing regulation • Monitoring using satellite data and field inspections	• Long-term ecosystem preservation • Prevention of erosion activation • Enhanced climate resilience

V. CONCLUSION AND PERSPECTIVES

This chapter proposed an integrated and dynamic framework for assessing and prioritizing soil erosion risk in the N'Fis watershed, combining RUSLE-based modeling, field-based validation, and multi-criteria decision analysis using the SMART methodology. This multi-level approach was designed to overcome the limitations of traditional static erosion assessments and to provide operational decision-support tools adapted to complex mountainous environments.

The application of the RUSLE model revealed a high spatial heterogeneity of soil erosion, with estimated soil loss rates ranging from less than 3 $t{\cdot}ha^{-1}{\cdot}yr^{-1}$ in well-protected areas to more than 22.7 $t{\cdot}ha^{-1}{\cdot}yr^{-1}$ in highly vulnerable zones. Moderate erosion rates (3–12.6 $t{\cdot}ha^{-1}{\cdot}yr^{-1}$) dominate approximately 43% of the watershed, while 26% of the basin experiences high erosion levels (12.6–22.7 $t{\cdot}ha^{-1}{\cdot}yr^{-1}$). Areas classified as having high to very high erosion risk account for nearly 15% of the total watershed area, mainly located in steep upstream sectors.

Rainfall erosivity values ranged between 41.4 and more than 86 $MJ{\cdot}mm{\cdot}ha^{-1}{\cdot}h^{-1}{\cdot}yr^{-1}$, with a mean value close to 65 $MJ{\cdot}mm{\cdot}ha^{-1}{\cdot}h^{-1}{\cdot}yr^{-1}$, confirming the strong erosive potential of precipitation in the High Atlas region. Soil erodibility values varied from 0.015 to 0.034 $t{\cdot}ha{\cdot}h{\cdot}ha^{-1}{\cdot}MJ^{-1}{\cdot}mm^{-1}$, reflecting moderate to high soil susceptibility. The LS factor reached values exceeding 60 in the steepest zones, highlighting the dominant role of topography in erosion dynamics.

Field surveys confirmed the model outputs while revealing important local contrasts. Sub-watersheds initially classified as moderate priority (e.g., C3 and F3) showed active erosion processes, sediment deposition, infrastructure exposure, and human settlements, justifying a higher operational priority. Conversely, some sub-watersheds ranked as high priority by RUSLE exhibited limited anthropogenic pressure and stable vegetation cover, indicating a lower immediate intervention urgency.

The implementation of the SMART multi-criteria decision analysis allowed the integration of erosion intensity, observed erosion dynamics, vegetation degradation, agricultural practices, and infrastructure exposure into a single objective prioritization index. This approach significantly reduced subjectivity and provided a transparent ranking of sub-watersheds according to their real vulnerability and socio-economic stakes.

From an operational perspective, the results clearly identify critical erosion hotspots contributing directly to sediment supply toward the Lalla Takerkoust dam, which is increasingly threatened by reservoir siltation. The prioritization framework developed in this study provides concrete guidance for targeting soil and water conservation measures, optimizing public investments, and enhancing watershed resilience under ongoing climate and land-use changes.

Beyond the N'Fis watershed, the proposed methodology is fully transferable to other Moroccan basins and offers a strong basis for the development of a national dynamic soil erosion monitoring and decision-support system. Such a system would contribute directly to improved land degradation control, dam protection strategies, and climate adaptation policies.

In conclusion, this study demonstrates that the combination of dynamic modeling, field expertise, and objective multi-criteria analysis constitutes a powerful and operational framework for sustainable soil and water conservation planning in semi-arid mountainous environments.

REFERENCES

Arnold, J. G., Moriasi, D. N., Gassman, P. W., Abbaspour, K. C., White, M. J., Srinivasan, R., & van Griensven, A. (2012). SWAT: Model use, calibration, and validation. *Transactions of the ASABE*, *55*(4), 1491–1508. DOI: 10.13031/2013.42256

Arnold, J. G., Srinivasan, R., Muttiah, R. S., & Williams, J. R. (1998). Large area hydrologic modeling and assessment: Part I—Model development. *Journal of the American Water Resources Association*, *34*(1), 73–89. DOI: 10.1111/j.1752-1688.1998.tb05961.x

Baghdadi, N., Cerdan, O., Zribi, M., Auzet, V., Darboux, F., El Hajj, M., & Bou Kheir, R. (2007). Operational performance of current SAR sensors in mapping soil surface characteristics: Application to hydrological and erosion modelling. *Hydrological Processes*, *22*(1), 9–20. DOI: 10.1002/hyp.6609

Boardman, J., Poesen, J., & Evans, R. (2003). Socio-economic factors in soil erosion and conservation. *Environmental Science & Policy*, *6*(1), 1–6. DOI: 10.1016/S1462-9011(02)00120-X

Boix-Fayos, C., Martínez-Mena, M., Arnau-Rosalén, E., Calvo-Cases, A., Castillo, V., & Albaladejo, J. (2006). Measuring soil erosion by field plots: Understanding the sources of variation. *Earth-Science Reviews*, *78*(3–4), 267–285. DOI: 10.1016/j.earscirev.2006.05.005

Borrelli, P., Robinson, D. A., Fleischer, L. R., Lugato, E., Ballabio, C., Alewell, C., Meusburger, K., Modugno, S., Schütt, B., Ferro, V., Bagarello, V., Oost, K. V., Montanarella, L., & Panagos, P. (2017). An assessment of the global impact of 21st century land use change on soil erosion. *Nature Communications*, *8*(1), 2013. DOI: 10.1038/s41467-017-02142-7 PMID: 29222506

Brown, L. C., & Foster, G. R.. (1987). Storm erosivity using idealized intensity distributions. *Transactions of the ASAE. American Society of Agricultural Engineers*, *30*(2), 379–386. DOI: 10.13031/2013.31957

Cerdan, O., Govers, G., Le Bissonnais, Y., Van Oost, K., Poesen, J., Saby, N., Gobin, A., Vacca, A., Quinton, J., Auerswald, K., Klik, A., Kwaad, F. J. P. M., Raclot, D., Ionita, I., Rejman, J., Rousseva, S., Muxart, T., Roxo, M. J., & Dostal, T. (2010). Rates and spatial variations of soil erosion in Europe: A study based on erosion plot data. *Geomorphology*, *122*(1–2), 167–177. DOI: 10.1016/j.geomorph.2010.06.011

Desmet, P. J. J., & Govers, G. (1996). A GIS procedure for automatically calculating the USLE LS factor on topographically complex landscape units. *Journal of Soil and Water Conservation*, *51*(5), 427–433. DOI: 10.1080/00224561.1996.12457102

Djoukbala, O., Hasbaia, M., Benselama, O., & Mazour, M. (2019). Comparison of erosion prediction models USLE, MUSLE and RUSLE in a Mediterranean watershed. *Modeling Earth Systems and Environment, 5*(2), 725–743. DOI: 10.1007/s40808-018-0562-6

FAO & ITPS. (2015). *Status of the World's Soil Resources (SWSR) – Main Report.* Food and Agriculture Organization of the United Nations.

Ghorbanzadeh, O., Blaschke, T., Gholamnia, K., & Aryal, J. (2019). Evaluation of different machine learning methods and deep-learning CNNs for landslide detection. *Remote Sensing, 11*(2), 196. DOI: 10.3390/rs11020196

Gitas, I. Z. (2009). Multi-temporal soil erosion risk assessment using a modified USLE raster model. *EARSeL eProceedings.*

Goodwin, P., & Wright, G. (2014). *Decision analysis for management judgment* (4th ed.). Wiley.

Hengl, T., de Jesus, J. M., Heuvelink, G. B. M., Gonzalez, M. R., Kilibarda, M., Blagotić, A., & Kempen, B. (2017). SoilGrids250m: Global gridded soil information based on machine learning. *PLoS One, 12*(2), e0169748. DOI: 10.1371/journal.pone.0169748 PMID: 28207752

Itzkin, A., Scholes, M. C., Clifford-Holmes, J. K., Rowntree, K., Van der Waal, B., & Coetzer, K. (2021). A social-ecological systems understanding of drivers of degradation. *Sustainability (Basel), 13*(2), 516. DOI: 10.3390/su13020516

Jensen, J. R. (2000). *Remote sensing of the environment: An earth resource perspective.* Prentice Hall.

Kouli, M., Soupios, P., & Vallianatos, F. (2009). Soil erosion prediction using RUSLE in GIS. *Environmental Geology, 57*, 483–497. DOI: 10.1007/s00254-008-1318-9

Labbaci, A., Moukrim, S., Lahssini, S., Laaribya, S., Alaoui, H. M., & Hallam, J. (2021). Estimation of land degradation loss by water erosion: case of the site of biological and ecological interest of ain Asmama (western high Atlas, Morocco). *Adv. Sci. Technol. Eng. Syst. J, 6*, 241-247. doi:10.25046/aj060326

Markhi, A., Laftouhi, N.-E., Soulaimani, A., & Fniguire, F. (2015). Quantification et evaluation de l'erosion hydrique en utilisant le modèle rusle et déposition intégrés dans un sig. application dans le bassin versant n'fis dans le haut atlas de marrakech (MAROC). *European Scientific Journal, ESJ, 11*(29). Retrieved from https://eujournal.org/index.php/esj/article/view/6342

Meusburger, K., Steel, A., Panagos, P., Montanarella, L., & Alewell, C. (2012). Spatial and temporal variability of rainfall erosivity. *Hydrology and Earth System Sciences*, *16*(1), 167–177. DOI: 10.5194/hess-16-167-2012

Morgan, R. P. C., Quinton, J. N., Smith, R. E., Govers, G., Poesen, J. W. A., Auerswald, K., Chisci, G., Torri, D., & Styczen, M. E. (1998). The European Soil Erosion Model (EUROSEM): A dynamic approach for predicting sediment transport from fields and small catchments. *Earth Surface Processes and Landforms*, *23*(6), 527–544. DOI: 10.1002/(SICI)1096-9837(199806)23:6<527::AID-ESP868>3.0.CO;2-5

Naipal, V., Reick, C. H., Van Oost, K., & Pongratz, J. (2015). Improving the global applicability of RUSLE. *Geoscientific Model Development*, *8*(9), 2893–2911. DOI: 10.5194/gmd-8-2893-2015

Nearing, M. A. (1997). A single, continuous function for slope steepness influence. *Soil Science Society of America Journal*, *61*(4), 917–919. DOI: 10.2136/sssaj1997.03615995006100030029x

Nearing, M. A., Jetten, V., Baffaut, C., Cerdan, O., Couturier, A., & Van Oost, K. (2017). Modeling response of soil erosion to changes in precipitation and cover. *Catena*, *61*(2-3), 131–154. doi:10.1016/j.catena.2005.03.007

OpenStreetMap Contributors. (2023). *OpenStreetMap*. https://www.openstreetmap.org

Padarian, J., Minasny, B., & McBratney, A. B. (2020). Machine learning and soil sciences: A review aided by machine learning tools. *Soil (Göttingen)*, *6*(1), 35–52. DOI: 10.5194/soil-6-35-2020

Panagos, P., Borrelli, P., & Meusburger, K. (2015a). A new European LS-factor. *Geosciences*, *5*(3), 117–126. DOI: 10.3390/geosciences5020117

Panagos, P., Borrelli, P., Poesen, J., Ballabio, C., Lugato, E., Meusburger, K., Montanarella, L., & Alewell, C. (2015b). The new assessment of soil loss in Europe. *Environmental Science & Policy*, *54*, 438–447. DOI: 10.1016/j.envsci.2015.08.012

Panagos, P., & Katsoyiannis, A. (2019). Soil erosion modelling: New challenges. *Environmental Research*, *172*, 470–474. DOI: 10.1016/j.envres.2019.02.043 PMID: 30844572

Rajabi, A. M., Yavari, A., & Cheshomi, A. (2022). Soil erosion assessment in semi-arid watersheds. *Natural Hazards*, *112*, 1685–1704. DOI: 10.1007/s11069-022-05244-8

Renard, K. G., Foster, G. R., Weesies, G. A., McCool, D. K., & Yoder, D. C. (1997). *Predicting soil erosion by water: A guide to conservation planning with the RUSLE*. USDA Handbook 703.

Schmitter, P., Dercon, G., Hilger, T., Hertel, M., Treffner, J., Lam, N., & Cadisch, G. (2010). Linking crop response with sediment deposition. *Agriculture, Ecosystems & Environment*, *140*, 34–45. DOI: 10.1016/j.agee.2010.11.009

Vrieling, A. (2006). Satellite remote sensing for water erosion assessment. *Catena*, *65*(1), 2–18. DOI: 10.1016/j.catena.2005.10.005

Wischmeier, W. H., & Smith, D. D. (1978). *Predicting rainfall erosion losses: a guide to conservation planning* (No. 537). Department of Agriculture, Science and Education Administration.

Zhou, P., Luukkanen, O., Tokola, T., & Nieminen, J. (2008). Effect of vegetation cover on soil erosion. *Catena*, *75*(3), 319–325. DOI: 10.1016/j.catena.2008.07.010

Zouagui, A., Sabir, M., Naimi, M., Chikhaoui, M., & Benmansour, M. (2018). Modélisation du risque d'érosion hydrique par l'équation universelle des pertes en terre dans le Rif Occidental: Cas du bassin versant de Moulay Bouchta (Maroc). *European Scientific Journal*, *14*(3), 1857–788. DOI: 10.19044/esj.2018.v14n3p524

Chapter 8
Spatial Bigdata and AI for Assessing the Impact of Climate Change on the Bioclimatic Zones in Mediterranean Context:
Case Study of Morocco

Hicham Mharzi Alaoui
http://orcid.org/0000-0003-2226-5133
National Center for Climate and Envoronnemental Risks, Morocco

Youness Bouziani
National Agency for Water and Forest, Morocco

Majdouline Belhouari
http://orcid.org/0009-0006-4241-5361
National Agency for Water and Forest, Morocco

Zineb Loukili
National Agency for Water and Forest, Morocco

Hicham Hajji
IAV Hassan II, Rabat, Morocco, Morocco

Assali Fouad
National Agency for Water and Forest, Morocco

ABSTRACT

Climate change is an increasing global threat to biodiversity and ecosystem functioning, affecting species distributions, ecological interactions, and habitat conditions. In Morocco, rising temperatures and increasing aridity pose major challenges to ecosystem conservation and adaptation. This study assesses the national-scale impacts of climate change on Moroccan ecosystems by integrating artificial intelligence

DOI: 10.4018/979-8-2600-0783-9.ch008

and cloud computing to analyze the evolution of Emberger's bioclimatic zones. The results indicate a marked aridification of bioclimatic zones, with projected losses of approximately 85% and 36% of humid and sub-humid zones, respectively, while arid and semi-arid zones are expected to expand by about 15% and 10%. These changes reflect heterogeneous vulnerabilities among bioclimatic zones and sub-zones, ranging from intra-zone thermal shifts to transitions toward new bioclimatic regimes. These findings support adaptive ecosystem management, biodiversity conservation, and the sustainability of ecosystem services under future climate scenarios.

I. INTRODUCTION

The history of life on Earth is linked to environmental changes occurring across multiple spatial and temporal scales (Davis & Shaw, 2001). This relationship critically assesses the species apacity to adapt their distribution in response to tectonic, oceanographic, and climatic events (Rosen, 1984). Furthermore, considering that climate change represents a pervasive and escalating global threat to biodiversity and ecosystems (Díaz et al., 2019), it affects species individually as well as their interactions with other organisms and their habitats. This, in turn, alters the structure and function of ecosystems, along with the goods and services that natural systems provide to society (Díaz et al., 2019).

Moreover, it has been observed that marine, freshwater, and terrestrial organisms modify their distribution to remain in their preferred environmental conditions (Chen et al., 2011; Lenoir & Svenning, 2015). Thus, these species are affected by the climate in many ways, including changes in their distribution range, changes in relative abundance within species' ranges, and more subtle shifts in the timing of biological processes and use of microhabitats (Bates et al., 2014; Williams et al., 2008). On the other hand, when the climate changes, species must either tolerate the change, move, adapt, or face extinction (Berg et al., 2010). Furthermore, when these shifts in the species distribution occur in foundational or habitat-organizing species, there can be widespread effects that propagate through entire communities (Zarnetske et al., 2012).

Therefore, understanding the direction and magnitude of ecological responses will allow human communities to better anticipate these changes and adapt, if necessary. Periodic assessments of the current and future impacts of climate change on ecosystems are essential for developing and updating natural resource management plans and evaluating adaptation measures (West et al., 2009).

In Morocco, the impacts of climate change, characterized by trends toward increased aridity and warming (MDCE, 2020), represent significant challenges for the conservation and adaptation of various ecosystems and their biodiversity.

Recent studies (Lahssini et al., 2015; Moukrim et al., 2018, 2022) have attempted to highlight the impact of climate change on terrestrial ecosystems, particularly forests, by studying the evolution of their potential distribution areas under different climate change scenarios. The approach used in these studies is based on Species Distribution Modeling (SDM) techniques with the Maxent algorithm. However, these fragmented studies will not provide a clear picture of the impacts of climate change on the ecosystems, particularly terrestrial ecosystems and their flora, fauna, mycology and micro-organisms, until we have completed a study of all the species of the animal and vegetal reigns in Morocco, or at least of the main ecosystems.

In this context, and to address the urgency of assessing the vulnerability of terrestrial ecosystems to climate change and provide decision-makers and managers with reliable decision-making tools, the present work aims to examine the impacts of climate change on these ecosystems at the national level using machine learning algorithms and spatial bigdata using google earth engine. This will be achieved through the characterization of the evolution of bioclimatic zones defined by Emberger, which serve as the essential ecological entry point for the study of ecosystems and their faunistic, floristic, and other components in Mediterranean regions in general and Morocco in particular. The selection of Emberger's bioclimatic classification is justified by its specific development for the Mediterranean context and its reliance on climatic thresholds, particularly the pluviothermic quotient (Q2), which are highly relevant for explaining vegetation distribution patterns in this region. Such ecological specificity is not captured with the same level of precision by more general global climate classification systems, such as the Köppen–Geiger classification (Emberger, 1930; Daget, 1977).

Moreover, the simplicity and accessibility of the Emberger Index have favored its widespread use in technical reports, forest management plans, and ecological restoration guidelines. This index can also support forest managers in identifying the most suitable species for reforestation and ecosystem restoration programs (Vessella & Schirone, 2022).

The present study aims to contribute to a better understanding of the vulnerability of ecosystems to climate change in Morocco. Although several studies have examined the impacts of climate change on individual species in Morocco, a systematic, nationwide, and reproducible assessment of the vulnerability of ecosystems as a whole remains limited. Moreover, existing bioclimatic maps are largely based on historical datasets and methodological approaches that are difficult to update (Emberger, 1964; Sauvage, 1963). Therefore, there is a clear need for a quantitative, forward-looking, and spatially explicit analytical framework capable of assessing how climate change may reshape Morocco's bioclimatic zones, which constitute the ecological basis for the distribution of its ecosystems.

More specifically, this study aims to develop a reproducible model of Morocco's bioclimatic zones at the national scale using a machine learning approach based on Emberger's classification. It also seeks to quantify the projected changes in these bioclimatic zones by 2050, particularly in terms of gains, losses, and transitions, under a high-emissions climate scenario RCP8.5. Finally, the study examines the implications of these projected shifts for Sustainable Forest Management (SFM), with particular emphasis on associated risks, priority areas for intervention, and potential adaptation strategies.

2. MATERIALS AND METHODS

2.1. Study Area

Due to its geographical location, serving as a true crossroad between Europe and Africa, covering the Mediterranean and the Atlantic, and the significance of its mountain ranges, with about ten peaks exceeding 4000 meters in altitude, Morocco is one of the most unique Mediterranean countries, characterized by its extreme ecological diversity, unmatched in terms of bioclimatic, morphological, floristic, and faunistic aspects (Benabid, 1985). This diversity has led to a trove of landscapes and natural environments, counting over forty major types of ecosystems providing habitats for nearly 7000 vascular plant species, 550 species of vertebrates, and thousands of invertebrates.This confers Morocco with an exceptional cultural heritage, ranking only second to Anatolia in the Mediterranean basin. The majority of this heritage comes from forest, pre-forest, pre-steppe, and steppe ecosystems, spread over more than 9 million hectares.

2.2. Methodology

2.2.1. Response Variables

The goal is to create a "large" sample that represents all bioclimatic zones and their variants for Moroccan territory, including geolocated points with their corresponding bioclimatic zones. To do so, we geo-referenced and digitized the map by Sauvage & Brignon (Emberger, 1964) republished in 1976. Subsequently, a random representative sample was extracted from the different bioclimatic zones, composed of 160,000 geolocated points, distributed proportionally to the area of each zone. Each of these bioclimatic zones is represented by a minimum of 1,000 points.

2.2.2. Explanatory Variables

The choice of explanatory variables, also known as independent variables, is a crucial step in modeling in general. Accordingly, we used the 19 bioclimatic variables from the WorldClim dataset at a spatial resolution of 30 arc-seconds. These variables, derived from monthly temperature and precipitation data, are widely used for SDM (Species Distribution Modeling) and ecological modeling in general. These variables characterize annual trends, seasonality, and extreme or limiting environmental factors. The data representing recent climatic conditions consist of the average for the reference period from 1970 to 2000. These variables include:

BIO1 = Annual mean temperature
BIO2 = Mean diurnal range (monthly mean of (tmax – tmin))
BIO3 = Isothermality (BIO2/BIO7) (* 100)
BIO4 = Temperature seasonality (standard deviation * 100)
BIO5 = Maximum temperature of the hottest month
BIO6 = Minimum temperature of the coolest month
BIO7 = Annual temperature range (BIO5-BIO6)
BIO8 = Mean temperature of the wettest quarter
BIO9 = Mean temperature of the driest quarter
BIO10 = Mean temperature of the warmest quarter
BIO11 = Mean temperature of the coolest quarter
BIO12 = Annual precipitation
BIO13 = Precipitation of the wettest month
BIO14 = Precipitation of the driest month
BIO15 = Precipitation seasonality (coefficient of variation)
BIO16 = Precipitation of the wettest quarter
BIO17 = Precipitation of the driest quarter
BIO18 = Precipitation of the warmest quarter
BIO19 = Precipitation of the coolest quarter

Additionally, the Emberger precipitation quotient and altitude were used along with the bioclimatic variables for model development.

Furthermore, the climate's evolution in the 21st century will depend on global anthropogenic greenhouse gas and aerosol emissions, as well as the intrinsic sensitivity of the climate system to these emissions, which remain relatively uncertain. In the 5th IPCC report (2014), four emission scenarios, called “Representative Concentration Pathways” (RCP), were chosen by the scientific community for climate modeling and the study of effects at various emission levels. For the present study, the RCP8.5 scenario was adopted, corresponding to a high greenhouse gas emissions trajectory leading to a global warming of more than +4.5°C by 2100. This methodological choice is employed here as a "stress test" to assess ecosystem responses under conditions of elevated climatic pressure, thereby identifying the upper range of potential vulnerability.

2.2.3. Modeling

For the modeling component of this study, the Random Forest algorithm was selected(Breiman, 2001), This machine learning approach is widely recommended for land cover classification and environmental modeling because of its strong predictive performance(Zhu et al., 2016), often outperforming or matching other classification techniques (Verikas et al., 2011).. Unlike traditional linear models, Random Forest is able to capture complex and non-linear relationships among variables, while remaining less complex to implement and more interpretable than other machine learning approaches such as neural networks (Santos et al., 2019).

This algorithm relies on the creation of a group of decision trees. Each tree is trained with a random sample of data. The final prediction is determined by a majority vote of the different trees. Moreover, the Random Forest algorithm classifier is relatively robust to noise, and the error rate for the overall accuracy of this algorithm remains almost the same as long as the noise does not exceed the 20% threshold (Rodriguez-Galiano et al., 2012). However, the Random Forest requires the adjustment of two parameters: the number of trees and the number of variables shared at each node. Adjusting these parameters affects its ability to analyze data while avoiding overfitting. Prediction will be more reliable and stable by adding trees without increasing the risk of errors. Additionally, for variable ranking, the algorithm considers not only the importance of each variable separately but also interactions with other variables (Chan & Paelinckx, 2008).

3. RESULTS

3.1. Predictive Quality of the Model

The accuracy of the bioclimatic zone modeling in Morocco was evaluated using two key parameters: the Out of Bag Error Estimate (OOBE) and the Overall Accuracy.

The OOBE value of 0.044 (an indicator of model precision) indicates that the overall error rate is very low, with only 4.4% of predictions being incorrect. This means that the model's predictions are correct 95.6% of the time. Such a low error rate demonstrates that the model has effectively learned from the training data, ensuring reliable predictions even for unseen data.

Regarding the Overall Accuracy, which is 0.97, this indicates that 97% of the geolocated points were correctly classified according to their respective bioclimatic zone. This high accuracy reflects the model's strong ability to correctly differentiate between the various bioclimatic zones and their variations.

3.2. Importance of Variables

The analysis of the importance of the variables (Figure. 1) identified the key factors influencing the distribution of bioclimatic zones and their variations in Morocco.

Among the *21 tested variables*, only *14 have a significant impact* on these distributions. The *six most influential variables*, in descending order of importance, are:

- **Emberger's Quotient (Q2)**: An indicator that simultaneously reflects heat and humidity levels.
- **Annual precipitation (BIO12)**: Determines soil water retention capacity and supports vegetation growth.
- **Temperature seasonality (BIO4)**: Captures extreme climatic variations marking seasonal transitions, which influence species diversity.
- **Elevation**: Directly affects temperature and humidity, playing a crucial role in biome distribution and ecosystem structure.
- **Maximum temperature of the warmest month (BIO5)**: Essential for assessing extreme thermal conditions that impact plant and animal physiology.
- **Annual temperature range (BIO7)**: Measures thermal variations throughout the year, which are critical for species adaptation to seasonal climatic changes.

Figure 1. Principal components

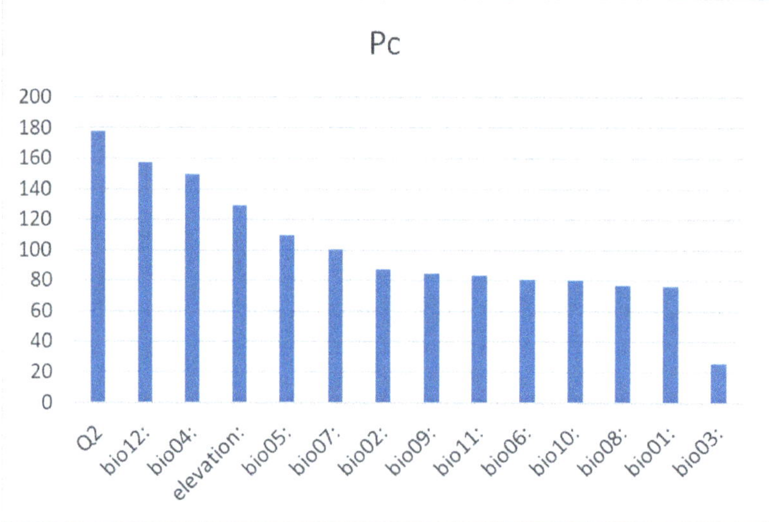

3.3. Recent Distribution of Bioclimatic Zones in Morocco

The results of the modeling of the recent distribution of bioclimatic zones (Figure. 2, Table 1) indicate a dominance of the Saharan bioclimate, which covers 60% of the national territory. This zone is primarily characterized by the fresh (90%) and temperate (8%) variants.

The arid and semi-arid zones rank second, each occupying 15% of Morocco's surface area. Within these zones, the temperate variant is the most dominant, followed by the cold and fresh variants in equal proportions.

The subhumid bioclimate accounts for 7% of the territory, with an almost equal distribution of the temperate, fresh, and cold variants, while the hot variant represents only 3%. The humid bioclimatic zone is the least extensive, covering just 2% of the country, with a predominance of the cold (50%), fresh (27%), and temperate (22%) variants.

Finally, the high mountain bioclimatic zone is the least represented, covering only 1% of the national territory.

Figure 2. Map of the recent distribution of bioclimatic zones in Morocco

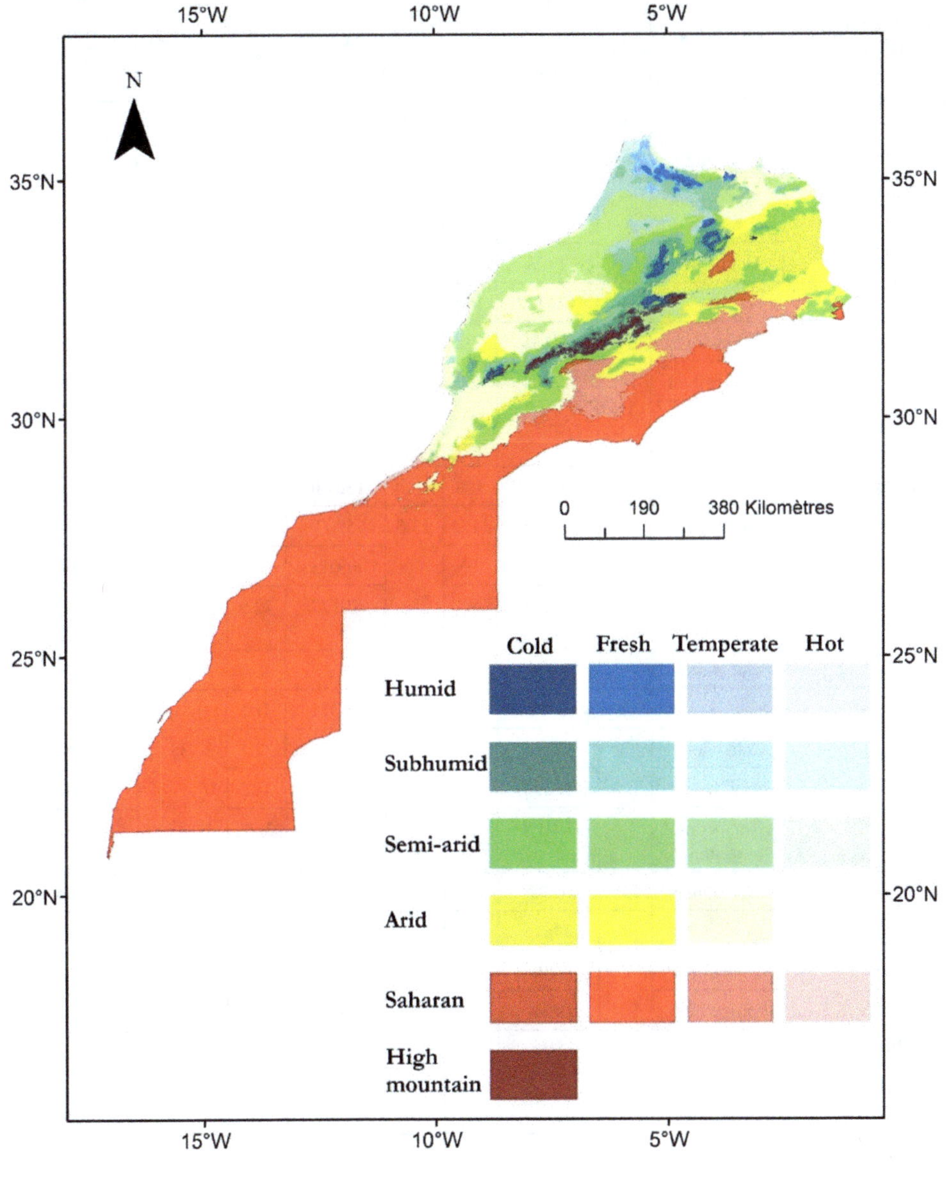

Table 1. Current and future status and evolution of bioclimatic zones in Morocco

Bioclimatic Zones (BZ)	Variants	Current BZ			BZ in 2050			Evolution	
		% [1]	Area (Km²)	% [2]	% [3]	Area (Km²)	% [4]	Change %[5]	Absolute change %[6]
Saharan	Cold winter	60	3640	0.9	62.9	1300	0.3	5	-64
	fresh winter		383836	90		394483	88.0		3
	Temperate winter		35523	8		50728	11.0		43
	Hot winter		1728	0.4		281	0.1		-84
Sub-total			**424727**	**100**		**446793**	**100**		
Arid	Cold winter	16	33151	29	18.4	22520	17.2	13	-32
	Fresh winter		27694	24		37240	28.5		34
	Temperate winter		54623	47		70986	54.3		30
Sub-total			**115468**	**100**		**130746**	**100**		-
Semi Arid	Hot winter	15	15925	15	13.6	5907	6.1	-7	-63
	Fresh winter		20668	20		16013	16.6		-23
	Temperate winter		58298	56		70177	72.9		20
	Hot winter		8998	9		4225	4.4		-53
Sub-total			**103890**	**100**		**96321**	**100**		-
Subhumid	Cold winter	7	17113	34	4.5	10358	32.1	-35	-39
	Fresh winter		18023	36		10085	31.2		-44
	Temperate winter		13103	26		11867	36.7		-9
	Hot winter		1377	3		0	0		-100
Sub-total			**49616**	**100**		**32309**	**100**		-
Humid	Cold winter	2	5290	49	0.3	854	40	-80	-84
	Fresh winter		2868	27		1109	51		-61
	Temperate winter		2366	22		196	9		-92
	Hot winter		258	2		0	0		-100
Sub-total			**10782**	**100**		**2159**	**100**		-
High mountain	High mountain	1	6367	100	0.4	2521	100	-60	-60
Total		**100**	**710850**	-	**100**	**710850**	-	-	-

1 &3: Percentage of national territory.2 &4: Percentage of bioclimatic stage. 5: 100*(4-2)/2. 6: 100*(5-3)/3.

3.4. Future Distribution of Bioclimatic Zones in Morocco:

The results of climate projections under the RCP8.5 scenario for the evolution of bioclimatic zones in 2050 (Figure. 3, Table. 1) show that, compared to the current state, Morocco would lose 80% of the humid bioclimatic zone, with the disappearance of its hot variant and significant regressions exceeding 60% for the other variants. In second place, the high mountain zone would decrease by 60%. The subhumid zone would fall by 35%, with the extinction of its hot variant and a 40% loss for both the cold and fresh variants. The arid bioclimatic zone would increase by 13%, with a 15% retreat of the cold variant and an expansion of the fresh and temperate variants by about 30% each. For the semi-arid zone, it would decrease of around 7%, with regressions for all variants except the temperate variant, which would increase by 20%. Finally, the Saharan zone would gain 5%, with significant regressions affecting the cold (-64%) and hot (-84%) variants, and increases for the temperate (43%) and fresh (3%) variants.

Figure 3. Map of the distribution of bioclimatic zones in Morocco in 2050 under the RCP8.5 scenario

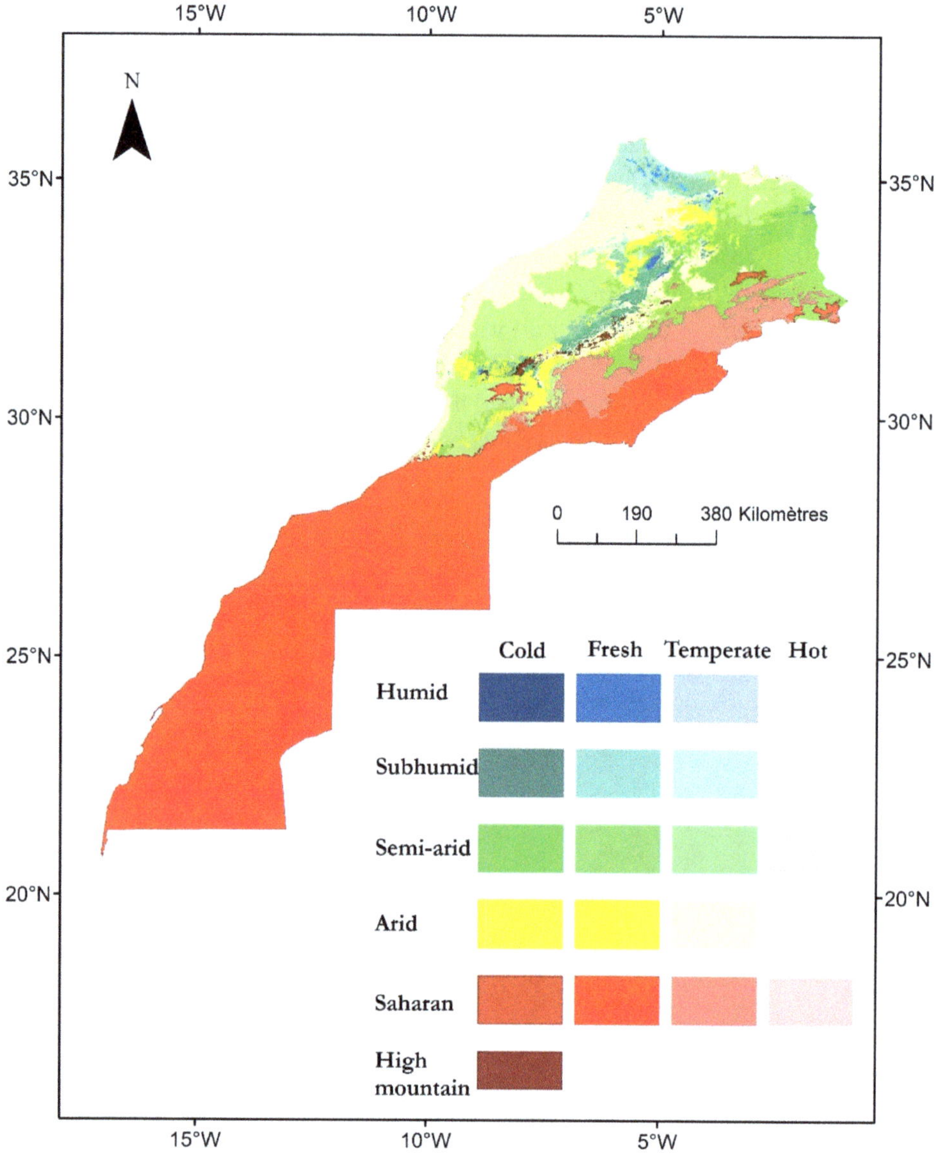

To highlight the details of the evolution of bioclimatic zones, a contingency matrix was developed by overlaying the maps of recent and future bioclimatic zones. Figure.4 illustrate the evolution of each bioclimatic zone by 2050.

For the evolution/variability of each bioclimatic zone (Figure.4), we present a focus on the predominant cases of stability and/or transformation by 2050:

Figure 4. Evolution of bioclimatic zones by 2050

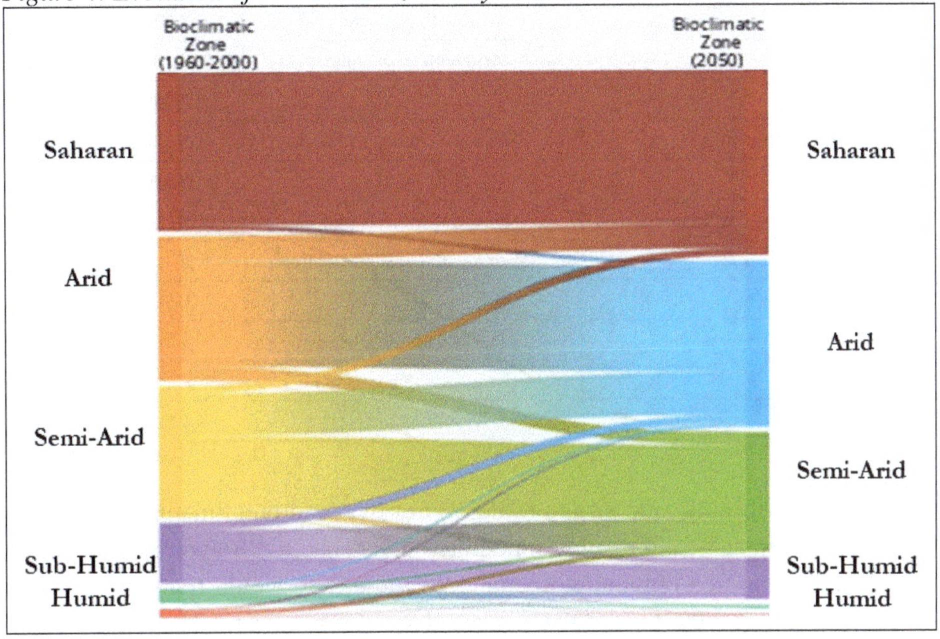

For the Saharan bioclimatic zone (Table 2), the cold variant would exhibit partial stability, with approximately 41% of its current geographical area remaining unchanged, while 32% and 18% would shift toward the temperate and fresh variants, respectively. The fresh variant would maintain almost its entire current distribution area, with approximately 99% remaining unchanged. The temperate variant would retain most of its current area (~86%), with 12% projected to shift toward the fresh variant. In contrast, the hot variant would predominantly transition toward the fresh variant (85%), while only about 10% of its current extent would remain unchanged.

Table 2. Major Evolutions/Variabilities of the Saharan Bioclimatic Zone by 2050

Bioclimatic Zone and variants		Area (Km²)	%
Current	**In 2050**		
Saharan with cold winter	Saharan with temperate winter	1171	32.17
	Arid with cold winter	1499	41.19
	Arid with fresh winter	642	17.64
Subtotal		**3640**	**91.01**
Saharan with fresh winter	Saharan with fresh winter	382051	99.53
Subtotal		**382051**	**99.53**
Saharan with temperate winter	Saharan with fresh winter	4298	12.09
	Saharan with temperate winter	30426	85.64
Subtotal		**35523**	**97.74**
Saharan with hot winter	Saharan with fresh winter	1463	84.68
	Saharan with hot winter	174	10.07
Subtotal		**1728**	**94.76**

For the Arid Bioclimatic Zone (Table. 3) For the Arid bioclimatic zone (Table 3), the cold variant would display moderate persistence, with approximately one third of its recent area remaining unchanged, while 37% would transition toward the fresh variant and 9% toward the semi-arid temperate variant. The fresh variant would retain 41% of its current extent, whereas 33% and 19% would shift toward the Saharan and arid temperate variants, respectively. The temperate variant would remain largely stable, with approximately three-quarters of its recent spatial distribution maintained, while 9% would convert to the semi-arid temperate variant.

Table 3. Major Evolutions/Variabilities of the Arid Bioclimatic Zone by 2050

Bioclimatic Zone and variants		Area (Km2)	%
Current	**En 2050**		
Arid with cold winter	Arid with cold winter	11354	34.25
	Arid with fresh winter	12236	36.91
	Semi-arid with temperate winter	2847	8.58
Subtotal		**33151**	**79.74**
Arid with fresh winter	Saharan with temperate winter	9139	32.99
	Arid with fresh winter	11479	41.45
	Aride with temperate winter	5257	18.98
Subtotal		**27694**	**93.43**

continued on following page

Table 3. Continued

Bioclimatic Zone and variants		Area (Km2)	%
Current	**En 2050**		
Arid with temperate winter	Arid with temperate winter	40063	73.34
	Semi-arid with temperate winter	5051	9.24
Subtotal		**54623**	**82.59**

Regarding the Semi-arid bioclimatic zone (Table 4), the cold variant would show limited persistence, with approximately 16% of its recent territory remaining unchanged, while 44% and 16% would shift toward the Arid zone (including cold, cool, and temperate variants) and the Saharan zone (with a temperate winter variant), respectively. The fresh variant would retain 30% of its current extent, while 7% and 44% would transition to the temperate variant and the Arid zone (with temperate and fresh variants), respectively. The temperate variant would remain relatively stable, with about 64% of its recent area maintained, whereas 25% would shift toward the Arid zone (fresh and temperate variants). In contrast, the hot variant would retain approximately one third of its recent area, while 35% would transition toward the temperate semi-arid variant, and about one-fifth would shift toward the temperate arid variant.

Table 4. Major Evolutions/Variabilities of the Semi-Arid Bioclimatic Zone by 2050

Bioclimatic Zone and variants		Area (Km²)	%
Current	**In 2050**		
Semi-arid with cold winter	Saharan with temperate winter	2605	16.35
	Arid with cold winter	2881	18.08
	Arid with fresh winter	2658	16.69
	Arid with temperate winter	1448	9.08
	Semi-arid with cold winter	2570	16.13
Subtotal		**15925**	**76.36**
Semi-arid with fresh winter	Arid with fresh winter	2770	13.40
	Arid with temperate winter	6351	30.72
	Semi-arid with fresh winter	6119	29.60
	Semi-arid with temperate winter	1427	6.90
Subtotal	-	20668	80.64

continued on following page

Table 4. Continued

Bioclimatic Zone and variants		Area (Km²)	%
Current	**In 2050**		
Semi-arid with temperate winter	Arid with fresh winter	4757	8.16
	Arid with temperate winter	10060	17.25
	Semi-arid with temperate H	37527	64.37
Subtotal		**58298**	**89.78**
Semi-arid with hot winter	Arid with temperate winter	1776	19.73
	Semi-arid with temperate winter	3187	35.41
	Semi-arid with hot winter	2814	31.27
Subtotal		**8139**	**85.97**

Regarding the Subhumid bioclimatic zone (Table 5), the cold variant would maintain approximately 31% of its recent area, while 41% would transition toward the semi-arid zone (with temperate and fresh variants) and 14% would shift toward the cold arid zone. The fresh variant would retain about 30% of its current extent, whereas 14% would evolve toward the temperate variant, and 28% would transition toward the temperate arid zone and the fresh semi-arid zone. The temperate variant would preserve approximately 27% of its recent area, while a substantial proportion (66%) would shift toward the temperate and fresh semi-arid zones. In contrast, the hot variant would predominantly transition toward the temperate variant (81%), while 15% would migrate toward the semi-arid zone (with hot and temperate variants).

Table 5. Major Evolutions/Changes in the Subhumid Bioclimatic Zone by 2050

Bioclimatic Zone and variants		Area(km²)	%
Current	**In 2050**		
Sub-humid with cold winter	Arid with cold winter	2402	14.03
	Semi-arid with fresh winter	1953	11.41
	Semi-arid with temperate winter	4773	27.89
	Subhumid with cold winter	5306	31.00
Subtotal		**17113**	**84.34**
Sub-humide with fresh winter	Arid with temperate winter	2485	13.78
	Semi-arid with fresh winter	2497	13.85
	Subhumid with fresh winter	5039	27.95
	Subhumid with temperate winter	2516	13.96
Subtotal		15900	70

continued on following page

Table 5. Continued

Bioclimatic Zone and variants		Area(km²)	%
Current	**In 2050**		
Sub-humid with temperate winter	Semi-arid with fresh winter	1361	10.38
	Semi-arid with temperate winter	7316	55.83
	Sub-humid with temperate winter	3563	27.19
Subtotal		13103	93.40
Sub-humid with hot winter	Semi-arid with temperate winter	98	7.14
	Semi-arid with hot winter	112	8.13
	Subhumid with temperate winter	1120	81.36
Subtotal		**1377**	**96.64**

For the Humid bioclimatic zone (Table 6), the cold variant would experience significant changes, losing 27% of its recent domain to the subhumid cold variant, while 38% would transition toward the semi-arid zone (with temperate and cold variants) and 16% toward the arid cold variant. The fresh variant would show limited stability, with about 16% remaining unchanged, whereas 60% would shift toward the subhumid zone (with fresh and temperate winters) and 10% toward the arid temperate zone. The temperate variant would undergo substantial contraction, with approximately 72% and 19% of its recent territory transitioning toward the subhumid zone with temperate and fresh variants, respectively. Finally, the hot variant would almost entirely transition toward the subhumid zone, with approximately 94% evolving toward the temperate variant.

Table 6. Major evolutions/variabilities of the Humid Bioclimatic Zone at the horizon 2050

Bioclimatic Zone and variants		Area (Km²)	%
Recent	**in 2050**		
Humid with cold winter	Arid with cold winter	860	16.25
	Semi-arid with cold winter	834	15.77
	Semi-arid with temperate winter	1158	21.88
	Subhumid with cold winter	1435	27.12
Subtotal		5290	81.04

continued on following page

Table 6. Continued

Bioclimatic Zone and variants		Area (Km²)	%
Recent	**in 2050**		
Humid with fresh winter	Arid with temperate winter	288	10.04
	Subhumid with temperate winter	1343	46.84
	Subhumid with temperate winter	365	12.72
	Humid with temperate winter	473	16.48
Subtotal		2868	86.09
Humid with temperate winter	Subhumid with fresh winter	445	18.81
	Subhumid with temperate winter	1706	72.09
Subtotal		2366	90.91
Humid with hot winter	Subhumid with temperate winter	244	94.47
Subtotal		**244**	**94.47**

Finally, for the High Mountain bioclimatic zone (Table. 7): It would retain a quarter of its current area and lose 57% and 11% to the semi-arid temperate and cold subhumid zones, respectively.

Table 7. Major Evolutions/ variabilities of the High Mountain Bioclimatic Zone by 2050

Bioclimatic Zone and variants		Area(km²)	%
Recent	**In 2050**		
High mountain	Semi-arid with temperate winter	3645	57.24
	Subhumid with cold winter	699	10.98
	High mountain	1574	24.71
Subtotal	-	**6367**	**92.94**

4. DISCUSSIONS

4.1. Impacts of Climate Change on Bioclimatic Ecosystems and Their Services

The results of this study highlight the significant vulnerability of Morocco's bioclimatic zones and sub-zones to climate change. The projected transformations vary in intensity depending on the spatial distribution of climatic shifts, including processes of humidification or aridification, as well as warming or cooling trends. These changes may affect ecosystems at different magnitudes depending on their geographic location and ecological characteristics. In this context, terrestrial ecosystems, particularly forest, pre-forest, pre-steppe, steppe, and Saharan ecosystems, together with their associated biodiversity, appear likely to face increasing environmental pressures. These ecosystems support diverse biological communities composed of flora, fauna (including invertebrates, mammals, birds, reptiles, and amphibians), fungi, and microorganisms, as well as significant genetic diversity. The projected climatic disturbances may therefore influence not only the dominant structuring species but also the broader ecological networks and habitats that sustain these ecosystems.

These findings are consistent with previous studies that have already documented the vulnerability of several Moroccan ecosystems to climate change. Research focusing on emblematic species such as *Abies marocana*, *Argania spinosa*, *Quercus suber*, *Quercus rotundifolia*, *Tetraclinis articulata*, *Juniperus phoenicea*, *Juniperus oxycedrus*, *Juniperus thurifera*, and *Pinus halepensis* has highlighted the high sensitivity of these species to projected climatic changes, particularly in relation to shifts in their suitable ecological niches and distribution ranges (Benabou et al., 2022; Elmalki et al., 2022; Moukrim et al., 2018, 2022). Moreover, recent research examining the potential impacts of climate change on Morocco's bioclimatic zones—although not including the High Mountain zone or the full diversity of bioclimatic variants—has indicated that Humid and Sub-humid zones may be among the most affected by future climatic changes (Meliho et al., 2024). By focusing on the evolution of bioclimatic zones at the national scale, the present study complements these species-centered approaches by providing an ecosystem-level perspective. This broader framework allows projected climatic trends to be linked with potential transformations in ecological conditions that structure ecosystems, thereby offering a more integrated basis for anticipating ecological responses and supporting the development of adaptive conservation and management strategies.

The impacts of climate change on these ecosystems may manifest through multiple ecological processes. Among the most significant are physiological stresses affecting components of the biosphere—including fauna, flora, fungi, and microorganisms—as well as the proliferation of pathogens, pests, and invasive species. Climatic disturbances may also disrupt key biological processes such as flowering, fruiting, germination, phenology, and growth, thereby affecting the reproductive and developmental capacities of many species. In some cases, these pressures may result in the decline, migration, or even extinction of certain species. At broader ecological scales, such disruptions may alter trophic interactions and lead to disturbances or breakdowns in food chains and ecological networks.

Beyond gradual changes in average climatic conditions, climate extremes play a crucial role in shaping ecosystem responses. Extreme climatic events often act as primary drivers of ecological stress and may amplify the impacts associated with long-term climate trends. For instance, increases in mean temperature projected by the model may be accompanied by a disproportionate rise in the frequency and intensity of heatwaves, associated with maximum temperature indicators such as BIO5. Such extreme conditions may particularly affect ecosystems adapted to cooler environments, notably those found in Humid and High Mountain zones. Similarly, projected reductions in annual precipitation (BIO12) may lead to an increased frequency and severity of ecological droughts. These conditions may further elevate wildfire risks, particularly in Sub-humid and Semi-Arid regions where vegetation biomass remains sufficiently abundant to sustain fire propagation.

The ecological disturbances associated with these climatic changes may also significantly affect the ecosystem services provided by Moroccan ecosystems. A decline in biological and economic productivity may occur, impacting resources such as woody forest products, medicinal and aromatic plant biomass, forage resources, melliferous flora essential for beekeeping, and fungal resources. In addition, changes in ecosystem functioning may disrupt fundamental regulatory processes, including the hydrological cycle and major biogeochemical cycles such as those of carbon, nitrogen, phosphorus, sulfur, and oxygen. These alterations may weaken the protective functions of ecosystems, potentially leading to increased soil degradation, including impacts on agricultural soils, and affecting water resources and critical infrastructure such as roads, ports, airports, railways, and renewable energy facilities. Furthermore, ecosystem degradation may also diminish cultural and recreational ecosystem services, including those related to ecotourism and nature-based leisure activities.

The degradation of these ecosystems and the services they provide may also have cascading impacts on specialized ecosystems that are particularly sensitive to environmental variations. Such ecosystems include coastal and inland dune systems, saline and halophytic environments, and wetland ecosystems. Among these are riparian forests composed of species such as *Populus*, *Salix*, *Tamarix*, and *Fraxinus*, as well as herbaceous and shrub vegetation associated with wetlands, halophytic vegetation in salt flats and estuarine environments, and psammophilous species adapted to sandy substrates. These specialized ecosystems often depend on narrow ecological conditions and hydrological balances, making them especially vulnerable to climatic shifts and environmental degradation. As a result, climate change may lead to significant alterations in their structure, distribution, and ecological functioning.

Overall, the projected climatic changes highlight the need for integrated ecosystem-based approaches to conservation and land management. Understanding the dynamics of bioclimatic transformations and their ecological consequences is essential for developing adaptive strategies aimed at preserving biodiversity, maintaining ecosystem services, and strengthening the resilience of Morocco's natural ecosystems in the face of climate change.

4.2. Implications of Climate-Induced Bioclimatic Shifts for Adaptive Ecosystem Management

Given the magnitude of the projected impacts of climate change on ecosystems and the ecosystem services they provide, it becomes essential to translate scientific findings into operational management frameworks capable of supporting effective adaptation strategies. In this context, the results of the present study highlight the need to adapt and update existing ecosystem management approaches and tools in order to better respond to emerging climatic pressures.

The main challenge lies in transforming the projected bioclimatic shifts into a structured decision-support framework linking bioclimatic change, ecological risk, and management response. Such an approach makes it possible to identify priority areas for intervention, define appropriate management options, and establish monitoring systems capable of evaluating the effectiveness of implemented adaptation measures.

Although the present study primarily focuses on projected changes in bioclimatic zones, such shifts may have important implications for the ecosystems associated with these climatic conditions. Based on the projected bioclimatic trends identified in this study and the ecological characteristics commonly associated with the main Moroccan bioclimatic zones, the following table provides a conceptual synthesis of potential ecosystem services at risk, climate-related pressures, and possible ecosystem management responses.

Table 8. Conceptual synthesis of potential ecological implications of projected bioclimatic shifts in Morocco.

Bioclimatic Zone (Projected Trend)	Main Ecosystems Potentially Affected	Key Ecosystem Services Potentially at Risk	Major Climate-Related Risks	Possible Ecosystem Management Responses
Humid / High Mountain (Contraction)	Montane forests dominated by Cedrus atlantica, Abies marocana..	Water regulation, biodiversity conservation, soil protection, climate regulation	Increased temperature stress, reduced snow and moisture availability, habitat contraction and fragmentation	Prioritization of conservation areas and ecological refugia; strengthening of fire prevention measures; monitoring of ecosystem health
Sub-humid (Contraction and progressive aridification)	Cork oak forests (Quercus suber), maritime pine (Pinus pinaster), mixed Mediterranean woodlands	Timber and cork production, biodiversity support, carbon storage	Increased drought stress, higher wildfire risk, regeneration difficulties	Adaptation of silvicultural practices (e.g., density regulation); diversification of species composition; strengthening fire prevention strategies
Semi-arid (Shifts between variants)	Woodlands dominated by Tetraclinis articulata, Pinus halepensis, and Argania spinosa	Non-wood forest products (e.g., argan oil), grazing resources, soil stabilization	Increased water scarcity, ecosystem degradation, risk of desertification at ecological margins	Promotion of agroforestry and silvopastoral systems; soil and water conservation measures; restoration using drought-tolerant species
Arid / Saharan (Expansion)	Steppe and desert ecosystems dominated by Stipa tenacissima and scattered Acacia formations	Pastoral resources, erosion control, cultural ecosystem services	Extreme heat events, prolonged droughts, land degradation, sand encroachment	Management of grazing pressure; implementation of land restoration and desertification control measures; conservation of remnant woody vegetation

These potential implications highlight the importance of developing adaptive ecosystem management strategies capable of responding to the projected climatic shifts

Within this perspective, ecosystem management strategies should be grounded in the principles of Sustainable Forest Management (SFM), supported by a set of operational indicators that can be monitored at large spatial scales. The bioclimatic shifts identified in this study should therefore be related to field-based ecosystem health indicators, enabling a better understanding of ecosystem responses to climatic pressures.

Remote sensing technologies, combined with artificial intelligence approaches applied to geospatial data, provide particularly valuable tools to support such monitoring and analysis. Vegetation indices derived from satellite imagery, such as NDVI or EVI, can help detect early signs of physiological stress and declining ecosystem productivity. In addition, the use of high-resolution imagery, together with machine learning techniques, can enhance the detection and analysis of forest degradation, tree mortality, and structural changes within ecosystems.

The integration of these indicators into a national ecosystem monitoring and information system, combining remote sensing data, field observations, and AI-based analytical tools, would represent a critical step toward moving from forward-looking climate impact assessments to adaptive ecosystem management, supported by continuous observation and dynamic analysis of environmental changes.

Within this framework, several strategic orientations can be identified to strengthen ecosystem management in the context of climate change. These include the adaptation of regeneration and reforestation strategies, the optimization of silvicultural and rangeland management practices, the development of phytosanitary monitoring systems for the early detection of ecological stresses, the improvement of forest fire risk assessment and prediction, the strengthening of integrated watershed management and land degradation control, the reinforcement of biodiversity conservation and ecosystem restoration strategies, and the enhancement of governance mechanisms addressing the socio-economic dimensions of ecosystem services management.

4.3. Limitations of the Methodological Approach Adopted

Some limitations of this study should nevertheless be acknowledged. First, the projections rely on a single high-emissions scenario (RCP8.5), which was used here as a climate stress test to explore ecosystem responses under conditions of strong climatic pressure. A multi-scenario approach would allow a broader consideration of uncertainties associated with different emission pathways and could therefore be explored in future research.

Second, the bioclimatic zoning presented in this study reflects primarily the climatic potential for the distribution of different ecosystem types. However, it does not explicitly account for several factors that may influence the vegetation effectively observed on the ground, including anthropogenic pressures, land use and land cover dynamics, soil characteristics, and habitat fragmentation. Consequently, the actual distribution of forest ecosystems may differ from their potential bioclimatic envelope, as it is also shaped by these non-climatic drivers.

Third, the spatial resolution of the climatic data (approximately 1 km) is appropriate for analyses conducted at the national scale but may not capture certain local microclimatic variations, particularly in topographically complex mountain environments.

Finally, as with any bioclimatic classification approach, the analysis does not explicitly incorporate biotic interactions that may influence ecosystem responses to climate change. Integrating such ecological processes in future work could further improve the understanding of ecosystem dynamics under changing climatic conditions.

5. CONCLUSION

The present study demonstrates that Morocco's bioclimatic landscape is likely to undergo substantial transformations over the coming decades as a consequence of climate change. These projected changes have important implications for forest ecosystems and highlight the necessity of integrating climate considerations into long-term forest governance. Within the framework of Sustainable Forest Management (SFM), the results emphasize the importance of shifting toward more anticipatory, adaptive, and spatially differentiated management approaches capable of addressing emerging ecological challenges.

Based on the findings of this study, three strategic priorities emerge for guiding future forest management and policy development.

First, spatial prioritization of conservation efforts should be strengthened. The analysis indicates that Humid and High Mountain bioclimatic zones may be particularly vulnerable to future climatic shifts. These areas represent critical ecological refuges that support high levels of endemic biodiversity and contribute significantly to the maintenance of essential ecosystem services, including water regulation and habitat preservation. Consequently, forest management strategies in these zones should prioritize enhanced conservation measures, reinforced fire prevention systems, and protective interventions aimed at preserving ecosystem integrity and resilience.

Second, forest management practices must evolve to accommodate ecological transitions. In Sub-humid and Semi-arid zones, which are expected to experience increasingly warmer and drier conditions, the central challenge lies in adapting management approaches to changing environmental constraints. Rather than focusing exclusively on maintaining current ecological conditions, management strategies should increasingly support adaptive transitions within forest ecosystems. This may include revising species selection in reforestation programs by prioritizing drought-tolerant species and provenances, adapting silvicultural practices to improve resistance to drought and wildfire risks, and promoting agroforestry systems that enhance landscape multifunctionality while diversifying ecological and socio-economic benefits.

Third, monitoring systems should be strengthened through the integration of geospatial artificial intelligence. Considering the spatial scale and accelerating pace of environmental change, conventional monitoring approaches may not provide sufficient capacity to capture the complexity of ongoing dynamics. In this context, the integration of remote sensing technologies with machine learning methods offers considerable potential for improving environmental monitoring. Geospatial artificial intelligence can support the development of dynamic and spatially explicit monitoring systems capable of tracking changes in forest conditions, identifying emerging risks, and evaluating the effectiveness of adaptation measures. Such systems would provide valuable decision-support tools for policymakers and forest managers.

Overall, the integration of climate-informed planning, adaptive forest management strategies, and advanced geospatial technologies will be essential for enhancing the resilience of Morocco's forest ecosystems. These approaches can contribute to strengthening the implementation of Sustainable Forest Management while supporting broader national and international commitments related to biodiversity conservation, climate change adaptation, and sustainable land management.

Future Research Directions

While this study provides valuable insights into the potential evolution of Morocco's bioclimatic zones, several research avenues remain open for further investigation. Future studies could focus on integrating higher-resolution climate projections and more detailed ecological datasets to refine spatial assessments of vulnerability and ecosystem responses. Additionally, the incorporation of species distribution models and ecological niche modeling could improve the understanding of how specific forest species may respond to changing climatic conditions.

Another promising direction involves the development of advanced geospatial artificial intelligence frameworks capable of combining multi-source data, including satellite imagery, drone observations, climatic variables, and ground-based forest inventories. Such integrated systems could significantly enhance early detection of ecosystem stress, wildfire risk, and vegetation shifts.

Finally, interdisciplinary research linking ecological modeling with socio-economic and governance dimensions would provide a more comprehensive understanding of adaptation pathways. Considering the importance of forests for local livelihoods and rural economies, future work should also explore participatory approaches and policy mechanisms that support climate-resilient forest management while ensuring sustainable use of natural resources.

REFERENCES

Bates, A. E., Pecl, G. T., Frusher, S., Hobday, A. J., Wernberg, T., Smale, D. A., Sunday, J. M., Hill, N. A., Dulvy, N. K., Colwell, R. K., Holbrook, N. J., Fulton, E. A., Slawinski, D., Feng, M., Edgar, G. J., Radford, B. T., Thompson, P. A., & Watson, R. A. (2014). Defining and observing stages of climate-mediated range shifts in marine systems. *Global Environmental Change, 26*, 27–38. DOI: 10.1016/j.gloenvcha.2014.03.009

Benabid, A. (1985). Les écosystèmes forestiers marocains: Diversité, bigéographie et dégradation. *Forêt Méditerranéenne*, *7*(1), 53–64.

Benabou, A., Moukrim, S., Lahssini, S., El Aboudi, A., Menzou, K., Elmalki, M., El Madihi, M., & Rhazi, L. (2022). Impact of climate change on potential distribution of Quercus suber in the conditions of North Africa. *Biosystems Diversity*, *30*(3), 289-294. doi:10.15421/012231

Berg, M. P., Kiers, E. T., Driessen, G., Van Der Heijden, M., Kooi, B. W., Kuenen, F., Liefting, M., Verhoef, H. A., & Ellers, J. (2010). Adapt or disperse: Understanding species persistence in a changing world. *Global Change Biology*, *16*(2), 587–598. DOI: 10.1111/j.1365-2486.2009.02014.x

Breiman, L. (2001). Random Forests. *Machine Learning*, *45*(1), 5–32. DOI: 10.1023/A:1010933404324

Chan, J. C.-W., & Paelinckx, D. (2008). Evaluation of Random Forest and Adaboost tree-based ensemble classification and spectral band selection for ecotope mapping using airborne hyperspectral imagery. *Remote Sensing of Environment*, *112*(6), 2999–3011. DOI: 10.1016/j.rse.2008.02.011

Chen, I.-C., Hill, J. K., Ohlemüller, R., Roy, D. B., & Thomas, C. D. (2011). Rapid Range Shifts of Species Associated with High Levels of Climate Warming. *Science*, *333*(6045), 1024–1026. DOI: 10.1126/science.1206432 PMID: 21852500

Daget, P. (1977). Le bioclimat méditerranéen: Caractères généraux, modes de caractérisation. *Vegetatio*, *34*(1), 1–20. DOI: 10.1007/BF00119883

Davis, M. B., & Shaw, R. G. (2001). Range Shifts and Adaptive Responses to Quaternary Climate Change. *Science*, *292*(5517), 673–679. DOI: 10.1126/science.292.5517.673 PMID: 11326089

Díaz, S. M., Settele, J., Brondízio, E., Ngo, H., Guèze, M., Agard, J., Arneth, A., Balvanera, P., Brauman, K., & Butchart, S. (2019). *The global assessment report on biodiversity and ecosystem services: Summary for policy makers.*

Elmalki, M., Ichen, A., Aarab, M., Mounir, F., El Mderssa, M., Oularbi, Y., Tajdi, A., Moukrim, S., & Ouzzaouit, L. A. (2022). Impact of climate change on forest resources: Case of Quercus rotundifolia, Tetraclinis articulata, Juniperus phoenicea, J. oxycedrus, J. thurifera and Pinus halepensis. *Biosystems Diversity, 30*(3), 255-262.

Emberger, L. (1930). La végétation de la région méditerranéenne. Essai d'une classification des groupements végétaux. *Revue Générale de Botanique, 42*, 641–662.

Emberger, L. (1964). La position phytogéographique du Maroc dans l'ensemble méditerranéen. *Al Awamia, 12*, 1–15.

IPCC. (2014). *Climate Change 2014: Synthesis Report. Contribution of Working Groups I, II and 3 to the Fifth Assessment Report of the Intergovernmental Panel on Climate Change* [Core Writing Team, R.K. Pachauri and L.A. Meyer (eds.)]. IPCC, Geneva, Switzerland, 151 pp.

Lahssini, S., Hajib, S., Lahlaoi, H., Mharzi Alaoui, H., & Khattabi, A. (2015). Modelling Spatial Distribution of the Carob Tree (Ceratonia siliqua L.) in Azilal Province, Morocco. *Journal of Geography and Geology, 7*(4), 33. DOI: 10.5539/jgg.v7n4p33

Lenoir, J., & Svenning, J.-C. (2015). Climate-related range shifts – a global multidimensional synthesis and new research directions. *Ecography, 38*(1), 15–28. DOI: 10.1111/ecog.00967

MDCE. (2020). *Troisième communication nationale à la CCNUCC. Ministère délégué auprès du Ministre de l'Énergie, des Mines, de l'Eau et de l'Environnement, Chargé de l'Environnement, Maroc.*

Meliho, M., Khattabi, A., Zine El Abidine, A., & Orlando, C. A. (2024). Assessing the impact of climate change on bioclimatic zones in Morocco. *Modeling Earth Systems and Environment, 10*, 3285–3298. doi:10.1007/s10661-024-13077-0

Moukrim, S., Lahssini, S., Mharzi Alaoui, H., Rifai, N., Arahou, M., & Rhazi, L. (2018). Modélisation de la distribution spatiale des espèces endémiques pour leur conservation: Cas de l'Argania spinosa (L.) Skeels. *Revue d'Écologie, 73*(2), 153–166. DOI: 10.3406/revec.2018.1923

Moukrim, S., Lahssini, S., Rhazi, M., Menzou, K., El Madihi, M., Rifai, N., Bouziani, Y., Azedou, A., Boukhris, I., & Rhazi, L. (2022). Climate Change Impact on Potential Distribution of an Endemic Species Abies marocana Trabut. *Ekologia (Bratislava), 41*(4), 329–339. DOI: 10.2478/eko-2022-0034

Rodriguez-Galiano, V. F., Ghimire, B., Rogan, J., Chica-Olmo, M., & Rigol-Sanchez, J. P. (2012). An assessment of the effectiveness of a random forest classifier for land-cover classification. *ISPRS Journal of Photogrammetry and Remote Sensing*, *67*, 93–104. DOI: 10.1016/j.isprsjprs.2011.11.002

Rosen, B. R. (1984). Reef coral biogeography and climate through the Late Cainozoic: Just islands in the sun or a critical pattern of islands? *Geological Journal*, *11*(Special issue), 201–262.

Santos, F., Graw, V., & Bonilla, S. (2019). A geographically weighted random forest approach for evaluate forest change drivers in the Northern Ecuadorian Amazon. *PloS one*, *14*(12), e0226224. doi:10.1371/journal.pone.0226224

Sauvage, C. (1963). *Étages bioclimatiques*. Atlas du Maroc, Section II: Physique du Globe et Météorologie, Notice explicative, Comité de Géographie du Maroc, Rabat.

Verikas, A., Gelzinis, A., & Bacauskiene, M. (2011). Mining data with random forests: A survey and results of new tests. *Pattern Recognition*, *44*(2), 330–349. DOI: 10.1016/j.patcog.2010.08.011

Vessella, F., & Schirone, B. (2022). Forest Conservation and Restoration Using the Emberger Index: Cork Oak as Study Case. *Forests*, *13*(2), 252. DOI: 10.3390/f13020252

West, J. M., Julius, S. H., Kareiva, P., Enquist, C., Lawler, J. J., Petersen, B., Johnson, A. E., & Shaw, M. R. (2009). U.S. Natural Resources and Climate Change: Concepts and Approaches for Management Adaptation. *Environmental Management*, *44*(6), 1001–1021. DOI: 10.1007/s00267-009-9345-1 PMID: 19636606

Williams, S. E., Shoo, L. P., Isaac, J. L., Hoffmann, A. A., & Langham, G. (2008). Towards an Integrated Framework for Assessing the Vulnerability of Species to Climate Change. *PLoS Biology*, *6*(12), e325. DOI: 10.1371/journal.pbio.0060325 PMID: 19108608

Zarnetske, P. L., Skelly, D. K., & Urban, M. C. (2012). Biotic Multipliers of Climate Change. *Science*, *336*(6088), 1516–1518. DOI: 10.1126/science.1222732 PMID: 22723403

Zhu, Z., Gallant, A. L., Woodcock, C. E., Pengra, B., Olofsson, P., Loveland, T. R., Jin, S., Dahal, D., Yang, L., & Auch, R. F. (2016). Optimizing selection of training and auxiliary data for operational land cover classification for the LCMAP initiative. *ISPRS Journal of Photogrammetry and Remote Sensing*, *122*, 206–221. DOI: 10.1016/j.isprsjprs.2016.11.004

Compilation of References

Aafi, A. (2007). *Etude de la diversité floristique de l'écosystème de Chêne-liège de la forêt de la Mamora*. Thèse Doct. Es-Sc. Agronomiques, IAV Hassan II, Rabat, 190 p.

Aafi, A., El Kadmiri, A. A., Benabid, A., & Rochdi, M. (2005). Richesse et diversité floristique de la suberaie de la Mamora (Maroc) [Richness and floristic diversity of the cork oak forest of Mamora (Morocco)]. *Acta Botanica Malacitana*, *30*, 127–138. DOI: 10.24310/abm.v30i0.7187

Abel-Schaad, D., Iriarte, E., López-Sáez, J. A., Pérez-Díaz, S., Sabariego Ruiz, S., Cheddadi, R., & Alba-Sánchez, F. (2018). Are *Cedrus atlantica* forests in the Rif Mountains of Morocco heading towards local extinction? *The Holocene*, *28*(6), 1023–1037. DOI: 10.1177/0959683617752842

Achhal, A., Akabli, A., Barbero, M., Benabid, A., M'Hirit, O., Peyre, C., Quézel, P., & Rivas-Martinez, S. (1979). A propos de la valeur bioclimatique et dynamique de quelques essences forestières au Maroc. *Ecologia Mediterranea*, *5*(1), 211–249. DOI: 10.3406/ecmed.1979.960

Achour, H., Toujani, A., Rzigui, T., & Faïz, S. (2018). Forest cover in Tunisia before and after the 2011 Tunisian Revolution: A spatial analysis approach. *Journal of Geovisualization and Spatial Analysis*, *2*(2), 10. DOI: 10.1007/s41651-018-0017-7

AfDB. (2025). Tunisia launches $24 million project to restore forests and revitalize rural economies. African Development Bank. https://www.afdb.org/en/news-and-events/press-releases/tunisia-launches-24-million-project-restore-forests-and-revitalize-rural-economies-83225

African Development Bank (AfDB). (2025). *Tunisia launches $24 million project to restore forests and revitalize rural economies*. African Development Bank. https://www.afdb.org/en/news-and-events/press-releases/tunisia-launches-24-million-project-restore-forests-and-revitalize-rural-economies-83225

Agence Nationale des Eaux et Forêts. (ANEF). (2022). *National Water and Forests Agency in Morocco*. http://www.eauxetforets.gov.ma/Biodiversite/GestionBiodiversite/Pages/Gestion-Biodiversite.aspx

AGIF. (2020). *Landscape fire governance framework*. Agência para a Gestão Integrada de Fogos Rurais.

Agrawal, A., Chhatre, A., & Hardin, R. (2008). Changing governance of the world's forests. *Science*, *320*(5882), 1460–1462. DOI: 10.1126/science.1155369 PMID: 18556552

Aguilera, E., Díaz-Gaona, C., García-Laureano, R., Reyes-Palomo, C., Guzmán, G. I., Ortolani, L., Sánchez-Rodríguez, M., & Rodríguez-Estévez, V. (2020). Agroecology for adaptation to climate change and resource depletion in the Mediterranean region. A review. *Agricultural Systems*, *181*, 102809. Advance online publication. DOI: 10.1016/j.agsy.2020.102809

Aissi, A., Beghami, Y., Lepais, O., & Vela, E. (2020). Analyse morphologique et taxonomique du complexe Quercus faginea (Fagaceae) en Algérie. *Botany*, *99*(2), 99–113. DOI: 10.1139/cjb-2020-0075

Alami, M.. (2024). Genomic scan of endangered Moroccan endemic trees: *Argania spinosa* as a case of adaptive capacity under Mediterranean stressors. *Environmental Advances*, *16*, 100339.

Alaoui, A., Laaribya, S., & Ayan, S. (2020). The evolution of the forest cover with the effect of anthropic pressure: The case study of Sehoul cork-oak forest in Morocco, North Atlantic. *Kastamonu Üniversitesi Orman Fakültesi Dergisi*, *20*(1), 62–73. DOI: 10.17475/kastorman.705872

Alasli, M. (2019). Static risk mapping of forest fires in the case of the Province of Chefchaouen (Morocco). *Proceedings of the ICA*. Copernicus GmbH. DOI: 10.5194/ica-proc-2-2-2019

Alexandrian, D., Esnault, F., & Calabri, G. (1999). Forest fires in the Mediterranean area. *Unasylva*, *50*(197), 35–41.

Alipour, M., La Puma, I., Picotte, J., Shamsaei, K., Rowell, E., Watts, A., Kosovic, B., Ebrahimian, H., & Taciroglu, E. (2023). A multimodal data fusion and deep learning framework for large-scale wildfire surface fuel mapping. *Fire (Basel, Switzerland)*, *6*(2), 36. DOI: 10.3390/fire6020036

Allen, C. D., Breshears, D. D., & McDowell, N. G. (2015). On underestimation of global vulnerability to tree mortality and forest die-off from hotter drought in the Anthropocene. *Ecosphere*, *6*(8), 1–55. DOI: 10.1890/ES15-00203.1

Allen, C. D., Macalady, A. K., Chenchouni, H., Bachelet, D., McDowell, N., Vennetier, M., Kitzberger, T., Rigling, A., Breshears, D. D., Hogg, E. H. T., Gonzalez, P., Fensham, R., Zhang, Z., Castro, J., Demidova, N., Lim, J.-H., Allard, G., Running, S. W., Semerci, A., & Cobb, N. (2010). A global overview of drought and heat-induced tree mortality reveals emerging climate change risks for forests. *Forest Ecology and Management, 259*(4), 660–684. DOI: 10.1016/j.foreco.2009.09.001

Allouche, O., Tsoar, A., & Kadmon, R. (2006). Assessing the accuracy of species distribution models : Prevalence, kappa and the true skill statistic (TSS): Assessing the accuracy of distribution models. *Journal of Applied Ecology, 43*(6), 1223–1232. DOI: 10.1111/j.1365-2664.2006.01214.x

Alonso-Forn, D., Sancho-Knapik, D., Ferrio, J. P., Peguero-Pina, J. J., Bueno, A., Onoda, Y., Cavender-Bares, J., Niinemets, Ü., Jansen, S., Riederer, M., Cornelissen, J. H. C., Chai, Y., & Gil-Pelegrín, E. (2020). Revisiting the Functional Basis of Sclerophylly within the Leaf Economics Spectrum of Oaks: Dierent Roads to Rome. *Current Forestry Reports, 6*(4), 260–281. DOI: 10.1007/s40725-020-00122-7

Ameztegui, A., Coll, L., De Cáceres, M., & Morán-Ordóñez, A. (2024). Disturbance impacts on Mediterranean forests across climate and management scenarios. *Journal of Environmental Management, 371*, 123193. DOI: 10.1016/j.jenvman.2024.123193 PMID: 39536583

Anderegg, W. R., Hicke, J. A., Fisher, R. A., Allen, C. D., Aukema, J., Bentz, B., Hood, S., Lichstein, J. W., Macalady, A. K., McDowell, N., Pan, Y., Raffa, K., Sala, A., Shaw, J. D., Stephenson, N. L., Tague, C., & Zeppel, M. (2015). Tree mortality from drought, insects, and their interactions in a changing climate. *The New Phytologist, 208*(3), 674–683. DOI: 10.1111/nph.13477 PMID: 26058406

Andersan, H. W., Hoover, Marvin D., Reinhart, K. G. (1976). Forests and water effects of forest management on floods, sedimentation and water supply [Report]. - US: USDA Forest service general technical report. *Pacific Southwest Forest and Range Exp*. Stn., Berkeley, Calif. PSW-18, 115 p., illus.

Andréassian, V. (2004). Waters and forests: From historical controversy to scientific debate. *Journal of Hydrology (Amsterdam), 291*(1-2), 1–27. DOI: 10.1016/j.jhydrol.2003.12.015

ANEF. (2019). *Bilan des activités 2018*. Ministry of Agriculture.

ANEF. (n.d.). *Incendies: stratégie de gestion*. http://www.eauxetforets.gov.ma/Pages/Incendies-strategie-de-gestion.aspx

Aponte, C., García, L., & Marañón, T. (2013). Tree species effects on nutrient cycling and soil biota: A feedback mechanism favouring species coexistence. *Forest Ecology and Management*, *309*, 36–46. DOI: 10.1016/j.foreco.2013.05.035

Arar, A., Nouidjem, Y., Bounar, R., Tabet, S., & Kouba, Y. (2020). Modeling of the current and future potential distribution of Atlas cedar (*Cedrus atlantica*) forests revealed shifts in the latitudinal, longitudinal and altitudinal range towards more humid conditions. *Ecological Questions*, *31*(3), 49–62. DOI: 10.12775/EQ.2020.022

Aria, M., & Cuccurullo, C. (2017). bibliometrix: An R-tool for comprehensive science mapping analysis. *Journal of Informetrics*, *11*(4), 959–975. DOI: 10.1016/j.joi.2017.08.007

Arnold, J. G., Moriasi, D. N., Gassman, P. W., Abbaspour, K. C., White, M. J., Srinivasan, R., & van Griensven, A. (2012). SWAT: Model use, calibration, and validation. *Transactions of the ASABE*, *55*(4), 1491–1508. DOI: 10.13031/2013.42256

Arnold, J. G., Srinivasan, R., Muttiah, R. S., & Williams, J. R. (1998). Large area hydrologic modeling and assessment: Part I—Model development. *Journal of the American Water Resources Association*, *34*(1), 73–89. DOI: 10.1111/j.1752-1688.1998.tb05961.x

Assali, F., D'Avezac de Castera, H., Mharzi Alaoui, H., Taberkant, S., & Nabaha, A. (2019). *Integrated forest fires management in Morocco: New national strategy and action plan 2019–2028*. https://doi.org/DOI: 10.37002/biodiversidadebrasileira.v9i1.1119

Assali, F., Mharzi Alaoui, H., Rouchdi, M., & Badraoui, M. (2016). Modelling and mapping forest fire ignition risk in North-West Morocco (Chefchaouen–Ouazzane Region). [La Terre et la Vie]. *Revue d'Écologie*, *71*, 111–128. DOI: 10.3406/revec.2016.1834

Aubert, P.-M. (2010). *Action publique et société rurale dans la gestion des forêts marocaines: Changement social et efficacité environnementale* (Thèse de doctorat en sciences sociales pour l'environnement). AgroParisTech.

Aubert, P.-M. (2013). Les évolutions de la politique forestière au Maroc: Entre réappropriation du modèle forestier français et idéalisation de la tribu. *Revue Forestière Française*, *65*(4), 305–316. DOI: 10.4267/2042/53627

Aussenac, G. (1973). Climat, microclimat et production ligneuse. *Annales des Sciences Forestieres*, *30*(3), 239–258. https://www.afs-journal.org/articles/forest/pdf/1973/03/AFS_0003-4312_1973_30_3_ART0003.pdf. DOI: 10.1051/forest/19730303

Ayache, A., Hellal, B., & Ayad, N.. (2020). Modelling of the changes in land cover in the forest of Sidi Bel'Abbes between 1987 and 2018 and the impact of fires on soil evolution in western Algeria. *Authorea*. DOI: 10.22541/au.159672708.89027787

Baghdadi, N., Cerdan, O., Zribi, M., Auzet, V., Darboux, F., El Hajj, M., & Bou Kheir, R. (2007). Operational performance of current SAR sensors in mapping soil surface characteristics: Application to hydrological and erosion modelling. *Hydrological Processes, 22*(1), 9–20. DOI: 10.1002/hyp.6609

Banerjee, S., Niyogi, R., Sarkar, M. S., & John, R. (2022). Assessing the vulnerability of protected areas in the eastern Himalayas based on their biological, anthropogenic, and environmental aspects. *Trees People 8*. 10.1016/j

Banque mondiale, Direction générale des forêts & Délégation nationale aux risques majeurs. (2023). *Note sur les forêts algériennes: Gestion durable des forêts pour lutter contre les feux de forêts*. World Bank.

Bar-Ilan, J. (2010). Citations to the "Introduction to informetrics" indexed by WOS, Scopus and Google Scholar. *Scientometrics, 82*(3), 495–506. DOI: 10.1007/s11192-010-0185-9

Bates, A. E., Pecl, G. T., Frusher, S., Hobday, A. J., Wernberg, T., Smale, D. A., Sunday, J. M., Hill, N. A., Dulvy, N. K., Colwell, R. K., Holbrook, N. J., Fulton, E. A., Slawinski, D., Feng, M., Edgar, G. J., Radford, B. T., Thompson, P. A., & Watson, R. A. (2014). Defining and observing stages of climate-mediated range shifts in marine systems. *Global Environmental Change, 26*, 27–38. DOI: 10.1016/j.gloenvcha.2014.03.009

Bazza, M., Kay, M., & Knutson, C. (2018). Drought characteristics and management in North Africa and the Near East. FAO Water Reports No. 45. Food and Agriculture Organization of the United Nations.

Bednar, J. (2008). *The robust federation: Principles of design*. Cambridge University Press., https://books.google.co.ma/books?id=cba5vpiFMiMC DOI: 10.1017/CBO9780511819445

Belghazi, B., Badouzi, M., Belghazi, T., & Moujjani, S. (2011). Semis et plantations dans la forêt de chêne-liège de la Maâmora (Maroc). *Forêt Méditerranéenne, 32*(3), 301–314.

Belgherbi, B., Benabdeli, K., & Mostefai, K. (2018). Mapping the risk of forest fires in Algeria: Application of the forest of Guetarnia in Western Algeria. *Ekologia (Bratislava), 37*(3), 289–300. DOI: 10.2478/eko-2018-0022

Belhadj-Khedher, C., Koutsias, N., Karamitsou, A., El-Melki, T., Ouelhazi, B., Hamdi, A., Nouri, H., & Mouillot, F. (2018). A revised historical fire regime analysis in Tunisia (1985–2010) from a critical analysis of the national fire database and remote sensing. *Forests*, *9*(2), 59. DOI: 10.3390/f9020059

Bellahirech, A., Branco, M., Catry, F., Bonifácio, L., Sousa, E., & Jamâa, M. (2019). Site- and tree-related factors affecting colonization of cork oaks Quercus suber L. by ambrosia beetles in Tunisia. *Annals of Forest Science*, *76*(2), 45. DOI: 10.1007/s13595-019-0815-1

Ben Jamâa, M. L., Sghaier, T., Mnara, S., Nouri, M., & Sellemi, H. (2005). Le dépérissement du chêne-liège dans la subéraie de Béllif (Tunisie): Caractérisation et évaluation de son impact sur l'accroissement du liège. Integrated Protection in Oak Forests. *IOBC/WPRS Bulletin*, *28*, 17–24.

Ben Salem, F., Tarhouni, M., Ouled Belgacem, A., & Neffati, M. (2007). Impact of Drought on Plant Cover Dynamics in Two Natural Areas of Southern Tunisia. *Journal of Biological Sciences (Faisalabad, Pakistan)*, *7*(8), 1539–1544. DOI: 10.3923/jbs.2007.1539.1544

Ben Yahia, K., Chaar, H., Bahri, S., Mhamdi, S., Soudani, K., Khouaja, A., & Hasnaoui, B. (2016). Phenological monitoring of Cork oak in Kroumirie (northwest Tunisia). *Journal of Chemical, Biological and Physical Sciences, Section B*, *6*(4), 1258–1281.

Benabid, A. (2000). *Flore et écosystèmes du Maroc : Evaluation et préservation de la biodiversité*. Ibis Press & Kalila Wa Dimna. https://agris.fao.org/agris-search/search.do?recordID=XF2015030117

Benabid, A. (2000). *Flore et écosystèmes du Maroc: Évaluation et préservation de la biodiversité* [Flora and ecosystems of Morocco: Assessment and preservation of biodiversity]. Ibis Press; Kalila Wa Dimna.

Benabid, A. (1985). Les écosystèmes forestiers marocains: Diversité, bigéographie et dégradation. *Forêt Méditerranéenne*, *7*(1), 53–64.

Benabid, A. (1985). Les écosystèmes forestiers, préforestiers et présteppiques du Maroc: Diversité, répartition biogéographique et problèmes posés par leur aménagement. *Forêt Méditerranéenne*, *7*, 53–64.

Benabid, A. (2000). *Flore et écosystème du Maroc « Evaluation et préservation de la biodiversité*. Ibis Press.

Benabid, A. (2000). *Flore et écosystèmes du Maroc: Diversité, phytogéographie et conservation*. Ibis Press.

Benabid, A., & Fennane, M. (1994). Connaissances sur la végétation du Maroc: Phytogéographie, phytosociologie et séries de végétation. *Lazaroa, 14*, 21–97.

Benabou, A., Moukrim, S., Lahssini, S., El Aboudi, A., Menzou, K., Elmalki, M., El Madihi, M., & Rhazi, L. (2022). Impact of climate change on potential distribution of Quercus suber in the conditions of North Africa. *Biosystems Diversity*, *30*(3), 289-294. doi:10.15421/012231

Benabou, A., Moukrim, S., Laaribya, S., Aafi, A., Chkhichekh, A., El Maadidi, T., & El Aboudi, A. (2022). Mapping Ecosystem Services of Forest Stands : Case Study of Maamora, Morocco. *Geography, Environment, Sustainability*, *15*(1), 141–149. DOI: 10.24057/2071-9388-2021-047

Benabou, A., Moukrim, S., Lahssini, S., El Aboudi, A., Menzou, K., Elmalki, M., El Madihi, M., & Rhazi, L. (2022). Impact of the climate change on potential distribution of Quercus suber in the conditions of North Africa. *Biosystems Diversity*, *30*(3), 289–294. DOI: 10.15421/012231

Benchekroun, F., & Buttoud, G. (1989). L'arganeraie dans l'économie rurale du sud-ouest marocain. *Forêt Méditerranéenne*, *11*, 127–136.

Benito Garzon, M., Sanchez de Dios, R., & Sainz Ollero, H. (2008). Effects of climate change on the distribution of Iberian tree species. *Applied Vegetation Science*, *11*(2), 169–178. DOI: 10.3170/2008-7-18348

Benito, B. M., Lorite, J., Pérez-Pérez, R., Gómez-Aparicio, L., & Peñas, J. (2014). Forecasting plant range collapse in a mediterranean hotspot: When dispersal uncertainties matter. *Diversity & Distributions*, *20*(1), 72–83. DOI: 10.1111/ddi.12148

Benzid, H., & Madani, N. (2025). Protection of mediterranean forest biodiversity: The case of Morroco-current status, legislation and opportunities. *Revista Catalana de Dret Ambiental*, *16*(1), 1–34. DOI: 10.17345/rcda4066

Benzyane, M. (2007). La gestion durable des ressources forestières au Maroc: Quelle stratégie? *Forêt Méditerranéenne*, *28*(1), 47–54.

Benzyane, M., Aafi, A., Sbay, H., El Antry, S., Yassine, M., Ilmen, R., & Ghailoule, D. (2010). *Les écosystèmes naturels marocains et les changements climatiques: la résilience écologique à l'épreuve*. Centre de Recherche Forestière.

Benzyane, M., Naggar, M., & Lahlou, B. (2002). L'aménagement des forêts sud-méditerranéennes: Quelle approche? *Forêt Méditerranéenne*, *23*(3), 201–210.

Berg, M. P., Kiers, E. T., Driessen, G., Van Der Heijden, M., Kooi, B. W., Kuenen, F., Liefting, M., Verhoef, H. A., & Ellers, J. (2010). Adapt or disperse: Understanding species persistence in a changing world. *Global Change Biology*, *16*(2), 587–598. DOI: 10.1111/j.1365-2486.2009.02014.x

Berkes, F. (2009). Evolution of co-management: Role of knowledge generation, bridging organizations and social learning. *Journal of Environmental Management*, *90*(5), 1692–1702. DOI: 10.1016/j.jenvman.2008.12.001 PMID: 19110363

Berndtsson, R., & Niemczynowicz, J. (1986). Spatial and Temporal Characteristics of Highintensive Rainfall in Northern Tunisia. *Journal of Hydrology (Amsterdam)*, *87*(3-4), 285–298. DOI: 10.1016/0022-1694(86)90020-X

Bhamra, J. K., Anantha Ramaprasad, S., Baldota, S., Luna, S., Zen, E., Ramachandra, R., Kim, H., Schmidt, C., Arends, C., Block, J., Perez, I., Crawl, D., Altintas, I., Cottrell, G. W., & Nguyen, M. H. (2023). Multimodal wildland fire smoke detection. *Remote Sensing (Basel)*, *15*(11), 2790. DOI: 10.3390/rs15112790

Bishop, C. M. (2006). *Pattern recognition and machine learning*. Springer.

Blagojević, B., Jonsson, R., Björheden, R., Nordfjell, T., & Lindroos, O. (2019). Multi-criteria decision analysis (MCDA) in forest operations – an introductional review. *Croatian Journal of Forest Engineering*, *40*(1), 191–205.

Boardman, J., Poesen, J., & Evans, R. (2003). Socio-economic factors in soil erosion and conservation. *Environmental Science & Policy*, *6*(1), 1–6. DOI: 10.1016/S1462-9011(02)00120-X

Boix-Fayos, C., Martínez-Mena, M., Arnau-Rosalén, E., Calvo-Cases, A., Castillo, V., & Albaladejo, J. (2006). Measuring soil erosion by field plots: Understanding the sources of variation. *Earth-Science Reviews*, *78*(3–4), 267–285. DOI: 10.1016/j.earscirev.2006.05.005

Bonan, G. B. (2008). Forests and climate change: Forcings, feedbacks, and the climate benefits of forests. *Science*, *320*(5882), 1444–1449. DOI: 10.1126/science.1155121 PMID: 18556546

Bongers, F. J., Olmo, M., Lopez-Iglesias, B., Anten, N. P. R., & Villar, R. (2017). Drought responses, phenotypic plasticity and survival of Mediterranean species in two different microclimatic sites. *Plant Biology*, *19*(3), 386–395. DOI: 10.1111/plb.12544 PMID: 28054449

Borreguero, I., Sánchez-Miranda, Á., Martínez-Sancho, E., Rubio-Casal, A., Sánchez-Salguero, R., & Matías, L. (2024). Negative resilience trends to more frequent extreme droughts in Mediterranean oak species. *bioRxiv*. https://doi.org/ DOI: 10.1101/2024.11.19.624310

Borrelli, P., Robinson, D. A., Fleischer, L. R., Lugato, E., Ballabio, C., Alewell, C., Meusburger, K., Modugno, S., Schütt, B., Ferro, V., Bagarello, V., Oost, K. V., Montanarella, L., & Panagos, P. (2017). An assessment of the global impact of 21st century land use change on soil erosion. *Nature Communications*, *8*(1), 2013. DOI: 10.1038/s41467-017-02142-7 PMID: 29222506

Bosman, J., Mourik, I., Rasch, M., & Sieverts, E. (2006). Scopus reviewed and compared: The coverage and functionality of the citation database Scopus. *Online Information Review*, *30*(1), 4–21.

Bou Dagher-Kharrat, M., de Arano, I. M., Zeki-Bašken, E., Feder, S., Adams, S., Briers, S., Fady, B., Lefèvre, F., Górriz-Mifsud, E., Mauri, E., Mavsar, R., Muys, B., Nocentini, S., Pettenella, D., Peñuelas, J., Sardans, J., Secco, L., Travaglini, D., Tripodi, G., Wunder, S., & Bozzano, M. (2022). *Plan stratégique de recherche pour la forêt méditerranéenne 2030*. Institut européen des forêts. DOI: 10.36333/rs5

Bouandas, A., Belhoucine-Guezouli, L., Oliva, F., Semir, B., Bendjebbar, K., Pardo, F., & Pujade-Villar, J. (2024). Morphological and Taxonomic Analysis of the Quercus faginea and Quercus canariensis (Fagaceae) Complexes in Algeria. *International Journal of Plant Biology*, *15*(4), 927–943. Advance online publication. DOI: 10.3390/ijpb15040066

Boubekraoui, H., Attar, Z., Maouni, Y., Ghallab, A., & Draoui, B. (2024b). Forest Loss Drivers and Landscape Pressures in a Northern Moroccan Protected Areas' Network: Introducing a Novel Approach for Conservation Effectiveness Assessment. *Conservation*, *4*(3), 452–485. DOI: 10.3390/conservation4030029

Boubekraoui, H., Maouni, Y., Ghallab, A., Draoui, B., & Attar, Z. (2024a). Deforestation drivers in northern Morocco: An exploratory spatial data analysis. *Environmental Research Communications*, *6*(7), 071005. DOI: 10.1088/2515-7620/ad5ad6

Boubekraoui, H., Maouni, Y., Ghallab, A., Draoui, M., & Maouni, A. (2023). Spatio-temporal analysis and identification of deforestation hotspots in the Moroccan Western Rif. *Trees, Forests and People*, *12*, 100388. DOI: 10.1016/j.tfp.2023.100388

Boudy, P. (1950). Economie forestière Nord-Africaine, TOME 2: Monographie et traitement des essences forestières. Ed. Larosa, Paris, 525p.

Boudy, P. (1950). *Economie forestière Nord-africaine-Tome 2 : Monographies et traitements des essences forestières*. Edition Larose.

Bourgoin, C., Oszwald, J., Bourgoin, J., Gond, V., Blanc, L., Dessard, H., Van Phan, T., Sist, P., Laderach, P., & Reymondin, L. (2022). Assessing the ecological vulnerability of forest landscape to agricultural frontier expansion in the central highlands of Vietnam. *International Journal of Applied Earth Observation and Geoinformation, 84*, •••. 10.1016/j

Bourougaaoui, A., Ben Jamâa, M. L., & Robinet, C. (2021). Has North Africa turned too warm for a Mediterranean forest pest because of climate change? *Climatic Change, 165*(3-4), 46. DOI: 10.1007/s10584-021-03077-1

Boussalim, Y., & Dallahi, Y. (2025). Predictive modeling of *Cedrus atlantica* natural regeneration: Grazing intensity and bioclimatic extremes as the primary drivers in the Ain Leuh Forest, Middle Atlas of Morocco. *Modeling Earth Systems and Environment, 11*(1), 17. DOI: 10.1007/s40808-024-02191-0

Bouta, A., Laaribya, S., Brache, O., & Dindaroglu, T. (2024). Geospatial modelling of forest fire risk and mitigation strategies in Moroccan Mediterranean regions. *International Journal of Global Warming*. Advance online publication. DOI: 10.1504/IJGW.2024.142423

Bowman, D. M. J. S., Kolden, C. A., Abatzoglou, J. T., Johnston, F. H., van der Werf, G. R., & Flannigan, M. (2020). Vegetation fires in the anthropocene. *Nature Reviews. Earth & Environment, 1*(10), 500–515. DOI: 10.1038/s43017-020-0085-3

Bowring, S. P. K., Li, W., Mouillot, F., Rosan, T. M., & Ciais, P. (2024). Road fragment edges enhance wildfire incidence and intensity, while suppressing global burned area. *Nature Communications, 15*(1), 9176. DOI: 10.1038/s41467-024-53460-6 PMID: 39448625

Bradshaw, A. D. (1965). Evolutionary significance of phenotypic plasticity in plants. *Advances in Genetics, 13*, 115–155. Advance online publication. DOI: 10.1016/S0065-2660(08)60048-6

Breiman, L. (2001). Random forests. *Machine Learning, 45*(1), 5–32. DOI: 10.1023/A:1010933404324

Breshears, D. D., Cobb, N. S., Rich, P. M., Price, K. P., Allen, C. D., Balice, R. G., Romme, W. H., Kastens, J. H., Floyd, M. L., Belnap, J., Anderson, J. J., Myers, O. B., & Meyer, C. W. (2005). Regional vegetation die-off in response to global-change-type drought. *Proceedings of the National Academy of Sciences of the United States of America, 102*(42), 15144–15148. DOI: 10.1073/pnas.0505734102 PMID: 16217022

Brisson, J. (2011). Neighborhood competition and crown asymmetry in Acer saccharum. *Canadian Journal of Forest Research, 31*(12), 2151–2159. DOI: 10.1139/x01-161

Brown, L. C., & Foster, G. R.. (1987). Storm erosivity using idealized intensity distributions. *Transactions of the ASAE. American Society of Agricultural Engineers, 30*(2), 379–386. DOI: 10.13031/2013.31957

Brunetti, M., Maugeri, M., Nanni, T., & Navarra, A. (2002). Droughts and extreme events in regional daily Italian precipitation series. *International Journal of Climatology, 22*(5), 543–558. DOI: 10.1002/joc.751

Bugalho, M. N., Caldeira, M. C., Pereira, J. S., Aronson, J., & Pausas, J. G. (2011). Mediterranean cork oak savannas require human use to sustain biodiversity and ecosystem services. *Frontiers in Ecology and the Environment, 9*(5), 278–286. DOI: 10.1890/100084

Bussotti, F., Pollastrini, M., Holland, V., & Brüggemann, W. (2015). Functional traits and adaptive capacity of European forests to climate change. *Environmental and Experimental Botany, 111*, 91–113. DOI: 10.1016/j.envexpbot.2014.11.006

Cabeza, L. F., Chàfer, M., & Mata, É. (2020). Comparative Analysis of Web of Science and Scopus on the Energy Efficiency and Climate Impact of Buildings. *Energies, 13*(2), 409. DOI: 10.3390/en13020409

Camarero, J. J., Gazol, A., Sangüesa-Barreda, G., Oliva, J., & Vicente-Serrano, S. M. (2015). To die or not to die: Early warnings of tree dieback in response to a severe drought. *Journal of Ecology, 103*(1), 44–57. DOI: 10.1111/1365-2745.12295

Cardil, A., Monedero, S., Selegue, P., Navarrete, M. Á., de-Miguel, S., Purdy, S., Marshall, G., Chavez, T., Allison, K., Quilez, R., Ortega, M., Silva, C. A., & Ramirez, J. (2023). Performance of operational fire spread models in California. *International Journal of Wildland Fire, 32*(11), 1492–1502. DOI: 10.1071/WF22128

Cash, D. W., Adger, W. N., Berkes, F., Garden, P., Lebel, L., Olsson, P., Pritchard, L., & Young, O. (2006). cale and cross-scale dynamics: Governance and information in a multilevel world. *Ecology and Society, 11*(2), 8. http://dx.doi.org/10.5751/ES-01759-110208. DOI: 10.5751/ES-01759-110208

Cerdan, O., Govers, G., Le Bissonnais, Y., Van Oost, K., Poesen, J., Saby, N., Gobin, A., Vacca, A., Quinton, J., Auerswald, K., Klik, A., Kwaad, F. J. P. M., Raclot, D., Ionita, I., Rejman, J., Rousseva, S., Muxart, T., Roxo, M. J., & Dostal, T. (2010). Rates and spatial variations of soil erosion in Europe: A study based on erosion plot data. *Geomorphology, 122*(1–2), 167–177. DOI: 10.1016/j.geomorph.2010.06.011

Chaabani, S. (2013). Influence du climat sur la croissance radicale de chene zeen en Kroumirie, projet de fin d'étude, Ecole supérieur d'Agriculture de Mograne, 57p.

Chan, J. C.-W., & Paelinckx, D. (2008). Evaluation of Random Forest and Adaboost tree-based ensemble classification and spectral band selection for ecotope mapping using airborne hyperspectral imagery. *Remote Sensing of Environment, 112*(6), 2999–3011. DOI: 10.1016/j.rse.2008.02.011

Cheddadi, R., François, L., Bar-Hen, A., Huang, K., Demarteau, M., Roujjati, A., Baali, A., Damnati, B., Taieb, M., & Laurent, J.-M. (2006). Impact du climat sur quelques espèces méditerranéennes : Du passé au futur. *Réponses adaptatives au changement global : résultats et prospective*, 91-96.

Cheddadi, R., Fady, B., François, L., Hajar, L., Suc, J.-P., Huang, K., Demarteau, M., Vendramin, G. G., & Ortu, E. (2009). Putative glacial refugia of *Cedrus atlantica* deduced from Quaternary pollen records and modern genetic diversity. *Journal of Biogeography*, *36*(7), 1361–1371. DOI: 10.1111/j.1365-2699.2008.02063.x

Cheddadi, R., Henrot, A.-J., François, L., Boyer, F., Bush, M., Carré, M., Coissac, E., De Oliveira, P. E., Ficetola, F., Hambuckers, A., Huang, K., Lézine, A.-M., Nourelbait, M., Rhoujjati, A., Taberlet, P., Sarmiento, F., Abel-Schaad, D., Alba-Sánchez, F., & Zheng, Z. (2017). Microrefugia, climate change, and conservation of *Cedrus atlantica* in the Rif Mountains, Morocco. *Frontiers in Ecology and Evolution*, *5*, 114. DOI: 10.3389/fevo.2017.00114

Chen, T., Kornblith, S., Norouzi, M., & Hinton, G. (2020). A simple framework for contrastive learning of visual representations. In *Proceedings of the 37th International Conference on Machine Learning* (pp. 1597–1607). doi:10.5555/3524938.3525087

Chen, I.-C., Hill, J. K., Ohlemüller, R., Roy, D. B., & Thomas, C. D. (2011). Rapid Range Shifts of Species Associated with High Levels of Climate Warming. *Science*, *333*(6045), 1024–1026. DOI: 10.1126/science.1206432 PMID: 21852500

Chen, T., & Guestrin, C. (2016). XGBoost: A scalable tree boosting system. In *Proceedings of the 22nd ACM SIGKDD International Conference on Knowledge Discovery and Data Mining* (pp. 785–794). DOI: 10.1145/2939672.2939785

Chriha, S., & Sghari, A. (2013). Les incendies de forêt en Tunisie. Séquelles irréversibles de la révolution de 2011. *Méditerranée. Revue géographique des pays méditerranéens/Journal of Mediterranean geography*, (121), 87-93.

Chuvieco, E., Mouillot, F., van der Werf, G. R., San Miguel, J., Tanase, M., Koutsias, N., García, M., Yebra, M., Padilla, M., Gitas, I., Heil, A., Hawbaker, T. J., & Giglio, L. (2019). Historical background and current developments for mapping burned area from satellite Earth observation. *Remote Sensing of Environment, 225*, 45–64. DOI: 10.1016/j.rse.2019.02.013

Cobo, M. J., López-Herrera, A. G., Herrera-Viedma, E., & Herrera, F. (2011). Science mapping software tools: Review, analysis, and cooperative study among tools. *Journal of the American Society for Information Science and Technology, 62*(7), 1382–1402. DOI: 10.1002/asi.21525

Cochard, H. & Granier A. (1999). Fonctionnement hydraulique des arbres forestiers. Rev. For. Fr. Li, 121-133 pp.

Cosens, B. A., Craig, R. K., Hirsch, S. L., Arnold, C. A. T., Benson, M. H., DeCaro, D. A., Garmestani, A. S., Gosnell, H., Ruhl, J. B., & Schlager, E. (2017). The role of law in adaptive governance. *Ecology and Society*, *22*(1), 30. Advance online publication. DOI: 10.5751/ES-08731-220130

Craig, R. K., Garmestani, A. S., Allen, C. R., Arnold, C. A., Birgé, H., DeCaro, D. A., Fremier, A. K., Gosnell, H., & Schlager, E. (2017). Balancing stability and flexibility in adaptive governance: An analysis of tools available in U.S. *Environmental Law. Ecology and Society*, *22*(2), 3. http://dx.doi.org/10.5751/ES-08983-220203. DOI: 10.5751/ES-08983-220203

Cui, G., Kwak, H., Choi, S., Kim, M., Lim, C.-H., Lee, W.-K., Kim, J.-S., & Chae, Y. (2016). Assessing vulnerability of forests to climate change in South Korea. *Journal of Forestry Research*, *27*(3), 489–503. 10.1007/s11676-015-0201-2. DOI: 10.1007/s11676-015-0201-2

Cutter, S. L., Boruff, B. J., & Shirley, W. L. (2003). Social vulnerability to environmental hazards. *Social Science Quarterly*, *84*(2), 242–261. 10.1111/1540-6237.8402002. DOI: 10.1111/1540-6237.8402002

D'Odorico, P., Bhattachan, A., Davis, K. F., Ravi, S., & Runyan, C. W. (2013). Global desertification: Drivers and feedbacks. *Advances in Water Resources*, *51*, 326–344. DOI: 10.1016/j.advwatres.2012.01.013

Daget, P. (1977). Le bioclimat méditerranéen: Caractères généraux, modes de caractérisation. *Vegetatio*, *34*(1), 1–20. DOI: 10.1007/BF00119883

Dale, V. H., Joyce, L. A., McNulty, S., Neilson, R. P., Ayres, M. P., Flannigan, M. D., Hanson, P. J., Irland, L. C., Lugo, A. E., Peterson, C. J., Simberloff, D., Swanson, F. J., Stocks, B. J., & Wotton, B. M. (2001). Climate Change and Forest Disturbances : Climate change can affect forests by altering the frequency, intensity, duration, and timing of fire, drought, introduced species, insect and pathogen outbreaks, hurricanes, windstorms, ice storms, or landslides. *Bioscience*, *51*(9), 723–734. DOI: 10.1641/0006-3568(2001)051[0723:CCAFD]2.0.CO;2

Dask-GeoPandas. (2024). *Dask-GeoPandas documentation*. https://dask-geopandas.readthedocs.io

David, T. S., Henriques, M. O., Kurz-Besson, C., Nunes, J., Valente, F., Vaz, M., Pereira, J. S., Siegwolf, R., Chaves, M. M., Gazarini, L. C., Gazarini, L. C., & David, J. S. (2007). Water-use strategies in two co-occurring Mediterranean evergreen oaks: Surviving the summer drought. *Tree Physiology*, *27*(6), 793–803. DOI: 10.1093/treephys/27.6.793 PMID: 17331898

Davis, M. B., & Shaw, R. G. (2001). Range Shifts and Adaptive Responses to Quaternary Climate Change. *Science*, *292*(5517), 673–679. DOI: 10.1126/science.292.5517.673 PMID: 11326089

De Riguo, D., & Caudullo, G. (2016). Quercus suber in Europe: distribution, habitat, usage and threats. Eds.), European Atlas of Forest, Tree Species. Publ. Off. EU, Luxembourg, p 165.

Département des eaux et Eaux et Forêts. (2020). *Stratégie nationale de gestion intégrée des incendies de forêt 2020–2030.*

Desmet, P. J. J., & Govers, G. (1996). A GIS procedure for automatically calculating the USLE LS factor on topographically complex landscape units. *Journal of Soil and Water Conservation*, *51*(5), 427–433. DOI: 10.1080/00224561.1996.12457102

Díaz, S. M., Settele, J., Brondízio, E., Ngo, H., Guèze, M., Agard, J., Arneth, A., Balvanera, P., Brauman, K., & Butchart, S. (2019). *The global assessment report on biodiversity and ecosystem services: Summary for policy makers.*

Diffenbaugh, N. S., Pal, J. S., Giorgi, F., & Gao, X. (2007). Heat stress intensification in the Mediterranean climate change hotspot. *Geophysical Research Letters*, *34*(11), 2007GL030000. Advance online publication. DOI: 10.1029/2007GL030000

Dinerstein, E., Olson, D., Joshi, A., Vynne, C., Burgess, N. D., Wikramanayake, E., Hahn, N., Palminteri, S., Hedao, P., Noss, R., Hansen, M., Locke, H., Ellis, E. C., Jones, B., Barber, C. V., Hayes, R., Kormos, C., Martin, V., Crist, E., & Saleem, M. (2017). An Ecoregion-Based Approach to Protecting Half the Terrestrial Realm. *Bioscience*, *67*(6), 534–545. DOI: 10.1093/biosci/bix014 PMID: 28608869

Direction de la Météorologie Nationale (DMN). (2020). *Synthèse du climat au Maroc en 2024*. Rabat. Disponible sur: https://www.marocmeteo.ma/fr/climat-2024

Djoukbala, O., Hasbaia, M., Benselama, O., & Mazour, M. (2019). Comparison of erosion prediction models USLE, MUSLE and RUSLE in a Mediterranean watershed. *Modeling Earth Systems and Environment*, *5*(2), 725–743. DOI: 10.1007/s40808-018-0562-6

Donthu, N., Kumar, S., Mukherjee, D., Pandey, N., & Lim, W. M. (2021). How to conduct a bibliometric analysis: An overview and guidelines. *Journal of Business Research*, *133*, 285–296. DOI: 10.1016/j.jbusres.2021.04.070

Dugré, M., Hayot-Sasson, V., & Glatard, T. (2023). Performance comparison of Dask and Apache Spark on HPC systems for neuroimaging. *Concurrency and Computation*, *e7635*(21), e7635. DOI: 10.1002/cpe.7635

Dukes, J. S., Chiariello, N. R., Cleland, E. E., Moore, L. A., Rebecca Shaw, M., Thayer, S., Tobeck, T., Mooney, H. A., & Field, C. B. (2005). Responses of grassland production to single and multiple global environmental changes. *PLoS Biology*, *3*(10), e319. DOI: 10.1371/journal.pbio.0030319 PMID: 16076244

Economique, C. Social et Environnemental (CESE). (2022). Les écosystèmes forestiers du Maroc: Risques, défis et opportunités. Rabat, Maroc.

El Amarty, F., Chakir, A., Purohit, S., Elhag, A., & Taleb, M. A. (2025). Comparative evaluation of machine learning models for forest fire prediction in Morocco. *Earth Systems and Environment*. Advance online publication. DOI: 10.1007/s41748-025-00823-8

El Khaddari, M., & El Mousadik, A.. (2019). Seed germination and juvenile growth of *Tetraclinis articulata* from five Moroccan provenances. *Plant Cell Biotechnology and Molecular Biology*, *20*(11–12), 556–565.

El Maachi, S. E., Saadane, R., Abdellah, A., et Jamal, D. (2024). Navigating change: enhancing transportation resilience in fire-prone Moroccan areas. DOI: 10.1109/WFPST58552.2024.00030

El Mazi, M., Bouhlal, A., Saber, E.-R., Hmamouchi, M., & Ait Haddou, Y. (2025). Effects of wildfire severity on the soil properties of Mediterranean forest ecosystems in Morocco. *Euro-Mediterranean Journal for Environmental Integration*. Advance online publication. DOI: 10.1007/s41207-025-00896-x

El Mazi, M., Boutallaka, M., Saber, E.-R., Chafik, N., & Hmamouchi, M. (2024). Forest fire risk modeling in Mediterranean forests of Morocco using remote sensing and geospatial tools. *Euro-Mediterranean Journal for Environmental Integration*. Advance online publication. DOI: 10.1007/s41207-024-00591-3

El Mderssa, M., Malki, F., Ikraoun, H., Nassiri, L., & Ibijbijen, J. (2022). Détermination des paramètres d'évaluation du stock de carbone dans les écosystèmes forestiers (*Cedrus atlantica* Manetti, cèdre de l'Atlas au Maroc): Méthodes spécifiques et génériques. *Bois et Forêts des Tropiques, 351*, 67–77. DOI: 10.19182/bft2022.351.a36330

ELI5. (2024). *ELI5 documentation.* https://eli5.readthedocs.io

Elia, M., D'Este, M., Ascoli, D., Giannico, V., Spano, G., Ganga, A., Colangelo, G., Lafortezza, R., & Sanesi, G. (2020). Estimating the probability of wildfire occurrence in Mediterranean landscapes using artificial neural networks. *Environmental Impact Assessment Review, 85*, 106474. DOI: 10.1016/j.eiar.2020.106474

Elith, J., & Graham, C. H. (2009). Do they? How do they? WHY do they differ? On finding reasons for differing performances of species distribution models. *Ecography, 32*(1), 66–77. DOI: 10.1111/j.1600-0587.2008.05505.x

Ellatifi, M. (2012). *L'économie de la forêt et des produits forestiers au Maroc: Bilan et perspectives* (Thèse de doctorat en économie). Université Montesquieu – Bordeaux IV.

Ellegaard, O., & Wallin, J. A. (2015). The bibliometric analysis of scholarly production: How great is the impact? *Scientometrics, 105*(3), 1809–1831. DOI: 10.1007/s11192-015-1645-z PMID: 26594073

Elmalki, M., Ichen, A., Aarab, M., Mounir, F., El Mderssa, M., Oularbi, Y., Tajdi, A., Moukrim, S., & Ouzzaouit, L. A. (2022). Impact of climate change on forest resources: Case of Quercus rotundifolia, Tetraclinis articulata, Juniperus phoenicea, J. oxycedrus, J. thurifera and Pinus halepensis. *Biosystems Diversity, 30*(3), 255-262.

ElMderssa, M., Atlas, S., Ettaqy, A., Mouridi, Z. E., Abbas, Y., & Ibijbijen, J. (2025). Comparative evaluation of three methods for soil organic carbon quantification in Morocco's Middle Atlas forests. *Discover Soil, 2*(1), 98. Advance online publication. DOI: 10.1007/s44378-025-00127-3

Emberger, L. (1925). Le domaine naturel de l'arganier. *Bulletin de la Société Botanique de France, 72*(4), 770–774. DOI: 10.1080/00378941.1925.10832788

Emberger, L. (1930). La végétation de la région méditerranéenne. Essai d'une classification des groupements végétaux. *Revue Générale de Botanique, 42*, 641–662.

Emberger, L. (1938). Contribution à la connaissance des cèdres et en particulier du déodar et du cèdre de l'Atlas. *Revue de Botanique Appliquée et d'Agriculture Coloniale, 18*(198), 77–92. DOI: 10.3406/jatba.1938.5828

Emberger, L. (1939). Aperçu général sur la végétation du Maroc. *Veröffentlichungen des Geobotanischen Instituts Rübel Zürich, 14*, 40–157.

Emberger, L. (1955). Une classification biogéographique des climats. *Recueil des Travaux du Laboratoire de Botanique et de Zoologie de la Faculté des Sciences de Montpellier, 7*, 3–43.

Emberger, L. (1964). La position phytogéographique du Maroc dans l'ensemble méditerranéen. *Al Awamia, 12*, 1–15.

Engler, R., Randin, C. F., Thuiller, W., Dullinger, S., Zimmermann, N. E., Araújo, M. B., Pearman, P. B., Le Lay, G., Piedallu, C., Albert, C. H., Choler, P., Coldea, G., De LAMO, X., Dirnböck, T., Gégout, J.-C., Gómez-García, D., Grytnes, J.-A., Heegaard, E., Høistad, F., … Guisan, A. (2011). 21st century climate change threatens mountain flora unequally across Europe. *Global Change Biology, 17*(7), 2330-2341. DOI: 10.1111/j.1365-2486.2010.02393.x

Ennajah, A. (2010). Croissance et productivité des forêts de chênes lièges « Quercus-suber L.» en Tunisie. Vulnérabilité aux changements climatiques. Thèse de Doctorat en Sciences Biologiques – faculté des Sciences de Tunis - 192 p.

EUFORGEN. Distribution map of cork oak (Quercus suber) (2009). www.euforgen .org

European Commission. (2019). *Peer review report Tunisia 2018*. Directorate-General for European Civil Protection and Humanitarian Aid Operations (DG ECHO).

Fady, B., Farsakoglou, A. M., Caron, M., Abulaila, K., Aleksic, J., Alipour, S., Balian, D., Bedair, H., Bogunić, F., Cheikh Albassatneh, M., David-Schwartz, R., Delgado Clavero, C., Dönmez, A. A., Fennane, M., Gateble, G., Gauquelin, T., Hachi Illoul, M., Khaldi, A., Mahfoud-Saad, I., & Bou Dagher Karrat, M. (2025). Native trees of the Mediterranean region: Distribution, diversity and conservation challenges. *Current Forestry Reports, 11*(1), 20. DOI: 10.1007/s40725-025-00252-w

FAO & ITPS. (2015). *Status of the World's Soil Resources (SWSR) – Main Report*. Food and Agriculture Organization of the United Nations.

FAO. (2006). *Fire management: voluntary guidelines. Principles and strategic actions* (Fire Management Working Paper 17). Food and Agriculture Organization of the United Nations.

FAO. (2016). *Global Forest Resources Assessment*. https://www.fao.org/3/a-i4808e .pdf

Feng, Q., Li, Y., Zhang, H., & Chen, X. (2024). Interdisciplinary perspectives on forest ecosystems and their functions in hydrology, climate and biodiversity. *Environmental Reviews*. Advance online publication. DOI: 10.1139/er-2024-0010

Fennane, M., & Ibn Tattou, M. (2012). Statistiques et commentaires sur l'inventaire actuel de la flore vasculaire du Maroc. *Bulletin de l'Institut Scientifique, Rabat, section Sciences de la Vie, 34*(1), 1-9.

Fennane, M., & Ibn Tattou, M. (2012). Statistiques et commentaires sur l'inventaire actuel de la flore vasculaire du Maroc [Statistics and comments on the current inventory of the vascular flora of Morocco]. *Bulletin de l'Institut Scientifique, Rabat. Section Sciences de la Vie, 34*(1), 1–9.

Fernandes, S., Niknejad, M., Barata, C., & Bernardino, A. (2023). Forest fires identification using self-supervised learning. In Rousseau, J.-J., & Kapralos, B. (Eds.), *Pattern Recognition, Computer Vision, and Image Processing* (LNCS Vol. 13644). Springer. https://doi.org/DOI: 10.1007/978-3-031-37742-6_17

Fick, S. E., & Hijmans, R. J. (2017). WorldClim 2 : New 1-km spatial resolution climate surfaces for global land areas. *International Journal of Climatology, 37*(12), 4302–4315. DOI: 10.1002/joc.5086

Fischer, J., & Lindenmayer, D. B. (2007). Landscape modification and habitat fragmentation: A synthesis. *Global Ecology and Biogeography, 16*(3), 265–280. DOI: 10.1111/j.1466-8238.2007.00287.x

Flannigan, M. D., Krawchuk, M. A., de Groot, W. J., Wotton, B. M., & Gowman, L. M. (2009). Implications of changing climate for global wildland fire. *International Journal of Wildland Fire, 18*(5), 483–507. DOI: 10.1071/WF08187

Folke, C., Carpenter, S., Walker, B., Scheffer, M., Elmqvist, T., Gunderson, L., & Holling, C. S. (2004). Regime shifts, resilience, and biodiversity in ecosystem management. *Annual Review of Ecology, Evolution, and Systematics, 35*(1), 557–581. DOI: 10.1146/annurev.ecolsys.35.021103.105711

Food and Agriculture Organization (FAO). (2020). *State of the World's Forests 2020: Forests, Biodiversity and People*. FAO.

Food and Agriculture Organization of the United Nations, & Programme des Nations Unies pour l'Environnement. (2020). *La situation des forêts du monde 2020: Forêts, biodiversité et activité humaine.* Food and Agriculture Organization of the United Nations. DOI: 10.4060/ca8642fr

Food and Agriculture Organization of the United Nations, & Silva Mediterranea. (2024, February). *Restoring Mediterranean forests — UN Decade Flagship.* Food and Agriculture Organization of the United Nations. https://www.fao.org/silva-mediterranea/un-decade-of-ecosystem-restoration/restoring-mediterranean-forests-flagship/en

Food and Agriculture Organization of the United Nations. (2000). FAO Forestry Paper: Vol. 140. *Thuriferous juniper stands in Morocco: Research and management.* Food and Agriculture Organization of the United Nations.

Food and Agriculture Organization of the United Nations. (2017). *Production d'outils d'aide à la décision et de gestion en matière de vulnérabilité des écosystèmes forestiers méditerranéens aux effets du changement climatique et valorisation des efforts d'optimisation des biens et services (Projet FFEM/Plan Bleu).* Food and Agriculture Organization of the United Nations.

Food and Agriculture Organization of the United Nations. (2017). *State of Mediterranean forests 2017.* Food and Agriculture Organization of the United Nations.

Food and Agriculture Organization of the United Nations. (2020). *Technical assistance for forest fire management in Algeria.* Food and Agriculture Organization of the United Nations. https://www.fao.org/in-action/nenfire/projects/algeria/fr/

Food and Agriculture Organization of the United Nations. (2023, November 10). *Innovative monitoring of forest and landscape restoration in the Mediterranean region.* Food and Agriculture Organization of the United Nations. https://www.fao.org/forest-resources-assessment/resources/news/detail/innovative-monitoring-of-forest-and-landscape-restoration-in-the-mediterranean-region/en

Food and Agriculture Organization of the United Nations. (2024). *La situation des forêts du monde 2024 – Innovations dans le secteur forestier pour un avenir plus durable.* Food and Agriculture Organization of the United Nations., DOI: 10.4060/cd1211fr

Food and Agriculture Organization of the United Nations. (2024). *Restoring the Mediterranean region: Status and challenges (Unasylva No. 255, Vol. 75, 2024/1).* Food and Agriculture Organization of the United Nations. DOI: 10.4060/cd1720en

Ford, A. E. S., Coughlan, M. R., & Meentemeyer, R. K. (2021). modelling human–fire interactions: Combining alternative perspectives and approaches. *Frontiers in Environmental Science*, *9*, 649835. DOI: 10.3389/fenvs.2021.649835

Francos, M., Colino-Prieto, F., & Sánchez-García, C. (2024). How Mediterranean ecosystem deals with wildfire impact on soil ecosystem services and functions: A review. *Land (Basel)*, *13*(4), 407. DOI: 10.3390/land13040407

Franklin, J. (2009). *Mapping species distributions : Spatial inference and prediction.* Cambridge University Press., DOI: 10.1017/CBO9780511810602

Friedman, J. H. (2001). Greedy function approximation: A gradient boosting machine. *Annals of Statistics*, *29*(5), 1189–1232. DOI: 10.1214/aos/1013203451

Frouja, O., Hammami, M., Dakhlaoui, S., Wannes, W., Hessini, K., Msaada, K., & Ahmed, B. (2022). Intra and interspecific variability of Quercus suber and Quercus canariensis, an intrinsic water-use efficiency approach for differentiation. *Environmental Research Communications*, *4*(10), 105002. Advance online publication. DOI: 10.1088/2515-7620/ac9729

Füssel, H.-M. (2010). *Review and Quantitative Analysis of Indices of Climate Change Exposure, Adaptive Capacity, Sensitivity, and Impacts.* Development and Climate Change. DOI: 10.1596/9193

Gaertner, M., Den Breyen, A., Hui, C., & Richardson, D. M. (2009). Impacts of alien plant invasions on species richness in Mediterranean-type ecosystems: A meta-analysis. *Progress in Physical Geography*, *33*(3), 319–338. DOI: 10.1177/0309133309341607

Gallo Granizo, C., Bacciu, V., Eden, J., Hernández Paredes, E., García Feced, C., Bedhief, S., Oom, D., Goldammer, J., Dieppois, B., Belen, İ., Mitri, G., Turhan, Ü., Ascoli, D., Alfonso, L., & San-Miguel-Ayanz, J. (2025). fire management in the Mediterranean region. In *The status of Mediterranean forests 2025* (pp. 83-103). Food and Agriculture Organization of the United Nations.

Gandomi, A., & Haider, M. (2015). Big data concepts, methods, and analytics. *International Journal of Information Management*, *35*(2), 137–144. DOI: 10.1016/j.ijinfomgt.2014.10.007

Gauquelin, T., Michon, G., Joffre, R., Duponnois, R., Genin, D., Fady, B., Bou Dagher-Kharrat, M., Derridj, A., Slimani, S., Badri, W., Alifriqui, M., Auclair, L., Simenel, R., Aderghal, M., Baudoin, E., Galiana, A., Prin, Y., Sanguin, H., Fernandez, C., & Baldy, V. (2018). Mediterranean forests, land use and climate change: A social-ecological perspective. *Regional Environmental Change*, *18*(3), 623–636. DOI: 10.1007/s10113-016-0994-3

Geist, H. J., & Lambin, E. F. (2002). Proximate causes and underlying driving forces of tropical deforestation: Tropical forests are disappearing as the result of many pressures, both local and regional, acting in various combinations in different geographical locations. *Bioscience*, *52*(2), 143–150. DOI: 10.1641/0006-3568(2002)052[0143:PCAUDF]2.0.CO;2

Ghanam, M. (2003, décembre). *La désertification au Maroc – Quelle stratégie de lutte?* 2nd FIG Regional Conference, Marrakech, Maroc.

Ghani, N. A., Teo, P. C., Ho, T. C., Choo, L. S., Kelana, B. W. Y., Adam, S., & Ramliy, M. K. (2022). Bibliometric analysis of global research trends on higher education internationalization using Scopus database: Towards sustainability of higher education institutions. *Sustainability (Basel)*, *14*(14), 8810. DOI: 10.3390/su14148810

Ghanmi, M., Aafi, A., Satrani, B., Aberchane, M., Khia, A., & Yakhlef, S. B. (2014). Aromatic and medicinal plants in North Africa: Opportunities, constraints and prospects. In Gurib-Fakim, A. (Ed.), *Novel Plant Bioresources: Applications in Food, Medicine and Cosmetics* (pp. 411–421). Wiley-Blackwell. DOI: 10.1002/9781118460566.ch29

Gharnit, Y., Outourakht, A., Boulli, A., & Hassib, A. (2023). Biodiversity, autecology and status of aromatic and medicinal plants in Geopark M'Goun (Morocco). *Annali di Botanica*, *13*, 39–54.

Ghefar, M., Boutkhil, M., & Bouazzaoui, A. (2024). Assessing the impact of anthropogenic activities on land use and land cover changes in the semi-arid and arid regions of Algeria. *Environmental Monitoring and Assessment*, *196*(4), 383. DOI: 10.1007/s10661-024-12524-2

Ghorbanzadeh, O., Blaschke, T., Gholamnia, K., & Aryal, J. (2019). Evaluation of different machine learning methods and deep-learning CNNs for landslide detection. *Remote Sensing*, *11*(2), 196. DOI: 10.3390/rs11020196

Giglio, L., Boschetti, L., Roy, D. P., Humber, M. L., & Justice, C. O. (2018). The Collection 6 MODIS burned area mapping algorithm and product. *Remote Sensing of Environment*, *217*, 72–85. DOI: 10.1016/j.rse.2018.08.005 PMID: 30220740

Gimeno, T. E., Pías, B., Lemos-Filho, J. P., & Valladares, F. (2009). Plasticity and stress tolerance override local adaptation in the responses of Mediterranean holm oak seedlings to drought and cold. *Tree Physiology*, *29*(1), 87–98. DOI: 10.1093/treephys/tpn007

Giorgi, F., & Lionello, P. (2008). Climate change projections for the Mediterranean region. *Global and Planetary Change, Mediterranean climate: trends, variability and change, 63*(2), 90-104. DOI: 10.1016/j.gloplacha.2007.09.005

Gitas, I. Z. (2009). Multi-temporal soil erosion risk assessment using a modified USLE raster model. *EARSeL eProceedings*.

Gleckler, P. J., Taylor, K. E., & Doutriaux, C. (2008). Performance metrics for climate models. *Journal of Geophysical Research*, *113*(D6), 2007JD008972. Advance online publication. DOI: 10.1029/2007JD008972

Goeking, S. A., & Tarboton, D. G. (2020). Forests and water yield: A synthesis of disturbance effects on streamflow and snowpack in western coniferous forests. *Journal of Forestry*, *118*(2), 172–192. DOI: 10.1093/jofore/fvz069

Gómez, M., González, M., Gómez-Aparicio, L., & Serrano, M. (2020). Coexistent Mediterranean woody species as a driving factor of Phytophthora cinnamomi infectivity and survival. *Annals of Applied Biology*, *177*(1), 41–50. DOI: 10.1111/aab.12599

González, M., Romero, M.-Á., García, L.-V., Gómez-Aparicio, L., & Serrano, M.-S. (2020). Unravelling the role of drought as predisposing factor for Quercus suber decline caused by Phytophthora cinnamomi. *European Journal of Plant Pathology*, *156*(4), 1015–1021. DOI: 10.1007/s10658-020-01951-9

Goodwin, P., & Wright, G. (2014). *Decision analysis for management judgment* (4th ed.). Wiley.

Gratani, L., Catoni, R., & Varone, L. (2016). Evergreen species response to Mediterranean climate stress factors. *IForest (Viterbo)*, *9*(6), 946–953. DOI: 10.3832/ifor1848-009

Graus, S., Ferreira, T. M., Vasconcelos, G., & Ortega, J. (2024). Changing Conditions: Global Warming-Related Hazards and Vulnerable Rural Populations in Mediterranean Europe. *Urban Science (Basel, Switzerland)*, *8*(2), 42. DOI: 10.3390/urbansci8020042

Grotta, A. T., Gartner, B. L., & Radosevich, S. R. (2004). Influence of species proportion and timing of establishment on stem quality in mixed red alder – Douglas-fir plantations. *Canadian Journal of Forest Research, 34*(4), 863–873. DOI: 10.1139/x03-259

Guion, A., Turquety, S., Polcher, J., Pennel, R., Bastin, S., & Arsouze, T. (2022). Droughts and heatwaves in the Western Mediterranean: Impact on vegetation and wildfires using the coupled WRF-ORCHIDEE regional model. *Climate Dynamics, 58*(9–10), 2881–2903. DOI: 10.1007/s00382-021-05938-y

Guisan, A., Thuiller, W., & Zimmermann, N. E. (2017). *Habitat Suitability and Distribution Models With Applications in R*. Cambridge University Press., DOI: 10.1017/9781139028271

Guisan, A., & Zimmermann, N. E. (2000). Predictive habitat distribution models in ecology. *Ecological Modelling, 135*(2), 147–186. DOI: 10.1016/S0304-3800(00)00354-9

Haddad, N. M., Brudvig, L. A., Clobert, J., Davies, K. F., Gonzalez, A., Holt, R. D., Lovejoy, T. E., Sexton, J. O., Austin, M. P., Collins, C. D., Cook, W. M., Damschen, E. I., Ewers, R. M., Foster, B. L., Jenkins, C. N., King, A. J., Laurance, W. F., Levey, D. J., Margules, C. R., & Townshend, J. R. (2015). Habitat fragmentation and its lasting impact on Earth's ecosystems. *Science Advances, 1*(2), e1500052. DOI: 10.1126/sciadv.1500052 PMID: 26601154

Hajek, P., Seidel, D., & Leuschner, C. (2015). Mechanical abrasion, and not competition for light, is the dominant canopy interaction in a temperate mixed forest. *Forest Ecology and Management, 348*, 108–116. DOI: 10.1016/j.foreco.2015.03.019

Hakam, O., Ongoma, V., Beniaich, A., Meskour, B., El Kadi, M. A., Brouziyne, Y., & Baali, A. (2024). Assessment of the impact of climate change on argan tree in the Mediterranean GIAHS site, Morocco: Current and future distributions. *Modeling Earth Systems and Environment, 10*(4), 5529–5552. DOI: 10.1007/s40808-024-02077-1

Hallegatte, S., Fay, M., Bangalore, M., Kane, T., & Bonzanigo, L. (2015). *Shock Waves : Managing the Impacts of Climate Change on Poverty*. World Bank Publications., DOI: 10.1596/978-1-4648-0673-5

Hamida, B., & Kouba, Y. (2025). Atlas cedar response to climate-driven disturbances at its southern range edge. *Environmental Monitoring and Assessment, 197*(6), 673. DOI: 10.1007/s10661-025-14140-0

Hanley, J. A., & McNeil, B. J. (1982). The meaning and use of the area under a receiver operating characteristic (ROC) curve. *Radiology, 143*(1), 29–36. DOI: 10.1148/radiology.143.1.7063747 PMID: 7063747

Hannah, L., Midgley, G., Andelman, S., Araújo, M., Hughes, G., Martinez-Meyer, E., Pearson, R., & Williams, P. (2007). Protected area needs in a changing climate. *Frontiers in Ecology and the Environment, 5*(3), 131–138. DOI: 10.1890/1540-9295(2007)5[131:PANIAC]2.0.CO;2

Hantson, S., Pueyo, S., & Chuvieco, E. (2015). Global fire size distribution is driven by human impact and climate. *Global Ecology and Biogeography, 24*(1), 77–86. DOI: 10.1111/geb.12246

Hastie, T., Tibshirani, R., & Friedman, J. (2009). *The elements of statistical learning: Data mining, inference, and prediction* (2nd ed.). Springer. DOI: 10.1007/978-0-387-84858-7

Hellmann, J. J., Byers, J. E., Bierwagen, B. G., & Dukes, J. S. (2008). Five potential consequences of climate change for invasive species. *Conservation Biology, 22*(3), 534–543. DOI: 10.1111/j.1523-1739.2008.00951.x PMID: 18577082

Hengl, T., de Jesus, J. M., Heuvelink, G. B. M., Gonzalez, M. R., Kilibarda, M., Blagotić, A., & Kempen, B. (2017). SoilGrids250m: Global gridded soil information based on machine learning. *PLoS One, 12*(2), e0169748. DOI: 10.1371/journal.pone.0169748 PMID: 28207752

Herzi, M. (2016). Etude de l'impact des légumineuses associées au chêne liège sur sa croissance, cas de cytise triflore Cytisus triflorus. L'Hérit., Mémoire de Mastère de Recherche, INAT, 36-37 p.

Hidalgo-Triana, N., Solakis, A., Casimiro-Soriguer, F., Choe, H., Navarro, T., Pérez-Latorre, A. V., & Thorne, J. H. (2023). The high climate vulnerability of western Mediterranean forests. *The Science of the Total Environment, 895*, 164983. DOI: 10.1016/j.scitotenv.2023.164983 PMID: 37353024

Holling, C. S. (1973). Resilience and stability of ecological systems. *Annual Review of Ecology and Systematics, 4*(1), 1–23. DOI: 10.1146/annurev.es.04.110173.000245

Hulley, G. C., Dousset, B., & Kahn, B. H. (2020). Rising trends in heatwave metrics across southern California. *Earth's Future, 8*(7), e2020EF001480. DOI: 10.1029/2020EF001480

Ilmen, R., & Benjelloun, H. (2013). Les écosystèmes forestiers marocains à l'épreuve des changements climatiques. *Forêt Méditerranéenne, 34*(3), 195–208.

Institut Royal des Études Stratégiques (IRES). (2024). *L'avenir de la forêt marocaine: Rapport de synthèse*. IRES.

Intergovernmental Panel on Climate Change (IPCC). (2021). *Climate Change 2021: The Physical Science Basis – Summary for Policymakers*. Cambridge University Press. https://www.ipcc.ch/report/ar6/wg1

Intergovernmental Panel on Climate Change (IPCC). (2021). *Climate Change 2021: The Physical Science Basis*. Contribution of Working Group I to the Sixth Assessment Report. Cambridge University Press.

IPCC. (2001). *Climate Change 2001: Impacts, Adaptation, and Vulnerability*. Cambridge University Press.

IPCC. (2007). [Synthesis Report. Contribution of Working Groups I, II and III to the Fourth Assessment Report of the Intergovernmental Panel on Climate Change.]. *Climatic Change*, •••, 2007.

IPCC. (2014). *Climate Change 2014: Synthesis Report. Contribution of Working Groups I, II and 3 to the Fifth Assessment Report of the Intergovernmental Panel on Climate Change* [Core Writing Team, R.K. Pachauri and L.A. Meyer (eds.)]. IPCC, Geneva, Switzerland, 151 pp.

IPCC. (2014). Climate Change 2014: Synthesis Report. Contribution of Working Groups I, II and III to the Fifth Assessment Report of the Intergovernmental Panel on Climate Change. IPCC, Geneva. AR5 Synthesis Report [— IPCC]. *Climatic Change*, •••, 2014.

IPCC. (2021). *Climate Change 2021: The Physical Science Basis*. Cambridge University Press.

IPCC. (2021). Summary for Policymakers. In Allan, R. P., Cassou, C., Chen, D., Cherchi, A., Connors, L., Doblas-Reyes, F. J., Douville, H., Driouech, F., Edwards, T. L., Fischer, E., Flato, G. M., Forster, P., AchutaRao, K. M., Adhikary, B., Aldrian, E., & Armour, K. (Eds.), *Climate Change 2021 : The Physical Science Basis. Contribution of Working Group I to the Sixth Assessment Report of the Intergovernmental Panel on Climate Change* (p. 33). Cambridge University Press., DOI: 10.1017/9781009157896.001

IPCC. (2023). *Climate Change 2022 – Impacts, Adaptation and Vulnerability : Working Group II Contribution to the Sixth Assessment Report of the Intergovernmental Panel on Climate Change* (1re éd.). Cambridge University Press. DOI: 10.1017/9781009325844

Itzkin, A., Scholes, M. C., Clifford-Holmes, J. K., Rowntree, K., Van der Waal, B., & Coetzer, K. (2021). A social-ecological systems understanding of drivers of degradation. *Sustainability (Basel), 13*(2), 516. DOI: 10.3390/su13020516

Jactel, H., Brockerhoff, E. G., & Duelli, P. (2005). A test of the biodiversity-stability theory: meta-analysis of tree species diversity effects on insect pest infestations, and re-examination of responsible factors. In Scherer-Lorenzen, M., Körner, C., & Schulze, E. D. (Eds.), *Forest Diversity and Function* (pp. 235–262). Springer. DOI: 10.1007/3-540-26599-6_12

Jadouli, A., & El Amrani, C. (2024). *Advanced wildfire prediction in Morocco: development of a geospatial machine learning framework*. IEEE., DOI: 10.1109/ACCESS.2024.3516784

Jadouli, A., & El Amrani, C. (2025). *Hybrid parallel architecture integrating FFN, CNN, and MLP models for wildfire prediction*. Lecture Notes in Networks and Systems., DOI: 10.1007/978-3-031-88653-9_16

Jain, P., Coogan, S. C. P., Subramanian, S. G., Crowley, M., Taylor, S., & Flannigan, M. D. (2020). A review of machine learning applications in wildfire science and management. *Environmental Reviews*, *28*(4), 478–505. DOI: 10.1139/er-2020-0019

Jensen, J. R. (2000). *Remote sensing of the environment: An earth resource perspective*. Prentice Hall.

Jha, S. K., Negi, A. K., Alatalo, J. M., Prabhu, V., Jha, M. B., & Kumar, H. (2022). Forest Degradation Index: A Tool for Forest Vulnerability Assessment in Indian Western Himalaya. *Sustainability (Basel)*, *14*(23), 15655. 10.3390/su142315655. DOI: 10.3390/su142315655

Jia, S., Zhang, X., & Yin, G. (2024). Hydrological responses to global warming intensification. *Nature*, *629*, 137–143. DOI: 10.1038/s41586-025-09047-2

Jones, M. W., Abatzoglou, J. T., Veraverbeke, S., Andela, N., Lasslop, G., Forkel, M., Smith, A. J. P., Burton, C., Betts, R. A., van der Werf, G. R., Sitch, S., Canadell, J. G., Santín, C., Kolden, C., Doerr, S. H., & Le Quéré, C. (2022). Global and regional trends and drivers of fire under climate change. *Reviews of Geophysics, 60*(3), Article e2020RG000726. https://doi.org/DOI: 10.1029/2020RG000726

Jumilah, J., Nadi, S., Eko, H., & Friny, N. (2025). Bibliometrics analysis of environmental sustainability issues: Trends, challenges, and future directions. *E3S Web of Conferences, 640*, 01012. https://doi.org/DOI: 10.1051/e3sconf/202564001012

Kahane, R., Hodgkin, T., Jaenicke, H., Hoogendoorn, C., Hermann, M., Hughes, J. D. A., Padulosi, S., & Looney, N. (2013). Agrobiodiversity for food security, health and income. *Agronomy for Sustainable Development*, *33*(4), 671–693. DOI: 10.1007/s13593-013-0147-8

Kanzler, M., Böhm, C., Mirck, J., Schmitt, D., & Veste, M. (2019). Microclimate effects on evaporation and winter wheat (*Triticum aestivum* L.) yield within a temperate agroforestry system. *Agroforestry Systems*, *93*(5), 1821–1841. DOI: 10.1007/s10457-018-0289-4

Karmoude, Y., Idbraim, S., Saidi, S., Masse, A., & Arbelo, M. (2025). Efficient argan tree deforestation detection using Sentinel-2 time series and machine learning. *Applied Sciences (Basel, Switzerland)*, *15*(6), 3231. DOI: 10.3390/app15063231

Kawamura, M., Thomas, C. D., Tsurumoto, A., Sasahara, H., & Kawaguchi, Y. (2000). Lotka's law and productivity index of authors in a scientific journal. *Journal of Oral Science*, *42*(2), 75–78. DOI: 10.2334/josnusd.42.75 PMID: 10989588

Keeley, J. E., & Pausas, J. G. (2022). Evolutionary ecology of fire. *Annual Review of Ecology, Evolution, and Systematics*, *53*(1), 203–225. DOI: 10.1146/annurev-ecolsys-102320-095612

Keren, S., Svoboda, M., Janda, P., & Nagel, T. A. (2020). Relationships between structural indices and conventional stand attributes in an old-growth forest in Southeast Europe. *Forests*, *11*(1), 4. DOI: 10.3390/f11010004

Kint, V., Hein, S., Campioli, M., & Muys, B. (2010). Modelling self-pruning and branch attributes for young Quercus robur L. and Fagus sylvatica L. trees. *Forest Ecology and Management*, *260*(11), 2023–2034. DOI: 10.1016/j.foreco.2010.09.008

Kissinger, G., Herold, M., & De Sy, V. (2012). *Drivers of Deforestation and Forest Degradation: A Synthesis Report for REDD+ Policymakers*. Lexeme Consulting.

Klausmeyer, K. R., & Shaw, M. R. (2009). Climate change, habitat loss, protected areas and the climate adaptation potential of species in Mediterranean ecosystems worldwide. *PLoS One*, *4*(7), 1–9. DOI: 10.1371/journal.pone.0006392 PMID: 19641600

Knutti, R., Abramowitz, G., Eyring, V., Gleckler, P. J., Hewitson, B., & Mearns, L. (2010). Good practice guidance paper on assessing and combining multi model climate projections. In T. F. Stocker, D. Qin, G.-K. Plattner, M. Tignor, & P. M. Midgley (Éds.), *Meeting Report of the IPCC Expert meeting on assessing and combining multi model climate projections* (p. 1-11). IPCC Working Group I Technical Support Unit, University of Bern. https://academicjournals.org/journal/JENE/article-abstract/C1CDB822968

Kouachi, M. E.. (2024). Forty-year fire history reconstruction from Landsat data in Algeria. *Remote Sensing*, *16*(13), 2500. DOI: 10.3390/rs16132500

Kouli, M., Soupios, P., & Vallianatos, F. (2009). Soil erosion prediction using RUSLE in GIS. *Environmental Geology*, *57*, 483–497. DOI: 10.1007/s00254-008-1318-9

Kriegler, E., O'Neill, B. C., Hallegatte, S., Kram, T., Lempert, R. J., Moss, R. H., & Wilbanks, T. (2012). The need for and use of socio-economic scenarios for climate change analysis : A new approach based on shared socio-economic pathways. *Global Environmental Change*, *22*(4), 807–822. DOI: 10.1016/j.gloenvcha.2012.05.005

Ksontini, F. (1996). *Contribution à l'étude de la diversité génétique des peuplements de chêne-liège en Tunisie: implications pour la conservation*. Mémoire de Mastère de Recherche, Institut National Agronomique de Tunisie.

Kumar, M., Kalra, N., Singh, H., Sharma, S., Singh Rawat, P., Kumar Singh, R., & Kumar Gupta, A. Kumar, P. Ravindranath, N. (2021). Indicator-based vulnerability assessment of forest ecosystem in the Indian Western Himalayas: an analytical hierarchy process integrated approach. *Ecol. Indic, 125*. 10.1016/j.ecolind.2021.107568

Kuter, N., Yenilmez, F., & Kuter, S. (2011). Forest fire risk mapping by kernel density estimation. *Croatian Journal of Forest Engineering*, *32*(2), 599–610.

Laaribya, S. (2011). *Dynamique des espaces forestiers et plan stratégique de gestion et de développement socioéconomique de la forêt de la Maâmora* (Thèse de doctorat en sciences de la vie et de l'environnement). Faculté des Sciences de Kénitra.

Laaribya, S., Alaoui, A., Ayan, S., & Dindaroglu, T. (2024). Changes in the potential distribution of Atlas cedar in Morocco in the twenty-first century according to the emission scenarios of RCP 4.5 and RCP 8.5. *Forestist*, *74*(1), 16–25. DOI: 10.5152/forestist.2023.0004

Laaribya, S., Alaoui, A., & Gmira, N. (2017). La forêt marocaine et le cas du développement durable de l'arganier *Argania spinosa* L. Skeels au Maroc. *Biodiversité Biologique et Conservation*, *10*(2), 1–7.

Labbaci, A., Moukrim, S., Lahssini, S., Laaribya, S., Alaoui, H. M., & Hallam, J. (2021). Estimation of land degradation loss by water erosion: case of the site of biological and ecological interest of ain Asmama (western high Atlas, Morocco). *Adv. Sci. Technol. Eng. Syst. J*, *6*, 241-247. doi:10.25046/aj060326

Lafond, V., Lagarrigues, G., Cordonnier, T., & Courbaud, B. (2014). Uneven-aged management options to promote forest resilience for climate change adaptation: Effects of group selection and harvesting intensity. *Annals of Forest Science*, *71*(2), 173–186. DOI: 10.1007/s13595-013-0291-y

Lahssini, S.. (2015). Modeling forest transition in the Maâmora landscape under climate change. *Annals of Forest Research*, *58*(2), 279–296.

Lahssini, S., Hajib, S., Lahlaoi, H., Mharzi Alaoui, H., & Khattabi, A. (2015). Modelling Spatial Distribution of the Carob Tree (Ceratonia siliqua L.) in Azilal Province, Morocco. *Journal of Geography and Geology*, *7*(4), 33. DOI: 10.5539/jgg.v7n4p33

Lahssini, S., Lahlaoi, H., Alaoui, H., Hlal, E., Bagaram, M., & Ponette, Q. (2015). Predicting cork oak suitability in Maâmora forest using random forest algorithm. *Journal of Geographic Information System*, *7*(2), 202–210. DOI: 10.4236/jgis.2015.72017

Lamhamedi, M. S.. (2015). Nursery technologies for drought-resistant reforestation in arid Morocco. *Revue Forestière Française*, *67*(2), 175–188.

Lamhamedi, M. S.. (2015). *Techniques culturales et formation en pépinière forestière*. ENFI/FAO.

Lamhamedi, M. S., Bakry, S. H., & Hamrouni, L. (2015, décembre). Mise en application de nouvelles innovations techniques, technologiques et biotechnologiques pour la restauration, la domestication et l'intensification de la culture de l'arganier. In *Actes du 3e Congrès International de l'Arganier* (pp. 215–226). Agadir, Maroc.

Leite, C., Oliveira, V., Lauw, A., & Pereira, H. (2019). Cork rings suggest how to manage Quercus suber to mitigate the effects of climate changes. *Agricultural and Forest Meteorology*, *266–267*, 12–19. DOI: 10.1016/j.agrformet.2018.11.032

Lenoir, J., & Svenning, J.-C. (2015). Climate-related range shifts – a global multidimensional synthesis and new research directions. *Ecography*, *38*(1), 15–28. DOI: 10.1111/ecog.00967

Léo. Thorand. (2018). Résistance à l'embolie chez deux espèces en association: chêne liège/chêne zeen. Ecole Nationale Supérieure Agronomique de Toulouse. Stage d'été. 65p.

Limousin, J.-M., Rambal, S., Ourcival, J.-M., Rodríguez-Calcerrada, J., Pérez-Ramos, I. M., Rodríguez-Cortina, R., Misson, L., & Joffre, R. (2012). Morphological and phenological shoot plasticity in a Mediterranean evergreen oak facing long-term increased drought. *Oecologia, 169*(2), 565–577. DOI: 10.1007/s00442-011-2221-8 PMID: 22159896

Linares, J. C., Camarero, J. J., & Carreira, J. A. (2011). Competition modulates the adaptation capacity of forests to climatic stress: Insights from recent growth decline and death in relict stands of the Mediterranean fir Abies pinsapo. *Journal of Ecology*, *99*(2), 592–603.

Linares, J. C., Taïqui, L., & Camarero, J. (2011). Increasing drought sensitivity and decline of Atlas cedar (*Cedrus atlantica*) in the Moroccan Middle Atlas forests. *Forests*, *2*(3), 777–796. DOI: 10.3390/f2030777

Lindenmayer, D. B., & Likens, G. E. (2010). The science and application of ecological monitoring. *Biological Conservation*, *143*(6), 1317–1328. DOI: 10.1016/j.biocon.2010.02.013

Lionello, P. (Ed.). (2012). *The climate of the Mediterranean region : From the past to the future* (1st ed.). Elsevier.

Locatelli, B., Herawati, H., Brockhaus, M., Idinoba, M., & Kanninen, M. (2008). *Methods and tools for assessing the vulnerability of forests and people to climate change* (Working Paper No. 43).

Lopez-Tirado, J., & Hidalgo, P. J. (2016). Predictive modelling of climax oak trees in southern Spain: Insights in a scenario of global change. *Plant Ecology*, *217*(4), 451–463. DOI: 10.1007/s11258-016-0589-6

Lopez-Tirado, J., Vessella, F., Schirone, B., & Hidalgo, P. J. (2018). Trends in evergreen oak suitability from assembled species distribution models: Assessing climate change in south-western Europe. *New Forests*, *49*(4), 471–487. DOI: 10.1007/s11056-018-9629-5

Lybbert, T. J., Aboudrare, A., Chaloud, D., Magnan, N., & Nash, M. (2011). L'essor du marché de l'huile d'argan marocaine semble bénéficier à certains ménages ruraux tout en menaçant la forêt d'arganiers endémique. *Proceedings of the National Academy of Sciences of the United States of America*, *108*, 13963–13968. DOI: 10.1073/pnas.1106382108

M'hirit, O., Samih, A., & Malagnoux, M. (Eds.). (1994). *Le cèdre de l'Atlas (Cedrus atlantica Manetti): Présentation générale et état des connaissances*. Réseau Silva Mediterranea; Annales de la Recherche Forestière au Maroc. https://www.fao.org/3/AH002F/AH002F00.pdf

M'hirit, O., Benzyane, M., Benchekroun, F., El Yousfi, S. M., & Bendaoun, M. (1998). *L'arganier: Une espèce fruitière forestière à usage multiple*. Mardaga.

Maghreb Arab Presse / Agence Nationale des Eaux et Forêts (MAP/ANEF). (2024, 1 août). L'ANEF renforce la digitalisation du secteur forestier. *Maghreb Arab Presse*. https://www.mapnews.ma/fr/actualites/social/l%E2%80%99anef-renforce-la-digitalisation-du-secteur-forestier

Maliha, S. N., Deborah, J. C., William, G. K., & Samuel, S. (2008). Regional assessment of landscape and land use change in the Mediterranean region: Morocco case study (1981–2003). In Liotta, P. H. (Eds.), *Environmental change and human security* (pp. 143–165). Springer.

MAP/ANEF. (2024, August 1). L'ANEF renforce la digitalisation du secteur forestier. Maghreb Arab Presse. https://www.mapnews.ma/fr/actualites/social/l%E2%80%99anef-renforce-la-digitalisation-du-secteur-forestier

MAPMDREF. (2020, February 27). *Présentation de la stratégie « Forêts du Maroc 2020–2030 »*. Haut-Commissariat aux Eaux et Forêts et à la Lutte Contre la Désertification._https://www.agriculture.gov.ma/fr/actualites/presentation-de-la-strategie-forets-du-maroc-2020-2030-skhirat

Markhi, A., Laftouhi, N.-E., Soulaimani, A., & Fniguire, F. (2015). Quantification et evaluation de l'erosion hydrique en utilisant le modèle rusle et déposition intégrés dans un sig. application dans le bassin versant n'fis dans le haut atlas de marrakech (MAROC). *European Scientific Journal, ESJ, 11*(29). Retrieved from https://eujournal.org/index.php/esj/article/view/6342

Martín-Martín, A., Orduna-Malea, E., Thelwall, M., & Delgado López-Cózar, E. (2018). Google Scholar, Web of Science, and Scopus: A systematic comparison. *Journal of Informetrics, 12*(4), 1160–1177. DOI: 10.1016/j.joi.2018.09.002

Martin-StPaul, N., Delzon, S., & Cochard, H. (2017). Plant resistance to drought depends on timely stomatal closure. *Ecology Letters, 20*(11), 1437–1447. DOI: 10.1111/ele.12851

Matesanz, S., & Valladares, F. (2014). Ecological and evolutionary responses of Mediterranean plants to global change. *Environmental and Experimental Botany, 103*, 53–67. DOI: 10.1016/j.envexpbot.2013.09.004

Mauri, E., Hernández Paredes, E., Núñez Blanco, I., & García Feced, C. (2023). *Key recommendations on wildfire prevention in the Mediterranean*. European Forest Institute., DOI: 10.36333/rs6

McCormick, J. (2020, October 1). California firefighters tap AI for an edge in battling wildfires. *The Wall Street Journal.*

MDCE. (2020). *Troisième communication nationale à la CCNUCC. Ministère délégué auprès du Ministre de l'Énergie, des Mines, de l'Eau et de l'Environnement, Chargé de l'Environnement, Maroc.*

Mechergui, K., Jaouadi, W., Altamimi, A., Naghmouchi, S., & Ammari, Y. (2021). Effect of climate change on the spatial distribution and cork production of Quercus suber L., the risk of exclusion by the Aleppo pine expansion, and management practices to protect Q. suber habitat: A review. *Notulae Botanicae Horti Agrobotanici Cluj-Napoca, 49*(1), 12218. DOI: 10.15835/nbha49112218

Mechergui, K., Jaouadi, W., & Ammari, Y. (2022). Le chêne zéen (Quercus canariensis Willd.) en Tunisie: Écologie, production, état de dégradation et les pratiques d'aménagement. *Revue Marocaine des Sciences Agronomiques et Vétérinaires, 10*(1). https://www.agrimaroc.org/index.php/Actes_IAVH2/article/view/1075

Médail, F., & Diadema, K. (2006). Biodiversité végétale méditerranéenne. *Annales de Géographie, 115*(651), 618–640. DOI: 10.3406/geo.2006.21290

Medail, F., & Quezel, P. (1999). Biodiversity hotspots in the Mediterranean Basin : Setting global conservation priorities. *Conservation Biology, 13*(6), 1510–1513. DOI: 10.1046/j.1523-1739.1999.98467.x

Meddour-Sahar, O. (2015). Wildfires in Algeria: Problems and challenges. *iForest - Biogeosciences and Forestry, 8*(6), 818–826. https://doi.org/DOI: 10.3832/ifor1279-007

Meddour-Sahar, O., González-Cabán, A., Meddour, R., & Derridj, A. (2012). Wildfire management policies in Algeria: Present and future needs. In *Proceedings of the Fourth International Symposium on Fire Economics, Planning, and Policy: Climate Change and Wildfires* (pp. 382–395).

MedECC. (2019). *Report on climate and environmental changes in the Mediterranean Region.* Union for the Mediterranean.

MedECC. (2020). *Climate and Environmental Change in the Mediterranean Basin – Current Situation and Risks for the Future. First Mediterranean Assessment Report (MAR1).* Union for the Mediterranean, Plan Bleu, UNEP/MAP. DOI: 10.5281/zenodo.4768833

Mediavilla, S., & Escudero, A. (2004). Stomatal responses to drought of mature trees and seedlings of two co-ccurring Mediterranean oaks. *Forest Ecology and Management, 187*(2-3), 281–294. DOI: 10.1016/j.foreco.2003.07.006

Meliho, M., Khattabi, A., Zine El Abidine, A., & Orlando, C. A. (2024). Assessing the impact of climate change on bioclimatic zones in Morocco. *Modeling Earth Systems and Environment, 10*, 3285–3298. doi:10.1007/s10661-024-13077-0

Merow, C., Smith, M. J., & Silander, J. A.Jr. (2013). A practical guide to MaxEnt for modeling species' distributions : What it does, and why inputs and settings matter. *Ecography, 36*(10), 1058–1069. DOI: 10.1111/j.1600-0587.2013.07872.x

Messier, C., Puettmann, K., Chazdon, R., Andersson, K. P., Angers, V. A., Brotons, L., Filotas, E., Tittler, R., Parrott, L., & Levin, S. A. (2015). From management to stewardship: Viewing forests as complex adaptive systems in an uncertain world. *Conservation Letters, 8*(5), 368–377. DOI: 10.1111/conl.12156

Meusburger, K., Steel, A., Panagos, P., Montanarella, L., & Alewell, C. (2012). Spatial and temporal variability of rainfall erosivity. *Hydrology and Earth System Sciences, 16*(1), 167–177. DOI: 10.5194/hess-16-167-2012

Mharzi Alaoui, H., Assali, F., Hajji, H., Lahssini, S., Aadel, T., & Moukrim, S. (2019). Machine learning approach for predicting seasonal risk of forest fires in Morocco. *Biodiversidade Brasileira, 9*(1). doi:10.37002/biodiversidadebrasileira.v9i1.1116

Mharzi Alaoui, H., Assali, F., Rouchdi, M., Tahiri, D., Lahssini, S., Aafi, A., & Moukrim, S. (2017). Cartographie de l'intensité du feu et de la sensibilité aux feux de cimes en forêts méditerranéennes. Cas de la province de Chefchaouen au nord du Maroc. [Terre et Vie]. *Revue d'Écologie, 72*(4), 387–409. DOI: 10.3406/revec.2017.1900

Miller, R. G. (1974). The jackknife-a review. *Biometrika, 61*(1), 1–15. DOI: 10.1093/biomet/61.1.1

Mina, M., Messier, C., Duveneck, M. J., Fortin, M. J., & Aquilu'e, N. (2022). Managing for the unexpected: Building resilient forest landscapes to cope with global change. *Global Change Biology, 28*(14), 4323–4341. DOI: 10.1111/gcb.16197 PMID: 35429213

Miranda, A., Syphard, A. D., Berdugo, M., Carrasco, J., Gómez-González, S., Ovalle, J. F., Delpiano, C. A., Vargas, S., Squeo, F. A., Miranda, M. D., Dobbs, C., Mentler, R., Lara, A., & Garreaud, R. (2023). Widespread synchronous decline of Mediterranean-type forest driven by accelerated aridity. *Nature Plants, 9*(11), 1810–1817. DOI: 10.1038/s41477-023-01541-7 PMID: 37845335

MITECO. (2022). *Strategic guidelines for wildland fire management.* Ministerio para la Transición Ecológica y el Reto Demográfico.

Moreira, F., Ascoli, D., Safford, H., Adams, M. A., Moreno, J. M., Pereira, J. M. C., Catry, F. X., Armesto, J., Bond, W., González, M. E., Curt, T., Koutsias, N., McCaw, L., Price, O., Pausas, J. G., Rigolot, E., Stephens, S., Tavşanoğlu, C., Vallejo, V. R., & Fernandes, P. M. (2020). Wildfire management in Mediterranean-type regions: Paradigm change needed. *Environmental Research Letters, 15*(1), 011001. DOI: 10.1088/1748-9326/ab541e

Moretti, V., Di Bartolomei, R., Sorgi, T., Aromolo, R., & Salvati, L. (2015). Soil water deficit and climate conditions during the dry season along the coastal-inland gradient in Castelporziano forest, central Italy. *Rendiconti Lincei. Scienze Fisiche e Naturali, 26*(S3), 283–288. DOI: 10.1007/s12210-014-0336-7

Morgan, R. P. C., Quinton, J. N., Smith, R. E., Govers, G., Poesen, J. W. A., Auerswald, K., Chisci, G., Torri, D., & Styczen, M. E. (1998). The European Soil Erosion Model (EUROSEM): A dynamic approach for predicting sediment transport from fields and small catchments. *Earth Surface Processes and Landforms, 23*(6), 527–544. DOI: 10.1002/(SICI)1096-9837(199806)23:6<527::AID-ESP868>3.0.CO;2-5

Moukrim, S., Lahssini, S., Mharzi Alaoui, H., Rifai, N., Arahou, M., & Rhazi, L. (2018). Modélisation de la distribution spatiale des espèces endémiques pour leur conservation: Cas de l'Argania spinosa (L.) Skeels. *Revue d'Écologie, 73*(2), 153–166. DOI: 10.3406/revec.2018.1923

Moukrim, S., Lahssini, S., Naggar, M., Lahlaoi, H., Rifai, N., Arahou, M., & Rhazi, L. (2019b). Local community involvement in forest rangeland management: Case study of compensation on forest area closed to grazing in Morocco. *The Rangeland Journal, 41*(1), 43–53. DOI: 10.1071/RJ17119

Moukrim, S., Lahssini, S., Rhazi, M., Menzou, K., El Madihi, M., Rifai, N., Bouziani, Y., Azedou, A., Boukhris, I., & Rhazi, L. (2022). Climate Change Impact on Potential Distribution of an Endemic Species *Abies marocana* Trabut. *Ekologia (Bratislava), 41*(4), 329–339. DOI: 10.2478/eko-2022-0034

Moukrim, S., Lahssini, S., Rhazi, M., Mharzi, A. H., Benabou, A., Wahby, I., El Madihi, M., Arahou, M., & Rhazi, L. (2019). Climate change impacts on potential distribution of multipurpose agro-forestry species : Argania spinosa (L.) Skeels as case study. *Agroforestry Systems, 93*(4), 1209–1219. DOI: 10.1007/s10457-018-0232-8

Moukrim, S., Lahssini, S., Rifai, N., Menzou, K., Mharzi-Alaoui, H., Labbaci, A., Rhazi, M., Wahby, I. W., El Madihi, M., & Rhazi, L. (2020). Modélisation de la distribution potentielle de Cedrus atlantica Manetti au Maroc et impacts du changement climatique. *Bois et Forêts des Tropiques*, *344*(2), 3–16. DOI: 10.19182/bft2020.344.a31888

Moukrim, S., Tellal, J., Lahssini, S., Boudjaj, A., Rhazi, M., Madihi, M. E., Bammou, M. B., Moukrim, M., Saber, E.-R., Wahby, I., Rhazi, L., & Grillas, P. (2025). Long-Time Dynamics in a North African Coastal Ramsar Site, the Sidi Boughaba Lake. *Ekologia (Bratislava)*, *44*(2), 147–159. DOI: 10.2478/eko-2025-0017

Muller, S. D., Daoud-Bouattour, A., Fauquette, S., Bottollier-Curtet, M., Rifai, N., Robles, M., Saber, E.-R., El Madihi, M., Moukrim, S., & Rhazi, L. (2022). Holocene history of peatland communities of central Rif (Northern Morocco). *Geobios*, *70*, 35–53. DOI: 10.1016/j.geobios.2021.12.001

Naggar, M. (2020). La forêt et le nouveau modèle de développement au Maroc. *Forêt Méditerranéenne*, 2, 125–134. https://www.foret-mediterraneenne.org/upload/biblio/FORET_MED_2020_2_125-134.pdf

Naipal, V., Reick, C. H., Van Oost, K., & Pongratz, J. (2015). Improving the global applicability of RUSLE. *Geoscientific Model Development*, *8*(9), 2893–2911. DOI: 10.5194/gmd-8-2893-2015

Natividade, J. V. (1956). *Subériculture, édition française de l'ouvrage portugais «Subericultura»*. Ecole Nationale des Eaux et Forêts.

Navas, A., Machin, J., & Gaspar, L. (2009). Les sols dans le pays du Ziz (Sud-est marocain), caractéristiques et aspects de dégradation. *Applied Geography (Sevenoaks, England)*, *20*, 369–394.

Nearing, M. A. (1997). A single, continuous function for slope steepness influence. *Soil Science Society of America Journal*, *61*(4), 917–919. DOI: 10.2136/sssaj1997.03615995006100030029x

Nearing, M. A., Jetten, V., Baffaut, C., Cerdan, O., Couturier, A., & Van Oost, K. (2017). Modeling response of soil erosion to changes in precipitation and cover. *Catena*, *61*(2-3), 131–154. doi:10.1016/j.catena.2005.03.007

NFI. (2005). *National Forest Inventory*. High Commission for Water, Forests and Combating Desertification.

NFI. (2018). *National Agency for Water and Forests, Morocco. Map*. NFI.

Niemeijer, D., & de Groot, R. S. (2008). A conceptual framework for selecting environmental indicator sets. *Ecological Indicators*, *8*(1), 14–25. DOI: 10.1016/j.ecolind.2006.11.012

Nikinmaa, L., Lindner, M., Cantarello, E., Jump, A. S., Seidl, R., Winkel, G., & Muys, B. (2020). Reviewing the Use of Resilience Concepts in Forest Sciences. *Current Forestry Reports*, *6*(2), 61–80. DOI: 10.1007/s40725-020-00110-x PMID: 35747899

Niknejad, M., & Bernardino, A. (2021). Attention on classification for fire segmentation. In *Proceedings of the 20th IEEE International Conference on Machine Learning and Applications* (pp. 616–621). doi:10.48550/arXiv.2111.03129DOI: 10.1109/ICMLA52953.2021.00104

Nouri, J., Houcine, S., Foued, A., & Châabane, A. (2022). Dynamics of the natural regeneration of a natural zeen oak (Quercus canariensis Willd.) forests in Kroumirie, north-western Tunisia. *Arabian Journal of Geosciences*, *15*(22), 1661. DOI: 10.1007/s12517-022-10783-2

OpenStreetMap Contributors. (2023). *OpenStreetMap*. https://www.openstreetmap.org

Organisation de Coopération et de Développement Économiques (OCDE). (1993). *OECD core set of indicators for environmental performance reviews*. Environment Monographs No. 83. Paris: OECD Publications.

Ozenda, P. (1975). Sur les étages de végétation dans les montagnes du bassin méditerranéen. *Documents de Cartographie Écologique*, *16*, 1–32.

Padarian, J., Minasny, B., & McBratney, A. B. (2020). Machine learning and soil sciences: A review aided by machine learning tools. *Soil (Göttingen)*, *6*(1), 35–52. DOI: 10.5194/soil-6-35-2020

Panagos, P., Borrelli, P., & Meusburger, K. (2015a). A new European LS-factor. *Geosciences*, *5*(3), 117–126. DOI: 10.3390/geosciences5020117

Panagos, P., Borrelli, P., Poesen, J., Ballabio, C., Lugato, E., Meusburger, K., Montanarella, L., & Alewell, C. (2015b). The new assessment of soil loss in Europe. *Environmental Science & Policy*, *54*, 438–447. DOI: 10.1016/j.envsci.2015.08.012

Panagos, P., & Katsoyiannis, A. (2019). Soil erosion modelling: New challenges. *Environmental Research*, *172*, 470–474. DOI: 10.1016/j.envres.2019.02.043 PMID: 30844572

Pandey, R., Sharma, J., Singh, R., Rawat, M., Saklani, H., Tomar, P. K., Tiwari, L., Bhatt, I. D., Chand, T., Bala, N., Panwar, V. P., & Semeraro, T. (2022). Vegetation Characteristics Based Climate Change Vulnerability Assessment of Temperate Forests of Western Himalaya. *Forests, 13*(6), 848. 10.3390/f13060848. DOI: 10.3390/f13060848

Pan, Y., Birdsey, R. A., Fang, J., Houghton, R., Kauppi, P. E., Kurz, W. A., Phillips, O. L., Shvidenko, A., Lewis, S. L., Canadell, J. G., Ciais, P., Jackson, R. B., Pacala, S. W., McGuire, A. D., Piao, S., Rautiainen, A., Sitch, S., & Hayes, D. (2011). A large and persistent carbon sink in the world's forests. *Science, 333*(6045), 988–993. DOI: 10.1126/science.1201609 PMID: 21764754

Parmesan, C., & Yohe, G. (2003). A globally coherent fingerprint of climate change impacts across natural systems. *Nature, 421*(6918), 37–42. DOI: 10.1038/nature01286 PMID: 12511946

Pausas, J. G., & Fernández-Muñoz, S. (2012). Fire regime changes in the Western Mediterranean Basin: From fuel-limited to drought-driven fire regime. *Climatic Change, 110*(1-2), 215–226. DOI: 10.1007/s10584-011-0060-6

Pausas, J. G., & Keeley, J. E. (2014). Abrupt climate-independent fire regime changes. *Ecosystems (New York, N.Y.), 17*(6), 1109–1120. DOI: 10.1007/s10021-014-9773-5

Pausas, J. G., & Keeley, J. E. (2019). Wildfires as an ecosystem service. *Frontiers in Ecology and the Environment, 17*(5), 289–295. DOI: 10.1002/fee.2044

Peguero-Pina, J. J., Sancho-Knapik, D., Barrón, E., Camarero, J. J., Vilagrosa, A., & Gil-Pelegrín, E. (2014). Morphological and physiological divergences within Quercus ilex support the existence of different ecotypes depending on climatic dryness. *Annali di Botanica, 114*(2), 301–313. DOI: 10.1093/aob/mcu108 PMID: 24941998

Peguero-Pina, J. J., Sisó, S., Sancho-Knapik, D., Diaz-Espejo, A., Flexas, J., Galmés, J., & Gil-Pelegrín, E. (2016). Leaf morphological and physiological adaptations of a deciduous oak (Quercus faginea Lam.) to the Mediterranean climate: A comparison with a closely related temperate species (Quercus robur L.). *Tree Physiology, 36*(3), 287–299. DOI: 10.1093/treephys/tpv107 PMID: 26496958

Peñuelas, J., Rutishauser, T., & Filella, I. (2009). Phenology feedbacks on climate change. *Science, 324*(5929), 887–888. DOI: 10.1126/science.1173004 PMID: 19443770

Peñuelas, J., & Sardans, J. (2021). Global change and forest disturbances in the Mediterranean Basin: Breakthroughs, knowledge gaps, and recommendations. *Forests, 12*(5), 603. DOI: 10.3390/f12050603

Pérez-Ramos, I. M., Matías, L., Gómez-Aparicio, L., & Godoy, Ó. (2019). Functional traits and phenotypic plasticity modulate species coexistence across contrasting climatic conditions. *Nature Communications, 10*(1), 2555. Advance online publication. DOI: 10.1038/s41467-019-10453-0 PMID: 31186418

Petit, R. J., Hampe, A., & Cheddadi, R. (2005). Climate changes and tree phylogeography in the Mediterranean. *Taxon, 54*(4), 877–885. DOI: 10.2307/25065568

Phillips, S. J., Anderson, R. P., & Schapire, R. E. (2006). Maximum entropy modeling of species geographic distributions. *Ecological Modelling, 190*(3-4), 231–259. DOI: 10.1016/j.ecolmodel.2005.03.026

Pielke, R. J.Jr, Burgess, M. G., & Ritchie, J. (2022). Plausible 2005–2050 emissions scenarios project between 2 °C and 3 °C of warming by 2100. *Environmental Research Letters, 17*(2), 024027. DOI: 10.1088/1748-9326/ac4ebf

Pierce, D. W., Barnett, T. P., Santer, B. D., & Gleckler, P. J. (2009). Selecting global climate models for regional climate change studies. *Proceedings of the National Academy of Sciences of the United States of America, 106*(21), 8441–8446. DOI: 10.1073/pnas.0900094106 PMID: 19439652

Pina-Martins, F., Baptista, J., Pappas, G.Jr, & Paulo, O. (2018). New insights into adaptation and population structure of cork oak using genotyping by sequencing. *Global Change Biology, 25*(1), 337–350. DOI: 10.1111/gcb.14497 PMID: 30358018

Pons, J., & Pausas, J. G. (2006). Oak regeneration in heterogeneous landscape: The case of fragmented Quercus suber forests in the eastern Iberian Peninsula. *Forest Ecology and Management, 231*(1-3), 196–204. DOI: 10.1016/j.foreco.2006.05.049

Programme d'action national de lutte contre la désertification (PAN-LCD). (2022). *Programme d'action national de lutte contre la désertification – Version actualisée*. Rabat, Maroc.

Putz, F. E., Parker, G. G., & Archibald, R. M. (1984). Mechanical abrasion and intercrown spacing. *American Midland Naturalist, 112*(1), 24–28. DOI: 10.2307/2425452

Pyne, S. J. (2019). *Fire: A brief history* (2nd ed.). University of Washington Press. DOI: 10.1515/9780295746197

Qiao, Y., Jiang, W., Wang, F., Su, G., Li, X., & Jiang, J. (2023). FireFormer: An efficient transformer to identify forest fire from surveillance cameras. *International Journal of Wildland Fire, 32*(9), 1364–1380. DOI: 10.1071/WF22220

Quézel, P., & Médail, F. (2003). *Ecology and biogeography of Mediterranean forests*. Elsevier.

Rabhik., (2011). Ajustement de modèles hauteur – Circonférence-age pour le chêne zéen (Quercus canariensis Willd) dans la forêt d'Akfadou (TiziOuzou); effet de la densité et de la station, Thèse de magister en foresterie, Gestion et consèrvation des écosystèmes . Univ Tlemcen. 67p.

Rajabi, A. M., Yavari, A., & Cheshomi, A. (2022). Soil erosion assessment in semi-arid watersheds. *Natural Hazards, 112*, 1685–1704. DOI: 10.1007/s11069-022-05244-8

Ramírez-Valiente, J. A., Sánchez-Gómez, D., Aranda, I., & Valladares, F. (2010). Phenotypic plasticity versus local adaptation for leaf ecophysiological traits in thirteen contrasting cork oak populations under varying water availabilities. *Tree Physiology, 30*, 618–627. DOI: 10.1093/treephys/tpq013

Ramírez-Valiente, J., Valladares, F., Gil, L., & Aranda, I. (2009). Population differences in juvenile survival under increasing drought are mediated by seed size in cork oak (Quercus suber L.). *Forest Ecology and Management, 257*(8), 1676–1683. DOI: 10.1016/j.foreco.2009.01.024

Randerson, J. T., Chen, Y., van der Werf, G. R., Rogers, B. M., & Morton, D. C. (2012). Global burned area and biomass burning emissions from small fires. *Journal of Geophysical Research, 117*(G4), G04012. Advance online publication. DOI: 10.1029/2012JG002128

Reille, M. (1976). Analyse pollinique de sédiments postglaciaires dans le Moyen Atlas et le Haut Atlas marocains : Premiers résultats. *Ecologia Mediterranea, 2*(1), 154–172. DOI: 10.3406/ecmed.1976.925

Reis, F., Pereira, A. J., Tavares, R. M., Baptista, P., & Lino-Neto, T. (2021). Cork oak forests soil bacteria: Potential for sustainable agroforest production. *Microorganisms, 9*(9), 1973. DOI: 10.3390/microorganisms9091973 PMID: 34576868

Renard, K. G., Foster, G. R., Weesies, G. A., McCool, D. K., & Yoder, D. C. (1997). *Predicting soil erosion by water: A guide to conservation planning with the RUSLE.* USDA Handbook 703.

Reynolds, J. F., Stafford Smith, D. M., & Lambin, E. F.. (2007). Global desertification: Building a science for dryland development. *Science, 316*(5826), 847–851. DOI: 10.1126/science.1131634 PMID: 17495163

Rhanem, M. (2011). Aridification du climat régional et remontée de la limite inférieure du cèdre de l'Atlas (Cedrus atlantica Manetti) aux confins de la plaine de Midelt (Maroc). *Physio-Géo. Géographie physique et environnement, 5*(1), 143-165. DOI: 10.4000/physio-geo.1983

Rifai, N., Khattabi, A., Moukrim, S., Arahou, M., & Rhazi, L. (2018). Évaluation de la dynamique de l'occupation du sol dans la zone humide RAMSAR de Tahaddart (nord-ouest du Maroc). *Revue d'Ecologie (Terre et Vie), 73*(2), 142-152. https://hal.science/hal-03532629/document

Rocklin, M. (2015). Dask: Parallel computation with blocked algorithms and task scheduling. In *Proceedings of the 14th Python in Science Conference* (pp. 130–136). DOI: 10.25080/Majora-7b98e3ed-013

Rodriguez-Galiano, V. F., Ghimire, B., Rogan, J., Chica-Olmo, M., & Rigol-Sanchez, J. P. (2012). An assessment of the effectiveness of a random forest classifier for land-cover classification. *ISPRS Journal of Photogrammetry and Remote Sensing, 67*, 93–104. DOI: 10.1016/j.isprsjprs.2011.11.002

Rosen, B. R. (1984). Reef coral biogeography and climate through the Late Cainozoic: Just islands in the sun or a critical pattern of islands? *Geological Journal, 11*(Special issue), 201–262.

Roshani, S. H., Rahaman, M. H., Masroor, M., Sharma, Y., Sharma, A., & Saha, T. K. (2024). Vulnerability assessment of forest ecosystem based on exposure, sensitivity and adaptive capacity in the Valmiki Tiger Reserve, India: A geospatial analysis. *Ecological Informatics, 80*, 17. 10.1016/j.ecoinf.2024.102494

Ruffault, J., Curt, T., Moron, V., Trigo, R. M., Mouillot, F., Koutsias, N., Pimont, F., Martin-StPaul, N., Barbero, R., Dupuy, J.-L., Russo, A., & Belhadj-Khedher, C. (2020). Increased likelihood of heat-induced large wildfires in the Mediterranean Basin. *Scientific Reports, 10*(1), 13790. DOI: 10.1038/s41598-020-70069-z PMID: 32796945

San-Miguel-Ayanz, J., Durrant, T., Boca, R., Maianti, P., Liberta', G., Oom, D., Branco, A., De Rigo, D., Suarez-Moreno, M., Ferrari, D., Roglia, E., Scionti, N., Broglia, M., & Sedano, F. (2025). *Advance report on forest fires in Europe, Middle East and North Africa 2024* (JRC141505). Publications Office of the European Union. https://data.europea.eu/doi/10.2760/1264626, JRC141505

Santos, F., Graw, V., & Bonilla, S. (2019). A geographically weighted random forest approach for evaluate forest change drivers in the Northern Ecuadorian Amazon. *PloS one, 14*(12), e0226224. doi:10.1371/journal.pone.0226224

Sarris, D., Christopoulou, A., Angelonidi, E., Koutsias, N., Fulé, P. Z., & Arianoutsou, M. (2014). Increasing extremes of heat and drought associated with recent severe wildfires in southern Greece. *Regional Environmental Change, 14*(3), 1257–1268. DOI: 10.1007/s10113-013-0568-6

Sauvage, C. (1963). *Étages bioclimatiques.* Atlas du Maroc, Section II: Physique du Globe et Météorologie, Notice explicative, Comité de Géographie du Maroc, Rabat.

Savolainen, O., Pyhäjärvi, T., & Knürr, T. (2007). Gene flow and local adaptation in trees. *Annual Review of Ecology, Evolution, and Systematics, 38*(1), 595–619. DOI: 10.1146/annurev.ecolsys.38.091206.095646

Sayer, J., Sunderland, T., Ghazoul, J., Pfund, J. L., Sheil, D., Meijaard, E., Venter, M., Boedhihartono, A. K., Day, M., Garcia, C., van Oosten, C., & Buck, L. E. (2013). Ten principles for a landscape approach to reconciling agriculture, conservation, and other competing land uses. *Proceedings of the National Academy of Sciences of the United States of America, 110*(21), 8349–8356. DOI: 10.1073/pnas.1210595110 PMID: 23686581

Scheffer, M., Carpenter, S., Foley, J. A., Folke, C., & Walker, B. (2001). Catastrophic shifts in ecosystems. *Nature, 413*(6856), 591–596. DOI: 10.1038/35098000 PMID: 11595939

Scheffers, B. R., Meester, L. D., Bridge, T. C. L., Hoffmann, A. A., Pandolfi, J. M., Corlett, R. T., & Watson, J. E. M. (2016). The broad footprint of climate change from genes to biomes to people. *Science, 354*(6313), aaf7671. Advance online publication. DOI: 10.1126/science.aaf7671 PMID: 27846577

Schmitter, P., Dercon, G., Hilger, T., Hertel, M., Treffner, J., Lam, N., & Cadisch, G. (2010). Linking crop response with sediment deposition. *Agriculture, Ecosystems & Environment, 140*, 34–45. DOI: 10.1016/j.agee.2010.11.009

Scholes, R. J., & Biggs, R. (2005). A biodiversity intactness index. *Nature, 434*(7029), 45–49. DOI: 10.1038/nature03289 PMID: 15744293

Schultz, C. A., Timberlake, T. J., Wurtzebach, Z., McIntyre, K. B., Moseley, C., & Huber-Stearns, H. R. (2019). Policy tools to address scale mismatches: Insights from U.S. forest governance. *Ecology and Society, 24*(1), 13. https://www.jstor.org/stable/26796926. DOI: 10.5751/ES-10703-240121

Seddouki, M., Benayad, M., Aamir, Z., Tahiri, M., Maanan, M., & Rhinane, H. (2022). Using machine learning coupled with remote sensing for forest fire susceptibility mapping: Case study Tetouan Province, northern Morocco. *The International Archives of the Photogrammetry, Remote Sensing and Spatial Information Sciences, XLVIII-4*(W6), 333–2023.

Seidel, D., Leuschner, C., Müller, A., & Krause, B. (2011). Crown plasticity in mixed forests—Quantifying asymmetry as a measure of competition using terrestrial laser scanning. *Forest Ecology and Management, 261*(11), 2123–2132. DOI: 10.1016/j.foreco.2011.03.008

Serbouti, S., Ettaqy, A., Boukcim, H., Mderssa, M. E., El Ghachtouli, N., & Abbas, Y. (2023). Forests and woodlands in Morocco : Review of historical evolution, services, priorities for conservation measures and future research. *International Forestry Review, 25*(1), 121–145. DOI: 10.1505/146554823836838745

Serrano, M. S., Pérez, F. J., & Gómez-Aparicio, L. (2021). Disentangling the interactive effects of climate change and Phytophthora cinnamomi on coexisting Mediterranean tree species. *Agricultural and Forest Meteorology, 298–299*, 108295. Advance online publication. DOI: 10.1016/j.agrformet.2020.108295

Seto, K. C., Güneralp, B., & Hutyra, L. R. (2012). Global forecasts of urban expansion to 2030 and direct impacts on biodiversity and carbon pools. *Proceedings of the National Academy of Sciences of the United States of America, 109*(40), 16083–16088. DOI: 10.1073/pnas.1211658109 PMID: 22988086

Sharma, J., Upgupta, S., Jayaraman, M., Chaturvedi, R. K., Bala, G., & Ravindranath, N. H. (2017). Vulnerability of forests in India: A national scale assessment. *Environmental Management, 60*(3), 544–553. DOI: 10.1007/s00267-017-0894-4

Shibani, C., Pradeep, L., & Indra, P. (2025). Enhancing flood resilience through mixed forest cover: A case study of the Upper Sabarmati River Basin, Gujarat, India. *Journal of Information Systems Engineering and Management, 10* (33s) e-ISSN: 2468-4376. https://www.jisem-journal.com/. https://doi.org/DOI: 10.52783/jisem.v10i33s.5457

Silva Pedro, M., Rammer, W., & Seidl, R. (2015). Tree species diversity mitigates disturbance impacts on the forest carbon cycle. *Oecologia, 177*(3), 619–630. DOI: 10.1007/s00442-014-3150-0 PMID: 25526843

Silva, J., Araújo, S., Sales, H., Pontes, R., & Nunes, J. (2023). Quercus suber L. Genetic Resources: Variability and Strategies for Its Conservation. *Forests, 14*(9), 1925. DOI: 10.3390/f14091925

Simões, R., Rodrigues, A., Ferreira-Dias, S., Miranda, I., & Pereira, H. (2020). Chemical Composition of Cuticular Waxes and Pigments and Morphology of Leaves of Quercus suber Trees of Different Provenance. *Plants, 9*(9), 1165. Advance online publication. DOI: 10.3390/plants9091165 PMID: 32916803

Sinsin, T. E., Mounir, F., & El Aboudi, A. (2020). Analyse comparative des dynamiques spatio-temporelles dans les écosystèmes d'arganiers de plaine et de montagne, Maroc. *The International Journal of Environmental Studies*, *77*, 565–580. DOI: 10.1080/00207233.2020.1723960

Soheili, F., Heydari, M., Woodward, S., & Naji, H. R. (2023). Adaptive mechanism in Quercus brantii Lindl. leaves under climatic differentiation: Morphological and anatomical traits. *Scientific Reports*, *13*(1), 3580. DOI: 10.1038/s41598-023-30762-1 PMID: 36869142

Stiti, B., Piazzetta, R., Khaldi, A. (2014). Régénération de la suberaie tunisienne: état des lieux, contraintes et avancées techniques. *Forêt Méditerranéenne, XXXV* (2), pp.151-160. hal-03556655.

Stocker, T. F., & Qin, D. Plattner, G.-K., Alexander, L. V., Allen, Simon K., Bindoff, N. L., Bréon, F.-M., Church, J. A., Cubasch, U., Emori, S., & others. (2013). Technical summary. In T.F. Stocker, D. Qin, G.-K. Plattner, M. Tignor, S.K. Allen, J. Boschung, A. Nauels, Y. Xia, V. Bex, & P. M. Midgley (Éds.), *Climate change 2013 : The physical science basis. Contribution of Working Group I to the Fifth Assessment Report of the Intergovernmental Panel on Climate Change* (p. 33-115). Cambridge University Press.

Sun, S., Zhang, Y., Wang, N., Yang, W., Zhai, Y., Wang, H., Fan, P., You, C., Zheng, P., & Wang, R. (2024). Changing effects of energy and water on the richness distribution pattern of the Quercus genus in China. *Frontiers in Plant Science*, *15*, 1301395. Advance online publication. DOI: 10.3389/fpls.2024.1301395 PMID: 38298826

Taibi., (2012). Contribution à une étude descriptive des peuplements à chêne zéen (Quercus faginea Lamk) dans les forêts de Zariffet et Hafir (Wilaya de Tlemcen),70p.

Tedim, F., Leone, V., Amraoui, M., Bouillon, C., Coughlan, M. R., Delogu, G. M., Fernandes, P. M., Ferreira, C., McCaffrey, S., McGee, T. K., Parente, J., Paton, D., Pereira, M. G., Ribeiro, L. M., Viegas, D. X., & Xanthopoulos, G. (2018). Defining Extreme Wildfire Events: Difficulties, Challenges, and Impacts. *Fire (Basel, Switzerland)*, *1*(1), 9. DOI: 10.3390/fire1010009

Thorne, J. H., Choe, H., Stine, P. A., Chambers, J. C., Holguin, A., Kerr, A. C., & Schwartz, M. (2018). Climate Change Vulnerability Assessment of Forests in the Southwest USA. *Clim. Chang, 148*, 387–402. 10.1007/s10584-017-2010-4

Thuiller, W., Lavergne, S., Roquet, C., Boulangeat, I., Lafourcade, B., & Araujo, M. B. (2011). Consequences of climate change on the tree of life in Europe. *Nature*, *470*(7335), 531–534. DOI: 10.1038/nature09705 PMID: 21326204

Timzioura, R., Ezzine, S., Benomar, L., Lamhamedi, M. S., Ettaqy, A., Zine El Abidine, A., Zaher, H., Khasa, D. P., Pepin, S., & Abbas, Y. (2025). Bibliometric Analysis of Argan (Argania spinosa (L.) Skeels) Research : Scientific Trends and Strategic Directions for Climate-Resilient Ecosystem Management. *Forests*, *16*(6), 892. DOI: 10.3390/f16060892

Torralba, M., Fagerholm, N., Burgess, P. J., Moreno, G., & Plieninger, T. (2016). Do European agroforestry systems enhance biodiversity and ecosystem services? A meta-analysis. *Agriculture, Ecosystems & Environment*, *230*, 150–161. DOI: 10.1016/j.agee.2016.06.002

Touchan, R., Anchukaitis, K. J., Meko, D. M., Attalah, S., Baisan, C., & Aloui, A. (2008). Long-term context for recent drought in northwestern Africa. *Geophysical Research Letters*, *35*(13), L13705. Advance online publication. DOI: 10.1029/2008GL034264

Touhami, I., Chirino, E., Aouinti, H., Khorchani, E., Elaieb, M., Khaldi, A., & Nasr, Z. (2020). Decline and dieback of cork oak (Quercus suber L.) forests in the Mediterranean basin: A case study of Kroumirie, Northwest Tunisia. *Journal of Forestry Research*, *31*(5), 1–17. DOI: 10.1007/s11676-019-00974-1

Touhami, I., Rzigui, T., Zribi, L., Ennajah, A., Dhahri, S., Aouinti, H., Elaieb, M. T., Fkiri, S., Ghazghazi, H., Khorchani, A., Candelier, K., Khaldi, A., & Khouja, M. L. (2023). Climate change-induced ecosystem disturbance: A review on sclerophyllous and semi-deciduous forests in Tunisia. *Plant Biology*, *25*(4), 481–497. DOI: 10.1111/plb.13524

Trumbore, S., Brando, P., & Hartmann, H. (2015). Forest health and global change. *Science*, *349*(6250), 814–818. https://www.science.org/doi/10.1126/science.aac6759. DOI: 10.1126/science.aac6759 PMID: 26293952

Turco, M., Rosa-Cánovas, J. J., Bedia, J., Jerez, S., Montávez, J. P., Llasat, M. C., & Provenzale, A. (2018). Exacerbated fires in Mediterranean Europe due to anthropogenic warming projected with non-stationary climate-fire models. *Nature Communications*, *9*(1), 3821. DOI: 10.1038/s41467-018-06358-z PMID: 30279564

Turco, M., von Hardenberg, J., AghaKouchak, A., Llasat, M. C., Provenzale, A., & Trigo, R. M. (2017). On the key role of droughts in the dynamics of summer fires in Mediterranean Europe. *Scientific Reports*, *7*(1), 81. Advance online publication. DOI: 10.1038/s41598-017-00116-9 PMID: 28250442

United Nations Framework Convention on Climate Change (UNFCCC). (2015). *Paris Agreement*. UNFCCC.

United Nations. (1997). *Morocco's commitment to sustainable development: Implementation of Agenda 21 and the Rio Conventions*. United Nations.

Urbieta, I., Zavala, M., & Marañón, T. (2008). Human and non-human determinants of forest composition in southern Spain: Evidence of shifts towards cork oak dominance as a result of management over the past century. *Journal of Biogeography, 35*(9), 1688–1700. Advance online publication. DOI: 10.1111/j.1365-2699.2008.01914.x

Urbieta, M. S., Toril, E. G., Aguilera, Á., Giaveno, M. A., & Donatile, E. (2013). Cyanobacteria and photosynthetic species as part of the microbial community structure of biofilms in copahue geothermal springs (Neuquén, Argentina). *Advanced Materials Research, 825*, 11–14. . DOI: 10.4028/www.scientific.net/AMR.825.11

Valladares, F., Benavides, R., Rabasa, S., Díaz, M., Pausas, J. G., Paula, S., & Simonson, W. D. (2014). Global change and Mediterranean forests: Current impacts and potential responses. In Coomes, D. A., Burslem, D. F. R. P., & Simonson, W. D. (Eds.), *Forests and global change* (pp. 47–75). Cambridge University Press. DOI: 10.1017/CBO9781107323506.005

Van Der Werf, G. R., Randerson, J. T., Giglio, L., van Leeuwen, T. T., Chen, Y., Rogers, B. M., Mu, M., van Marle, M. J. E., Morton, D. C., Collatz, G. J., Yokelson, R. J., & Kasibhatla, P. S. (2017). Global fire emissions estimates during 1997–2016. *Earth System Science Data, 9*(2), 697–720. DOI: 10.5194/essd-9-697-2017

Van Eck, N. J., & Waltman, L. (2010). Software survey: VOSviewer, a computer program for bibliometric mapping. *Scientometrics, 84*(2), 523–538. DOI: 10.1007/s11192-009-0146-3 PMID: 20585380

Vanhove, M., Pina-Martins, F., Coelho, A., Branquinho, C., Costa, A., Batista, D., Silva, A., Sousa, P., Henriques, A., Marques, I., Belkadi, B., Knowles, L., & Paulo, O. (2021). Using gradient Forest to predict climate response and adaptation in Cork oak. *Journal of Evolutionary Biology, 34*(6), 910–923. DOI: 10.1111/jeb.13765 PMID: 33484040

Varela, M.-C. (2008). Dépérissement des peuplements de chêne-liège et changement climatique. *forêt méditerranéenne, XXIX*(2), 209-212.

Veldkamp, E., Schmidt, M., Markwitz, C., Beule, L., Beuschel, R., Biertümpfel, A., Bischel, X., Duan, X., Gerjets, R., Göbel, L., Grab, R., Guerra, V., Heinlein, F., Komainda, M., Langhof, M., Luo, J., Potthoff, M., van Ramshorst, J. G. V., Rudolf, C., & Wachendorf, M. (2023). Multifunctionality of temperate alley-cropping agroforestry outperforms open cropland and grassland. *Communications Earth & Environment, 4*(1), 20. Advance online publication. DOI: 10.1038/s43247-023-00680-1

Verikas, A., Gelzinis, A., & Bacauskiene, M. (2011). Mining data with random forests: A survey and results of new tests. *Pattern Recognition*, *44*(2), 330–349. DOI: 10.1016/j.patcog.2010.08.011

Vessella, F., López-Tirado, J., Simeone, M., Schirone, B., & Hidalgo, P. (2017). A tree species range in the face of climate change: Cork oak as a study case for the Mediterranean biome. *European Journal of Forest Research*, *136*(3), 555–569. DOI: 10.1007/s10342-017-1055-2

Vessella, F., & Schirone, B. (2022). Forest Conservation and Restoration Using the Emberger Index: Cork Oak as Study Case. *Forests*, *13*(2), 252. DOI: 10.3390/f13020252

Vicente, E., Carme, M., Silva, F., Stiti, B., Vizcaíno-Palomar, N., & Garzón, M. (2025). Phenotypic integration of post-germination traits in Quercus suber: Morphology is mediated by acorn mass and leaf physiology by populations' climate. *Annals of Botany*, *136*(1), 153–166. DOI: 10.1093/aob/mcaf068 PMID: 40220265

Vilà, M., Espinar, J. L., Hejda, M., Hulme, P. E., Jarošík, V., Maron, J. L., Pergl, J., Schaffner, U., Sun, Y., & Pyšek, P. (2011). Ecological impacts of invasive alien plants: A meta-analysis of their effects on species, communities and ecosystems. *Ecology Letters*, *14*(7), 702–708. DOI: 10.1111/j.1461-0248.2011.01628.x PMID: 21592274

Vila-Viçosa, C., Vázquez, F., Mendes, P., Río, S., Musarella, C., Cano-Ortiz, A., & Meireles, C. (2015). Syntaxonomic update on the relict groves of Mirbeck's oak (Quercus canariensis Willd. and Q. marianica C. Vicioso) in southern Iberia. *Plant Biosystems*, *149*(3), 512–526. DOI: 10.1080/11263504.2015.1040484

Vizinho, A., Príncipe, A., Vasconcelos, A., Rebelo, R., Branquinho, C., & Penha-Lopes, G. (2023). Using and Creating Microclimates for Cork Oak Adaptation to Climate Change. *Land (Basel)*, *12*(3), 531. Advance online publication. DOI: 10.3390/land12030531

Vrieling, A. (2006). Satellite remote sensing for water erosion assessment. *Catena*, *65*(1), 2–18. DOI: 10.1016/j.catena.2005.10.005

Weißhuhn, P., Müller, F., & Wiggering, H. (2018). Ecosystem Vulnerability Review: Proposal of an Interdisciplinary Ecosystem Assessment Approach. *Environmental Management*, *61*(6), 904–915. 10.1007/s00267-018-1023-8. DOI: 10.1007/s00267-018-1023-8

West, J. M., Julius, S. H., Kareiva, P., Enquist, C., Lawler, J. J., Petersen, B., Johnson, A. E., & Shaw, M. R. (2009). U.S. Natural Resources and Climate Change: Concepts and Approaches for Management Adaptation. *Environmental Management, 44*(6), 1001–1021. DOI: 10.1007/s00267-009-9345-1 PMID: 19636606

Williams, S. E., Shoo, L. P., Isaac, J. L., Hoffmann, A. A., & Langham, G. (2008). Towards an Integrated Framework for Assessing the Vulnerability of Species to Climate Change. *PLoS Biology, 6*(12), e325. DOI: 10.1371/journal.pbio.0060325 PMID: 19108608

Wischmeier, W. H., & Smith, D. D. (1978). *Predicting rainfall erosion losses: a guide to conservation planning* (No. 537). Department of Agriculture, Science and Education Administration.

Wright, I. J., Reich, P. B., Westoby, M., Ackerly, D. D., Baruch, Z., Bongers, F., Cavender-Bares, J., Chapin, T., Cornelissen, J. H. C., Diemer, M., Flexas, J., Garnier, E., Groom, P. K., Gulias, J., Hikosaka, K., Lamont, B. B., Lee, T., Lee, W., Lusk, C., & Villar, R. (2004). The world-wide leaf economics spectrum. *Nature, 428*(6985), 821–827. DOI: 10.1038/nature02403 PMID: 15103368

Wunder, S., Calkin, D. E., Charlton, V., Feder, S., Martínez de Arano, I., Moore, P., Rodríguez y Silva, F., Tacconi, L., & Vega-García, C. (2021). Resilient landscapes to prevent catastrophic forest fires: Socioeconomic insights towards a new paradigm. *Forest Policy and Economics, 128*, 102458. DOI: 10.1016/j.forpol.2021.102458

Yu, M., Masrur, A., & Blaszczak-Boxe, C. (2023). Predicting hourly PM2.5 concentrations in wildfire-prone areas using a spatio-temporal transformer model. *The Science of the Total Environment, 860*, 160446. DOI: 10.1016/j.scitotenv.2022.160446 PMID: 36436649

Zarnetske, P. L., Skelly, D. K., & Urban, M. C. (2012). Biotic Multipliers of Climate Change. *Science, 336*(6088), 1516–1518. DOI: 10.1126/science.1222732 PMID: 22723403

Zhou, G. Y., Morris, J. D., Yan, J. H., Yu, Z. Y., & Peng, S. L. (2002). Hydrological impacts of reafforestation with eucalyptus and indigenous specious: A case study in Southern China [Journal] // Elsevier Journal, *Forest ecology and management.* 209-222, ISSN 0378-1127, https://doi.org/DOI: 10.1016/S0378-1127(01)00694-6

Zhou, P., Luukkanen, O., Tokola, T., & Nieminen, J. (2008). Effect of vegetation cover on soil erosion. *Catena, 75*(3), 319–325. DOI: 10.1016/j.catena.2008.07.010

Zhou, Q., Keith, D. M., Zhou, X., Cai, M., Cui, X., Wei, X., & Luo, Y. (2018). Comparing the water holding characteristics of broad leaved, coniferous and mixed forest litter layers in a Karst Region. *Mountain Research and Development*, *38*(3), 220–229. DOI: 10.1659/MRD-JOURNAL-D-17-00002.1

Zhou, X., Abel, D. J., & Truffet, D. (1998). Data partitioning for parallel spatial join processing. *GeoInformatica*, *2*(2), 175–204. DOI: 10.1023/A:1009755931056

Zhu, Z., Gallant, A. L., Woodcock, C. E., Pengra, B., Olofsson, P., Loveland, T. R., Jin, S., Dahal, D., Yang, L., & Auch, R. F. (2016). Optimizing selection of training and auxiliary data for operational land cover classification for the LCMAP initiative. *ISPRS Journal of Photogrammetry and Remote Sensing*, *122*, 206–221. DOI: 10.1016/j.isprsjprs.2016.11.004

Zidane, I. E., Lhissou, R., Ismaili, M., Bouli, A., & Mabrouki, M. (2021). Characterization of fire severity in the Moroccan Rif using Landsat-8 and Sentinel-2 satellite images. *International Journal on Advanced Science, Engineering and Information Technology*, *11*(1), 72–83. DOI: 10.18517/ijaseit.11.1.10342

Zine El Abidine, A., Bouderrah, M., Lamhamedi, M., Mounir, F., & Ponette, Q. (2020). Choix des essences de reboisement pour la forêt de la Maâmora (Maroc) sur la base de la tolérance des plants juvéniles à la sécheresse. *Physio-Géo*, *15*, 133–160. DOI: 10.4000/physio-geo.11251

Zouagui, A., Sabir, M., Naimi, M., Chikhaoui, M., & Benmansour, M. (2018). Modélisation du risque d'érosion hydrique par l'équation universelle des pertes en terre dans le Rif Occidental: Cas du bassin versant de Moulay Bouchta (Maroc). *European Scientific Journal*, *14*(3), 1857–788. DOI: 10.19044/esj.2018.v14n3p524

Zuo, F., Yao, Q., Shi, L., Wang, Z., Bai, M., Fang, K., Guo, F., Yuan, L., & Zhang, W. (2024). Research on wildfire and soil water: A bibliometric analysis from 1990 to 2023. *Fire (Basel, Switzerland)*, *7*(12), 434. DOI: 10.3390/fire7120434

About the Contributors

Said Moukrim is an agricultural engineer and researcher specializing in ecology and natural resource management. He holds a PhD in Biological Sciences from Mohammed V University in Rabat, focusing on the impact of climate change on Moroccan forest ecosystems. His research applies species distribution modeling to assess ecological responses of key forest species. Currently a professor at Mohammed V University in Rabat, Dr. Moukrim previously worked at the Moroccan Department of Water and Forests at the Moroccan Department of Water and Forests. He has contributed to national and international studies on forest resilience. His work spans climate change adaptation, forest restoration, and land degradation. He has published in peer-reviewed journals and presented at national and international conferences. He also trains researchers and officials in forest management and climate resilience. He is also an active member of CEMED (Center of Experts 'Mastery' for Studies and Development), contributing to research and development initiatives.

Saïd Lahssini is a professor of higher education specializing in forest management and inventory. He holds a PhD in Forest Management from the Université catholique de Louvain, Belgium, and has nearly 24 years of experience in forestry research, digitalization, and training engineering. Currently, he teaches at the National School of Forestry Engineers (ENFI) in Morocco, where he leads courses on forest inventory, afforestation, growth modeling, and forest productivity. He has also contributed as a consultant for organizations such as FAO, GIZ, and UNDP, focusing on ecosystem restoration, biodiversity mapping, and climate change adaptation. His expertise includes geospatial technologies, data science applications in forestry, and the integration of digital tools for environmental monitoring. Dr. Lahssini has authored numerous scientific publications and actively participates in international conferences on forestry and environmental sustainability.

***

Abderrahman Aafi, a former student of Cidera, holds an Engineering Degree in Water and Forests from the National School of Forest Engineering (École Nationale Forestière d'Ingénieurs – Salé), obtained in 1987 (general forestry) and in 1995 (specialization: Ecology and Natural Resources Management), as well as a PhD in Agricultural Sciences from the Hassan II Agronomy and Veterinary Institute – Rabat (2007, option: Ecology). His professional career spans several decades (1987 to 2025), notably as a study engineer at the private consulting firm Maroc Développement (1987–1989), researcher and Head of the Ecology, Biodiversity and Soil Conservation Department at the National Center for Forestry Research (Rabat), part-time lecturer at the National School of Forest Engineering – Salé and at the Hassan II Agronomy and Veterinary Institute - Rabat (2007–2025), and former Director General of the National School of Forest Engineering – Salé (2015–2024). Mr. Aafi has participated in and coordinated numerous projects aimed at improving knowledge on biodiversity and natural ecosystems, as well as contributing to the socio-economic development of rural populations in forest and peri-forest areas, in partnership with FAO, GEF/UNDP, IPGRI, Diversitas, the EU, IRD – OSS/ROSELT. As the person in charge of the climate change dossier, he represented the Department of Water and Forests at several international meetings, notably in Germany and Denmark (COP 15, 2009). He is the author and co-author of numerous works and publications, including Maalamat Al Maghrib, covering forestry, agricultural, and environmental fields. In recognition of his substantial contributions, Mr. Abderrahman Aafi was awarded a royal distinction in 2022, receiving the National Order of Merit, Excellent Class.

Younes Abbas is a Professor at the Faculty of Polydisciplinary Studies of Beni Mellal, part of Sultan Moulay Slimane University. He currently serves as Head of the Department of Biology–Geology and previously acted as coordinator of academic programs within the department. Dr. ABBAS holds a PhD in Plant Ecophysiology and has more than 17 years of research experience, particularly through his work in collaboration with the Forest Research Center. His scientific interests focus on plant sciences and ecology, with a strong emphasis on forest ecosystems. His research is centered on plant ecophysiology and plant–microorganism interactions, with recent integration of advanced approaches such as metabolomics and genomics to better understand plant responses to environmental constraints. His work contributes to elucidating adaptive mechanisms in plants, particularly in Mediterranean and semi-arid forest ecosystems. Professor ABBAS has been involved in several national and international collaborative research projects, including FRQ–CNRST, PRAD, and ANPMA programs. He maintains active scientific collaborations with leading institutions such as Laval University (Canada), the Texas Biotechnology Center (USA), Valorhiz (France), and King Abdullah University of Science and

Technology (Saudi Arabia). He is the author of more than 100 peer-reviewed scientific publications in international journals. In addition to his research activities, he actively contributes to academic leadership, graduate supervision, and the development of interdisciplinary research programs addressing key environmental and socio-economic challenges.

Mohamed Ben Bammou holds a PhD in plant ecology, biodiversity, and environment, and is a lecturer at the Faculty of Sciences, Mohammed V University in Rabat. My research focuses on the impact of disturbances on the biodiversity of Mediterranean temporary ponds. I also work on carbon sequestration. Through experimental and field-based approaches, I am particularly interested in seed bank dynamics and the resilience of ecosystems to environmental pressures. Committed to both teaching and research, I contribute to the training of students in ecology and to the production of scientific knowledge aimed at reconciling biodiversity conservation with the sustainable management of natural ecosystems.

Mohamed El Mderssa is an Associate Professor at the Polydisciplinary Faculty of Beni Mellal, Sultan Moulay Slimane University, Morocco. He holds a PhD in ecology and environmental sciences and is a trained forest engineer. He is actively involved in higher education, research supervision, and participation in academic juries and scientific committees. His research focuses on forest ecology, soil science, carbon sequestration, and sustainable natural resource management. His work emphasizes the integration of field data, remote sensing, and artificial intelligence approaches to assess ecosystem services and support climate change adaptation and mitigation strategies in Mediterranean forest ecosystems. He maintains scientific collaborations with various national institutions as well as international academic partners within the framework of research projects focused on the sustainable management of forest resources.

Amel Ennajah is a dedicated research scientist specializing in forest ecology at Tunisia's National Research Institute for Rural Engineering, Water, and Forestry (INRGREF). Her research program centers on assessing and enhancing the resilience of Mediterranean forest ecosystems, particularly mixed-oak stands, to climate change and anthropogenic pressures. Through her work, which integrates field studies and ecological modeling, she aims to generate actionable science that informs adaptive forest management and sustainable conservation strategies. Her primary research examines the resilience, vulnerability, and adaptive capacity of mixed-oak landscapes, with a focus on informing evidence-based policy and sustainable management practices. She has also contributed to the characterization of native legumes such as Cytisus villosus and their potential role in restoring degraded silvopastoral systems through nitrogen fixation and soil improvement. She is actively involved in international research cooperation and in supervising graduate students in plant sciences and forest ecology. She actively disseminates her findings through peer-reviewed publications and contributions to scientific volumes on forestry.

Abderrahim Ettaqy holds a State Engineering Degree in Water and Forestry (2012) from the National School of Forest Engineers (Morocco) and a Ph.D. in Ecology and Plant Biodiversity (2024) from the Faculty of Science and Technics of Béni Mellal (Morocco). He currently leads the GIS, Cartography, and Inventory Unit at the Regional Directorate of Water and Forests of Béni Mellal, where he oversees forest management and biodiversity conservation projects. His research and professional activities have resulted in more than 20 publications, numerous conference presentations, lectures, and specialized training sessions. Dr. Ettaqy is actively involved in scientific initiatives related to hydrology, biodiversity monitoring, and watershed management, and regularly teaches courses in phytogeography, GIS, and watershed analysis. His expertise spans geospatial analysis, remote sensing, and ecological modeling. He is engaged in professional networks dedicated to environmental sustainability and has received awards recognizing his contributions to academic excellence and environmental research. His research interests include plant ecology, forest ecosystem sustainability, environmental geospatial modeling, remote sensing applications, and conservation strategies.

Sihem Jebari, original background is Agricultural Engineering. She has been working as a researcher at INRGREF (Tunisian National Research Institute for Rural Engineering, Water and Forestry) since 1999. Her research work is focused on semiarid and arid areas. It involves water erosion modeling, climate change impacts on water availability, agricultural environment degradation, reservoir siltation and transboundary water management.

Kamal Menzou is a teacher and researcher specializing in natural resource governance and environmental public policies. He holds a doctorate in Spatial Planning and Territorial Development from the National Institute of Spatial Planning and Urbanism in Rabat and a doctorate in Political Science from Mohammed V University in Rabat. Dr. Menzou has accumulated over 25 years of experience in the territorial management of forest areas. He also contributed to the management of international forestry programs with the EU, GIZ, AFD, FAO, etc., when he was head of the international cooperation department within the Moroccan forestry administration. He has contributed to studies on the management of forest areas in several regions of Morocco. He also has over 15 years of experience in teaching and supervising scientific research and served as President of the Third International

Colloquium of Political Science Researchers in Rabat in 2014. Currently, he teaches at the National School of Forestry Engineers (ENFI) and is the author and co-author of numerous scientific articles in national and international journals. He is also an active member of CEMED (Center of Experts 'Mastery' for Studies and Development), contributing to research and development initiatives.

Hicham Mharzi-Alaoui is the Head of the National Center for Climatic and Forest Risk Management in Rabat, Morocco. He holds a Ph.D. (2017) in Geoinformatics Sciences and Engineering from the Hassan II Institute of Agronomy and Veterinary Medicine. Following the completion of his doctorate, he joined several engineering schools as a Visiting Professor. He also served as a Postdoctoral Researcher at Mohammed VI Polytechnic University (UM6P), where he further developed innovative approaches to environmental monitoring and risk management. He has been actively involved in numerous national and international projects focused on the application of emerging technologies for climatic and forest risk management across multiple countries. His research interests include risk analysis, climate change, artificial intelligence, remote sensing, GIS-based analysis of ecological and environmental systems, land-use and land-cover change, environmental modeling, and human–environment interactions. He is the author of more than 40 scientific publications, including peer- reviewed journal articles, book chapters, and international reports. He is an active member of both the International Regional Science Association and its Moroccan section.

Mohammed El Madihi is an Associate Professor at the Faculty of Sciences, Chouaib Doukkali University (El Jadida, Morocco). He teaches general ecology and plant ecology at undergraduate and graduate levels. His research focuses on climate change impacts on ecosystems, with particular emphasis on Mediterranean wetlands, plant biodiversity, forest ecosystems, and temporary ponds. His work integrates ecological approaches to understand the structure, functioning, and resilience of plant communities under environmental stress. He has contributed to several national and international research projects related to biodiversity conservation and sustainable ecosystem management in Mediterranean environments.

Laila Rhazi is a professor at Mohammed V University of Rabat (Morocco), specializing in plant ecology and wetland biodiversity. Her research focuses on the structure, diversity, and dynamics of plant communities in Mediterranean temporary ponds under global change, including climate and anthropogenic pressures. She has developed extensive international collaborations on temporary pond ecosystems with partners across Europe (France, Belgium, Portugal, Spain, Italy, Finland), Africa (Tunisia, Algeria, Mauritania), and North America (USA, Canada). Her work contributes to advancing knowledge on wetland functioning and supports biodiversity conservation strategies in vulnerable ecosystems.

Rajaa Timzioura is a second-year PhD candidate specializing in climate change impacts on forest ecosystems. Her research examines the vulnerability of key Moroccan forest species to environmental stressors and focuses on the development of advanced, science-based adaptation strategies. She employs spatial modeling, species distribution modeling, and selective reforestation approaches to assess and mitigate forest decline. Her work contributes to strengthening the effectiveness and long-term sustainability of national reforestation and climate adaptation programs in Morocco.

Index

M

N

O

Q

R

S

V

W

X

www.ingramcontent.com/pod-product-compliance
Lightning Source LLC
LaVergne TN
LVHW081531300926
849161LV00006B/1379
9798260007839